D0363461

Two week

loan

BIOMEMBRANES
A Multi-Volume Treatise

Volume 4 • 1996

ENDOCYTOSIS AND EXOCYTOSIS

BIOMEMBRANES
A Multi-Volume Treatise

ENDOCYTOSIS AND EXOCYTOSIS

Editor: A. G. LEE
Department of Biochemistry
University of Southampton
Southampton, England

VOLUME 4 • 1996

 JAI PRESS INC.

Greenwich, Connecticut   *London, England*

CONTENTS

LIST OF CONTRIBUTORS

Anthony J. Baines

Research School of BioSciences and
  Biological Laboratory
University of Kent

H.D. Blackbourn

Department of Biochemistry
University of Cambridge

Eric J. Brown

Division of Infectious Diseases
Washington University School of
  Medicine

A. Lee Burns

Laboratory of Cell Biology and Genetics
National Institutes of Health

Gerhard A. Coetzee

Kenneth Norris Jr. Comprehensive
  Cancer Center
University of Southern California

Caroline A. Enns

Department of Cell Biology and Anatomy
Oregon Health Sciences University

Christian Fuhrer

Biozentrum
University of Basel
Switzerland

S.F.C. Hawkins

Department of Biochemistry
University of Cambridge

Nandini V.L. Hayes

Research School of BioSciences and
  Biological Laboratory
University of Kent

M. Fatima Horta   Department of Biochemistry-Immunology
                           Instituto de Ciencias Biologicas
                           Universidade Federal de Minas Gerais
                           Brazil

M.J.G. Hughes   Department of Biochemistry
                           University of Cambridge

A.P. Jackson   Department of Biochemistry
                           University of Cambridge

Chau-Ching Liu   Laboratory of Molecular Immunology
                              and Cell Biology
                           The Rockefeller University

Florin I. Niculescu   Department of Pathology
                           School of Medicine
                           University of Maryland

Adrian Ozinsky   Department of Medical Biochemistry
                           University of Cape Town Medical School
                           South Africa

Pedro M. Persechini   Instituto Biofisica
                           Universidade Federal do Rio de Jeneiro

Regina Pohlmann   Zentrum Biochemie and Molekulare
                              Zellbiologie
                           Georg-August-Universitat Gottingen
                           Germany

Harvey B. Pollard   Laboratory of Cell Biology and Genetics
                           National Institutes of Health

Horea G. Rus   Department of Pathology
                           School of Medicine
                           University of Maryland

Elizabeth A. Rutledge   Department of Cell Biology and Anatomy
                           Oregon Health Sciences University

Moon L. Shin   Department of Pathology
                           School of Medicine
                           University of Maryland

Martin Spiess   Biozentrum
                                  University of Basel
                                  Switzerland

Thomas H. Steinberg   Department of Biochemistry
                                  University of Cambridge

Deneys R. van der Westhuyzen   Department of Medical Biochemistry
                                  University of Cape Town Medical School
                                  South Africa

Anthony M. Williams   Department of Cell Biology and Anatomy
                                  Oregon Health Sciences University

John Ding-E Young   Laboratory of Molecular Immunology
                                     and Cell Biology
                                  The Rockefeller University

PREFACE

The quantity of information available about membrane proteins is now too large for any one person to be familiar with anything but a very small part of the primary literature. A series of volumes concentrating on molecular aspects of biological membranes therefore seems timely. The hope is that, when complete, these volumes will provide a convenient introduction to the study of a wide range of membrane functions.

Volume 4 of *Biomembranes* covers endocytosis, exocytosis and related processes. A major role of the plasma membrane is as a permeability barrier, keeping the inside of the cell inside and the outside, outside. Mechanisms must then exist to allow movement of material between the cell and its environment. One mechanism for export from the cell is by exocytosis, a process in which the membranes of secretory vesicles fuse with the plasma membrane releasing the contents of the vesicle into the extracellular medium. The process has been studied in particular depth for the release of neurotransmitters at the synapse. Import into the cell is possible by the process of receptor-mediated endocytosis in which selected plasma membrane proteins are internalized; when these proteins are receptors for macromolecules, the result is uptake of the macromolecule. Transferrin, the low density lipoprotein, and asialoglycoproteins are all taken up into cells in this way. Phagocytosis, the ingestion of cells and cell fragments by neutrophils and macrophages, also involves receptors—on the phagocytic membrane—of which the best studied are those for the Fc domain of IgG, for the third component of complement, and for the mannose/fructose carbohydrates. Protection of a host against infection can also be achieved by damaging the integrity of the plasma membrane of the invading organism. This is the strategy evolved by the cytotoxic T lymphocytes which produce a pore-forming toxin, perforin. Volume 4 of *Biomembranes* explores the structures and mechanisms involved in these biologically and medically important processes.

As editor, I wish to thank all the contributors for their efforts and the staff of JAI Press for their professionalism in seeing everything through to final publication.

A. G. Lee
Editor

RECEPTOR-MEDIATED ENDOCYTOSIS

A. P. Jackson, H. D. Blackbourn,
S. F. C. Hawkins, and M. J. G. Hughes

I. INTRODUCTION

Receptor-mediated endocytosis is a mechanism used by cells for the internalization of selected plasma membrane proteins. Many of these proteins are cell-surface

Biomembranes
Volume 4, pages 1–32.
Copyright © 1996 by JAI Press Inc.
All rights of reproduction in any form reserved.
ISBN: 1-55938-661-4.

1

receptors for extracellular macromolecules—typical examples are transferrin (Hopkins and Trowbridge, 1983), low-density lipoprotein (LDL) (Anderson et al., 1977), insulin (Terris et al., 1979), asialoglycoprotein in hepatocytes (Geuze et al., 1982), and antigen-bound immunoglobulin in B-lymphocytes (Watts et al., 1989; Brodsky, 1992). The process thus serves as a mechanism for the uptake of essential macromolecules and to connect the outside world with the internal compartments of the cell.

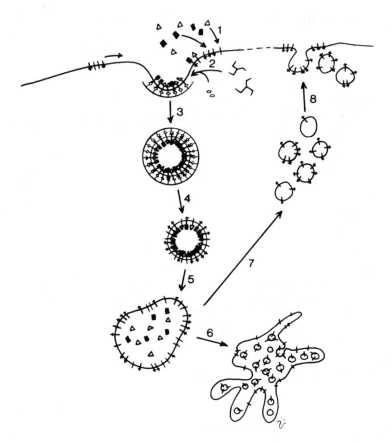

Figure 1. Receptor-mediated endocytosis. Extracellular ligands bind selected receptors at the plasma membrane (1). Clathrin and adaptors assemble at the under-surface of the plasma membrane to form the coated pit and trap receptors containing the appropriate internalization signal (2). The coated pit invaginates and detaches to form a coated vesicle (3), and is rapidly uncoated (4). Vesicles fuse with the endosome and receptors release their ligand in the mildly acidic endosomal environment (5). Some receptors are targeted for lysosomal degradation by their incorporation into multivesicular bodies of the late endosomes (6). Other receptors are recycled to the plasma membrane (7,8).

In the first step of the process, a ligand binds to a receptor on the plasma membrane. The receptor–ligand complexes become concentrated to the exclusion of other membrane proteins in specialized regions called coated pits. As the coated pit grows, it invaginates and detaches to form a coated vesicle. The rapid removal of the coat allows the vesicle to fuse with the endosome and further processing of receptor–ligand complex to proceed (Figure 1).

Over the last decade, there has been a large increase in our understanding of receptor-mediated endocytosis. In this chapter we will summarize the current state of knowledge from a variety of perspectives: how the major protein components interact, how specificity is built into the mechanism, and how the process has been adapted in a range of organisms.

II. STRUCTURAL COMPONENTS OF COATED VESICLES

Although this chapter describes their role in receptor-mediated endocytosis, coated vesicles play additional roles within the cell. Coated vesicles are responsible for the export of lysosomal enzymes from the trans-Golgi network and have been implicated in the early events of regulated secretion (Pearse, 1987; Brodsky, 1988). They are but one of a series of distinct transport organelles that transfer membrane and cargo between intracellular compartments (Pryer et al., 1992).

Under the electron microscope the coated vesicle reveals a remarkable closed hexagonal and pentagonal lattice-like structure (Figure 2). Geometric considerations predict that any closed sphere can be covered with just twelve pentagons and a variable number of hexagons. The predominant form in brain contains twelve pentagons and eight hexagons although other structures of different sizes are possible (Kanaseki and Kadota, 1969; Pearse, 1978).

The coated vesicle is composed of two discrete layers of coat proteins surrounding an inner layer of membrane-bound receptors. These two layers can be extracted with high concentrations of Tris buffer and separated by gel filtration (Figure 3). The outer layer is comprised of the protein clathrin, which is responsible for the characteristic polygonal lattice. The inner layer of adaptor proteins bind to the cytoplasmic tails of receptors and encourage clathrin assembly (Keen, 1990; Pearse and Robinson, 1990). In the cell, clathrin and adaptors exist in both assembled and unassembled pools, and there is rapid recycling between the two pools as the coated pits assemble, transform into coated vesicles, and uncoat. The relative size of the assembled and unassembled fractions can be used to estimate overall endocytic activity in a cell or tissue (Goud et al., 1985). Immunofluorescence microscopy reveals the assembled clathrin pool in a punctate distribution at the plasma membrane and around the Golgi apparatus representing the two main locations within the typical cell (Figure 4).

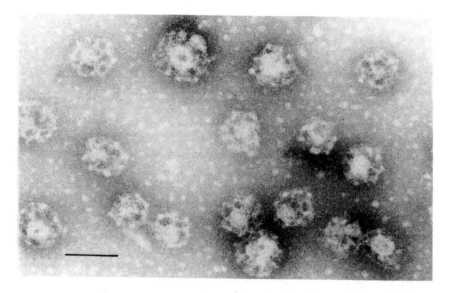

Figure 2. Pig brain coated vesicles negatively stained with uranyl acetate. The bar represents 0.1 μm. (Preparation and staining by A. C. Frank.)

A. The Clathrin Triskelions

Disassembled clathrin adopts a symmetrical three-legged structure called a triskelion and contains three molecules of clathrin heavy-chain (MW 180 kD) together with three molecules of clathrin light-chains (MW 20–30 kD) (Ungewickell and Branton, 1981). The triskelion leg comprises a single heavy-chain to which a light-chain is bound at the proximal domain (Figure 5a). The triskelion is nonplanar. Viewed from the convex face, with the vertex pointing "up" the arms are bent in a clockwise fashion (Kirchhausen et al., 1986). While predominantly rigid, the triskelion exhibits a limited degree of flexibility largely concentrated in three areas: the vertex, the elbow, and the terminal domain. The distance of the proximal arm from vertex to elbow is about 16–17 nm, which is approximately the distance between neighboring vertices in the assembled coated vesicle. All of these features ensure that, under the appropriate conditions, the required hexagons and pentagons of the coat form naturally from the assembly of triskelions. A combination of model building, proteolytic analysis, and high-resolution electron microscopy suggests how the triskelions fit together (Keen et al., 1979; Crowther and Pearse, 1981; Harrison and Kirchhausen, 1983; Vigers et al., 1986). Each vertex of the coat is composed of a triskelion vertex overlying three terminal domains, one from each of three separate triskelions. The terminal domains appear globular and point inwards. Each edge of the coat contains two antiparallel proximal domains

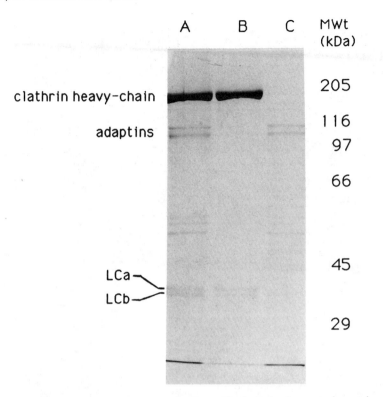

Figure 3. The protein components of: (A) purified pig brain coated vesicles, (B) purified triskelions, and (C), purified adaptors as detected by SDS-gel electrophoresis. (Preparation by A. C. Frank.)

and two antiparallel distal domains from separate triskelions. The continued accessibility of light-chains to proteolytic attack when assembled suggests that the proximal domains lie on the cytosolic face of the coat (Figure 5b and 5c).

The clathrin heavy-chain sequence has been determined for a variety of eukaryotic species (Kirchhausen et al., 1987a; Lemmon et al., 1991; O'Halloran and Anderson, 1992; Bazinet et al., 1993; Blackbourn and Jackson, 1996). It is unrelated to any other identified protein, but its properties suggest that it plays a predominantly structural role. The heavy-chain is highly conserved particularly along the proximal and distal arms. This should not be surprising in view of the extensive triskelion interactions noted above.

By contrast, the clathrin light-chains are far more divergent. Mammalian light-chains share only about 17% similarity with the yeast homolog (Silveira et al., 1990). Furthermore, vertebrate light-chains are polymorphic. Two isoforms designated LCa and LCb have been identified and sequenced (Jackson et al., 1987;

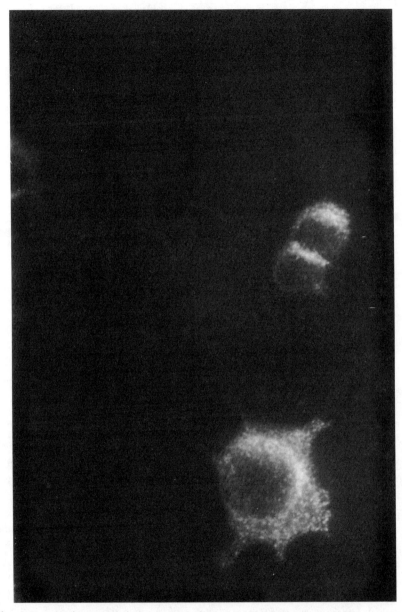

Figure 4. Distribution of clathrin within the rat pheochromocytoma cell line PC-12, detected by immunofluorescence microscopy using an anti-clathrin monoclonal antibody.

(a)

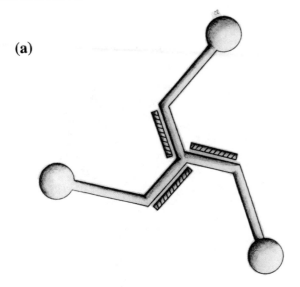

(b)

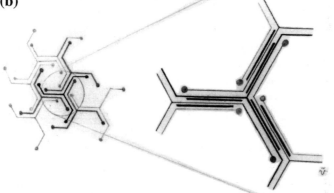

(*continued*)

Figure 5. (a) The clathrin triskelion. Each of the three legs is composed of a separate molecule of clathrin heavy-chain. The clathrin light-chain—represented by the stripped tube—lies along the heavy-chain proximal arm from the vertex to the elbow. The carboxyl-termini of the three heavy-chains meet at the triskelion vertex; the globular terminal domain lies at the corresponding amino-terminus. (b) Packing of triskelions into the assembled coat structure. (c) A coated vesicle showing the relative positions of two triskelions. The clathrin terminal domains are connected to the rest of the heavy-chain by a flexible hinge and point inwards in the assembled coat where they interact with the underlying adaptors. For clarity, the adaptors have not been illustrated.

(c)

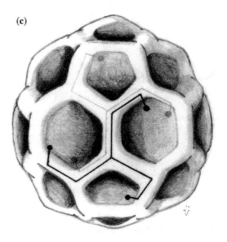

Figure 5. Continued.

Kirchhausen et al., 1987b; Jackson and Parham, 1988). Their sequences share some common features—most noticeably an alpha-helical region identified as the heavy-chain binding site, but they also contain isoform-specific regions (Brodsky et al., 1987). The proportion of LCa to LCb varies between tissues (Acton and Brodsky, 1990), although the polymorphism per se is not required for cell viability (Acton et al., 1993). Both light-chains bind to the same site on the heavy-chain and their distribution on triskelions is random (Winkler and Stanley, 1983; Kirchhausen et al., 1983). The light-chains bind the heavy-chain with their carboxy-termini towards the vertex (Brodsky et al., 1987). Two contrasting models have been proposed for the light-chain conformation when bound in triskelions. One assumes the light-chains bind in an extended manner via mutual alpha-helical regions and with the amino-terminal almost reaching the triskelion elbow (Kirchhausen et al., 1987b). More recently, Näthke et al. (1992) have proposed that the light-chains bind in a more compact U-shaped conformation in which both amino- and carboxy-termini point to the vertex (but see also Kirchhausen and Toyoda, 1993).

The function of the light-chains is not well understood, but there is some evidence that they play a regulatory role. For example, they bind calcium (Näthke et al., 1990) and calmodulin (Linden et al., 1981) and LCb can be phosphorylated at its amino-terminal (Hill et al., 1988). Interestingly, triskelions devoid of light-chains assemble more readily under physiological conditions (Ungewickell and Ungewickell, 1991) and light-chains confer calcium sensitivity on triskelion assembly (Lin et al., 1995). This indicates that light-chains may act as negative regulators of clathrin polymerization and prevent the premature or inappropriate assembly *in vivo*.

B. Adaptors

Underneath the outer clathrin coat lie the proteins of the adaptor complex. They can be separated into distinct classes, AP1 and AP2. Adaptors of each class are composed of two 100 kD molecules termed adaptins, together with smaller molecular weight components (Figure 6). The AP2 class is associated with plasma membrane-derived vesicles and the AP1 class is found in trans-Golgi network-derived coated vesicles (Pearse and Robinson, 1984). The receptor molecules differ between these two populations of coated vesicles, suggesting that the adaptors play an important role in the binding and selection of cargo. In addition, the adaptors interact with the clathrin lattice and stimulate its assembly *in vitro* (Keen et al., 1979). From plasma membrane derived coated vesicles, which presumably are engaged in receptor-mediated endocytosis, the AP2 adaptors are heterodimers of an α- and β-adaptin in association with a 50 kD medium chain and a 17 kD small chain. From Golgi-derived vesicles, the AP1 adaptors are composed of a γ- and β'-adaptin together with a 45 kD medium-chain and 19 kD small-chain (Ahle et al., 1988). The sequences of all major adaptins have been determined (Robinson, 1989; Kirchhausen et al., 1989; Robinson, 1990; Ponnambalam et al., 1990). In general they are highly conserved—in fact the amino acid sequence of β adaptin is identical between rat and human (Ponnambalam et al., 1990). If anything, the structural and packing constraints may be even more severe on the adaptins than on clathrin. There is a faint but significant homology between β adaptin and β COP, a component of the nonclathrin coated vesicles that mediate intra-Golgi transport—a strong hint that different transport organelles derive from a common ancestor and may still

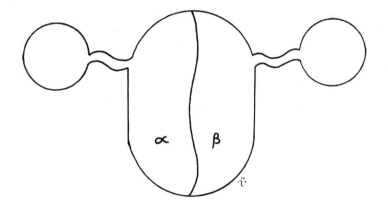

Figure 6. The plasma membrane adaptor complex adopts a symmetrical brick-like structure composed of α- and β-adaptins. The carboxy-terminal "ears" of each adaptin is connected to the main body by an extended flexible hinge. The 50- and 17-kD medium- and small-chain components bind within the main body, but their precise location is unknown and is not illustrated.

work in a fundamentally similar manner. With the exception of β and β', homology between each adaptin is weak suggesting that the molecules have become highly specialized since diverging from their shared ancestral protein, but some overall structural similarities have nevertheless been preserved. In particular, each adaptin contains a globular carboxy-terminal domain (representing about a third of the molecule) connected to the remaining amino-terminal region by a proline- and glycine-rich sequence of the type often found in extended flexible hinges. Electron micrographs of purified adaptors show a short brick-like structure with symmetrically protruding "ears" (Heuser and Keen, 1988). Controlled proteolysis confirms that these "ears" can easily be removed and correspond to the carboxy-terminal domains (Kirchhausen et al., 1989). However, the role of the "ears" is uncertain. They appear not to be required for adaptor binding to either clathrin or most receptors (Sorkin and Carpenter, 1993; Peeler et al., 1993), neither do they play a major role in targeting adaptors to membranes (Robinson, 1993). Yet the fact that the ears are connected to the adaptor body via a flexible hinge-like arm does suggest a binding domain of some kind. Recently, for example, the "ear" of α-adaptin has been shown to interact with the membrane protein eps15, protein previously identified as a substrate for the tyrosine kinase activity of the epidermal growth factor receptor (Benmerah et al., 1996). The clathrin-binding site is thought to reside predominantly within β-adaptin based on the ability of purified β-adaptin to interact with assembled triskelions (Ahle and Ungewickell, 1989). More recent experiments have narrowed the clathrin-binding site to a region of the β-adaptin hinge region (Shih et al., 1995). Similar experiments have indicated that the clathrin heavy-chain amino-terminal domain contains the adaptor-binding site. This is consistent with structural studies on coated vesicles showing the terminal domains pointing inward, although there may be additional interactions between adaptors and other regions of the triskelions (Keen et al., 1991).

The medium and small-chain components are located within the main brick-like body of the adaptor complex (Page and Robinson, 1995) and the sequences are known (Thurieau et al., 1988; Kirchhausen et al., 1991). The functions are still uncertain, but recent evidence implicates the medium chains as binding sites for internalized receptors containing tyrosine-based internalization signals (see below; Ohno et al., 1995).

C. Receptors

The fact that some receptors are selected with high efficiency to the exclusion of others implies an active sorting step during the assembly of the coated pit. The simplest way in which this can be envisioned is to have an "internalization signal" built into the receptor on its cytoplasmic domain. This is supported by studies of natural mutations in which the LDL receptor failed to be internalized. These mutations in homozygous form lead to familial hypercholesterolemia as the cholesterol-containing LDL particle cannot be removed from the bloodstream. Some of these mutations are caused by deletions or frameshifts of the entire cytoplasmic domain of the receptor. However, the most informative is the JD mutation in which a single tyrosine within the cytoplasmic domain is changed to cysteine (Goldstein et al., 1985). For a large number of receptors that enter coated pits, a tyrosine or other aromatic residue is necessary—even the influenza hemagglutinin, which does not normally enter coated pits, can be redirected there by introduction of a single tyrosine into its cytoplasmic tail (Lazarovits and Roth, 1988). However, the overall context is also critical. Further mutational analysis on LDL receptor, transferrin receptor, mannose-6-phosphate receptor, and immunoglobulin receptor has shown that the tyrosine must be surrounded by amino acids that favor the formation of tight turns in the peptide backbone (Vaux, 1992). This prediction has been supported by NMR studies on peptides containing the internalization sequence from the LDL receptor (Bansal and Gierasch, 1991).

Most receptors that have been analyzed in this manner enter coated pits whether or not they are loaded with ligand. Another class of receptors are not internalized until they bind their ligand; only then do they cluster into the pits with high efficiency. Growth factor receptors are good examples of this class and the epidermal growth factor (EGF) receptor in particular has been well studied.

The cytoplasmic tail of the EGF receptor contains a membrane-proximal tyrosine-kinase domain and a membrane distal regulatory domain. Binding of EGF induces receptor dimerization and tyrosine kinase activation, which in turn leads to both receptor autophosphorylation and the phosphorylation of other substrates (Sorkin and Waters, 1993). The regulatory domain contains at least three regions which act as internalization signals and which have sequence and structural similarities to the classic tyrosine-containing motifs (Chang et al., 1993). Somehow ligand binding must uncover or otherwise activate these signals. However, although the kinase activity is required for EGF receptor internalization, neither receptor dimerization nor the autophosphorylation is needed (Boll et al., 1995). Thus, the kinase activity may stimulate endocytosis by phosphorylating some additional target.

A different class of internalization signal has been identified by Letourner and Klausner (1992). This signal contains a dileucine motif near the C-terminus of the CD3 chains of the T cell receptor. Peptide binding studies indicate that the dileucine signals also bind adaptors, although it is not yet clear whether they recognize different adaptor subunits (Heilker et al., 1996).

The interaction between receptor and adaptors has been studied by a variety of means. For example, adaptors will interact with immobilized peptides containing the internalization motif (Pearse, 1988). More recently, Chang et al. (1993) have shown that soluble receptor tails will partially inhibit the binding of AP2 adaptors to stripped plasma membrane. These experiments also suggest that α-adaptin mediates the binding. In general, the binding revealed by these experiments is rather weak. This may not be a problem in the growing pit because as the triskelions assemble, they will effectively cross-link the underlying adaptors. The coated pit, relying as it does on multi-point attachments, will automatically build up high avidity interactions from relatively weak individual affinities.

The receptors which are internalized at the plasma membrane are also found in other intracellular compartments such as the endosome, Golgi apparatus, and trans-Golgi network. Why then does the AP2 adaptor only bind receptors when they are at the surface? The answer to this is not yet clear, but it seems likely that adaptors do not bind receptors directly. Rather it has been postulated that adaptors must first be "activated" at the appropriate membrane (Pearse and Robinson, 1990). This might be accomplished by a docking receptor which would first sequester the adaptor at the correct membrane, activate it, and deliver it to the receptor. Of course, this begs the question of how the pretargeting receptor is itself delivered to the correct location.

How selective are coated pits and vesicles? The exclusion of membrane proteins from coated pits has been simply ascribed to the physical "squeezing out" of receptors that lack an internalization signal (Bretscher et al., 1980). The receptors in a coated vesicle are indeed in close proximity. Nevertheless, some plasma membrane proteins that do not possess internalization signals, and even some proteins that lack cytoplasmic tails altogether can be detected in coated pits. It seems that these proteins enter coated pits inefficiently as part of the bulk-flow of membrane (Watts and Marsh, 1992). In some cases, even this background rate may be too high. For example, the mouse macrophage Fc receptor is efficiently inter-nalized while the lymphocyte isoform is excluded. This exclusion appears to be caused by the binding of the lymphocyte Fc receptor to the underlying cytoskeleton, which acts as a trap to retain the molecule at the cell surface (Miettinen et al., 1992). The original idea of the coated pit as a highly selective "molecular filter" is too simplistic. Rather we should envisage receptors interacting with the pits over a range of affinities which may themselves vary under different conditions.

D.   Coated Pit Assembly

In order to dissect coated pit formation in detail, an *in vitro* assay that will allow individual components to be characterized is essential. A number of such assays have been developed which differ in detail and, perhaps inevitably, in some of their conclusions (Smythe et al., 1989; Schmid, 1992, 1993; Anderson, 1993). Never-theless, a number of common themes have emerged from this work. It is generally agreed that adaptors and clathrin bind to a limited number of sites on the plasma membrane (Mahaffey et al., 1989). Moreover, triskelions and adaptors are found to be necessary, but not sufficient for the assembly of coated pits and their transfor-mation into coated vesicles; other cytosolic factors are also required. Interestingly, in some assays only triskelions extracted from cytosol will form coated pits—triskelions isolated from coated vesicles appear to be inactive (Smythe et al., 1992). The implication is that triskelions need to be primed before assembling at the plasma membrane. Schmid and colleagues have developed an elegant biochemical assay in which the endocytic cycle can be operationally divided into three phases: coated pit assembly, invagination, and budding, each step of which can be quantitatively analyzed (Carter et al., 1993). Under these conditions, ATP is required for both triskelion assembly and final budding. Multiple GTP-binding proteins are required, at least one of which appears to be a member of the heterotrimeric G-protein family. In this respect, the coated pit formation resembles the assembly of nonclathrin coated vesicle where small molecular weight GTPases act as "molecular switches" ensuring the correct sequential assembly of each component with each step being driven to completion by GTP hydrolysis (Goud and McCaffery, 1991). In particular, the assembly of the coatamer onto the donor membrane and its dissociation at the target membrane is regulated by a class of small molecular weight GTP/GDP-binding proteins called ADP-ribosylation factors (ARFs) of which there are many distinct isoforms, probably at least one for each subcellular compartment. At the donor membrane cytosolic GDP-bound ARF interacts with a membrane-bound guanine nucleotide exchange factor. This leads to the formation of GTP-bound ARF and its binding to the membrane which in turn drives the assembly of the coatamer. At the target membrane, a specific GTPase activating protein stimulates GTP hydrolysis by ARF. This leads to coat disassembly and the release of GDP-bound ARF (Donaldson and Klausner, 1994). The drug Brefeldin A inhibits the exchange factor and thus causes a profound disruption of coatamer-mediated membrane traffic. Significantly, Brefeldin A also causes the dissociation of AP1 adaptors from the *trans*-Golgi Network (Robinson and Kreis, 1992; Wong and Brodsky, 1992). This suggests that the assembly of AP1 adaptors is ARF-dependent and that the hypothetical docking protein which initially targets adaptors to their correct membrane is an ARF nucleotide exchange factor. Curiously, Brefeldin A does not inhibit the assembly of AP2 (Robinson and Kreis, 1992; Wong and Brodsky, 1992). However, this does not rule out an ARF-like molecule at the plasma membrane. For example, the non-hydrolyzable analogue GTPγS causes AP2 complexes to be mistargeted from the plasma membrane to an endosomal-like compartment (Seaman et al., 1993). This could be explained if the AP2 adaptors normally cycle between the plasma membrane and endosome but that GTPγS inhibits the GTPase activating protein.

Electron micrographs of plasma membranes attached to plastic consistently show flat sheets of hexagonal clathrin lattices (Heuser, 1980, 1989a). This may simply reflect nonproductive binding of coats to receptors tethered to the plastic substratum (frustrated endocytosis) as it is hard to see how lattice rearrangement can be accomplished from flat sheets. Furthermore, the triskelion is naturally non planar, so the incorporation of clathrin into the growing pit should automatically induce curvature (Kirchhausen and Harrison, 1981).

The final conversion of coated pits into free coated vesicles requires membrane fission. The molecular details are not completely understood but genetic analysis has identified a crucial role for another type of GTP-requiring protein at this step. In the temperature-sensitive *Drosophila* mutant *shibire*, the conversion of coated pits into coated vesicles is inhibited at the nonpermissive temperature (Kosaka and Ikeda, 1983a, 1983b). The defective gene is the *Drosophila* homologue of the GTPase dynamin (Van der Bliek and Meyerowitz, 1991; Chen et al., 1991). Over expression of human dynamin with mutations in the GTPase domain block the invaginations of coated pits (Van der Bliek et al., 1993) and incubation with GTPγS leads to the accumulation of highly invaginated coated pits connected to the plasma membrane by long thin collars containing rings of assembled dynamin (Hinshaw and Schmid, 1995; Takei et al., 1995). The hydrolysis of GTP may, therefore, normally drive a conformational change in assembled dynamin that leads to the final release of coated vesicles from the membrane.

E. Disassembly, Endosomal Fusion, and Sorting

How the final pinching off step is achieved is a mystery. It is possible that the increasing addition of triskelions alone will force opposing lipid bilayers to fuse. However, the fact that ATP hydrolysis is required might suggest a more active mechanism (Schmid, 1993).

Once released, coated vesicles have a half-life of only a few minutes. Uncoating is catalyzed by Hsc70, a constitutive member of the heat shock family of proteins (Ungewickell, 1985). The kinetics of uncoating are complex (Greene and Eisenberg, 1990). Earlier reports indicated a role for clathrin light-chains in uncoating as a substrate for Hsc70 (Schmid et al., 1984; DeLuca-Flaherty et al., 1990). More recently this has been challenged, since coats devoid of light-chains can still be uncoated by Hsc70 (Ungewickell et al., 1995). The uncoating enzyme shows remarkable specificity in that it fails to uncoat or even bind to nascent coated pits (Heuser and Steer, 1989).

When carried out *in vitro*, the uncoating enzyme only removes triskelions; adaptors remain attached to the vesicle. It is not known if this also occurs *in vivo*, but if it does then the adaptors could be acting as targeting molecules for the early endosomes. In general, it has been difficult to detect adaptors on the early endosome, although there is at least one such report (Guagliardi et al., 1990).

The internalization and trafficking of fluorescently-labeled transferrin has been followed in living cells by video microscopy and has revealed an endosomal system that is highly fluid and dynamic (Hopkins et al., 1990). Although the extent to which the early and late endosomes are discrete, stable organelles is still debated (Griffiths and Gruenberg, 1991; Murphy, 1991). A number of cell-free assays have been developed to study the fusion events within the early endosomal system (Gruenberg and Howell, 1989). Fusion of early endosomes and uncoated vesicles requires cytosolic components—ATP and GTP (Mayorga et al., 1989; Woodman and Warren, 1991). Of particular interest are the members of the rab family of small molecular weight GTPases. Their main role is probably to regulate vesicular transport between individual compartments, perhaps acting in a "proof-reading" capacity to ensure the correct vesicles are targeted to the correct location (Zerial and Stenmark, 1993). Over twenty such proteins have been described and distinct rab isoforms are localized to discrete intracellular compartments. Rab5 is localized to the plasma membrane, coated vesicles, and early endosomes. When over-expressed, rab5 causes an increase in the rate of receptor-mediated endocytosis and an enlargement of the early endosome (Bucci et al., 1992). The implication is that rab5 may control the rate limiting step in the fusion of endocytic vesicles and perhaps also an earlier step in the formation of coated vesicles. Fusion between individual endosomal elements occurs rapidly and serves to mix newly internalized ligand. Recent reconstitution assays have identified a limited number of proteins required for this purpose, one of which is annexin II, a protein previously implicated in other membrane fusion events (Emans et al., 1993).

The lumen of the early endosome is acidic due to the activity of a proton-translocating ATPase. Under these conditions, ligands such as LDL dissociate (Anderson et al., 1977). Iron is also released from transferrin although the protein component, apotransferrin, remains bound to the transferrin receptor (Hopkins and Trowbridge, 1983). Many receptors are recycled to the plasma membrane at this point (Figure 1). How this is achieved is still not clear, but may involve a separate recycling subcompartment of the endosome (Yamashiro et al., 1984). Rab4, a separate isoform, has been localized to the early endosome and its over-expression leads to a stimulation of transferrin recycling without stimulating endocytosis (van der Sluijs et al., 1992). Rab4 may therefore control the exit of recycling material from the endosome. It also implies that recycling to the plasma membrane is itself achieved by a distinct population of vesicular carriers.

Recycling appears to be the default pathway for endocytosed receptors and may be a bulk-flow process in which membrane is nonselectively returned to the cell surface (Dunn et al., 1989; Mayor et al., 1993). Some receptors, however, contain additional targeting information that prevents or diminishes recycling. An interesting example is the insulin-sensitive glucose transporter, GLUT4. In unstimulated adipocytes, the protein is located predominantly in specialized intracellular vesicles which only fuse with the plasma membrane following insulin activation. The cytoplasmic domain of GLUT4 contains a sequence clearly related to the tyrosine endocytic motif and will efficiently direct GLUT4 into coated pits. However, the sequence is sufficiently different also to act as a retention signal promoting intracellular sequestration rather than recycling (Piper et al., 1993).

Other receptors, such as those for growth factors, progress further along the endosomal pathway, and become concentrated within multivesicular bodies of the late endosomes before being delivered to lysosomes (Figure 1). This sorting step requires an active tyrosine kinase domain, since its deletion leads to the receptor recycling to the plasma membrane (Felder et al., 1990). One of the major substrates phosphorylated by the kinase is annexin I (Futter et al., 1993). These results offer the first hints as to how multivesicular bodies form and how sorting decisions are made within the endosomal system.

III. WIDER BIOLOGICAL RELEVANCE OF RECEPTOR-MEDIATED ENDOCYTOSIS

It would be surprising if such a fundamental mechanism as receptor-mediated endocytosis had not been adapted in a variety of ways during the course of evolution to solve different problems. One example is in signal transduction. Here the rapid internalization of growth factor receptor following stimulation serves as a mechanism to remove the receptor from the surface and attenuate the signal (Sorkin and Waters, 1993). Another example is afforded by the resident trans-Golgi network protein, TGN38. Despite its steady-state distribution, TGN38 contains within its cytoplasmic tail an efficient tyrosine-containing internalization signal for plasma membrane coated pits. The protein continually recycles from the trans-Golgi network to the cell surface, but is efficiently captured and reinternalized by coated pits. Additional context-dependent sequences in the cytoplasmic tail probably ensure it is re-routed to the trans-Golgi network from the endosome (Luzio and Banting, 1993). In this example, receptor-mediated endocytosis is, in effect, acting as a retrieval mechanism to maintain organelle integrity. The more general point is that receptor-mediated endocytosis cannot be divorced from other membrane traffic events.

Most of our detailed information on receptor-mediated endocytosis has come from relatively few species—typically mammalian cells in tissue culture. In the following sections we review our knowledge of the process in other systems.

A.   Receptor-mediated Endocytosis in Neurons

Neurons synthesize and package neurotransmitters in specialized organelles called synaptic vesicles. These small translucent vesicles have a distinctive protein composition and cluster underneath the plasma membrane of the synapse. They are part of the regulated secretory pathway and will release their content by exocytosis upon appropriate stimulation by secretogogue or nerve impulse. Similar organelles termed synaptic-like microvesicles (SLMV) occur in neuro-endocrine cells and have been studied as a more convenient model for synaptic vesicle biosynthesis (DeCamilli and Jahn, 1990). Intrinsic proteins of the SLMV, such as synaptophsin, are exported to the cell surface via the endoplasmic reticulum, Golgi apparatus, and the constitutive secretory pathway. At the plasma membrane, they are internalized by coated pits and sequestered into the early endosome. Synaptophysin typically recycles several times between the plasma membrane, coated pit, and endosome before being sequestered into SLMV (Reigner-Vigouroux et al., 1991). Targeting information to allow this probably resides within the synaptophysin molecule itself, although some proteins without targeting sequences may be delivered by binding other molecules that do (Bennett et al., 1992). Receptor-mediated endocytosis is therefore acting as a mechanism to collect proteins and concentrate them in the early endsome before further processing to the SLMV.

The broad details of this pathway are likely to be similar for true synaptic vesicles in neurons, but the unique anatomy of neurons means that they face a number of special problems in their membrane trafficking pathways. The neuron itself consists of a cell body with nucleus and intracellular organelles, an axon which can be less than a millimeter to several meters in length, and short branched dendrites ending in presynaptic nerve terminals. The site of synthesis of synaptic proteins is therefore physically separated from their site of use by considerable distances and synaptic vesicles must be efficiently recycled if the neuron is to remain functional (Figure 7). There is considerable evidence that receptor mediated endocytosis at the synapse is used to achieve this aim. First, coated pits are observed at the presynaptic membrane by quick freeze experiments immediately after stimulation (Heuser, 1989b). Second, coated vesicles from brain contain high levels of the major synaptic vesicle proteins (Maycox et al., 1992) and third, the major adult phenotype of the *shibire* mutation is a rapid and reversible temperature-induced paralysis. At the nonpermissive temperature, the neuromuscular junctions of these flies show a dramatic accumulation of coated pits at the presynaptic membrane and a concomitant decrease in the number of synaptic vesicles (Kosaka and Ikeda, 1983a). This is strong evidence that the coated pit is an *obligatory* intermediate in synaptic vesicle recycling.

The protein components of neuronal coated vesicles show a number of unique features that may reflect the specialized role in neurons. For example, the β-adaptin gene is alternatively spliced in neurons to generate an extra insert into the proline-rich hinge (Ponnambalam et al., 1990). Similarly, one α-adaptin isoform with an extended hinge sequence is also uniquely expressed in neurons (Robinson, 1989). These features occur in parts of the adaptins which are not directly involved in receptor binding, but they might impart a greater flexibility to the growing pit allowing faster assembly.

Several additional brain-specific coat components have been identified. One, AP-180 (Ahle and Ungewickell, 1986), is a monomeric clathrin-binding protein also referred to by independent work as AP3 (Keen and Black, 1986) and NP155 (Kohtz and Puszkin, 1988). AP-180 is a very potent promoter of clathrin cage assembly (Lindner and Ungewickell, 1992), although its sequence does not show any homology to adaptins (Morris et al., 1993). A second brain-specific clathrin-binding protein, termed auxilin, has also been described (Ahle and Ungewickell, 1990; Schröder et al., 1995). Auxilin binds to assembled clathrin coats and, in the presence of ATP, stimulates the binding of Hsc70. In this respect, auxilin acts in a similar manner to a class of chaperonins called DnaJ proteins which stimulate substrate binding and ATPase activity of Hsc70 in a variety of reactions. All DnaJ proteins contain a conserved motif called the J domain and this sequence is also present near the Carboxy-terminus of auxilin (Ungewickell et al., 1995).

In vertebrates, both isoforms of clathrin light-chains contain neuron-specific insertion sequences which arise by alternative mRNA splicing (Jackson et al., 1987;

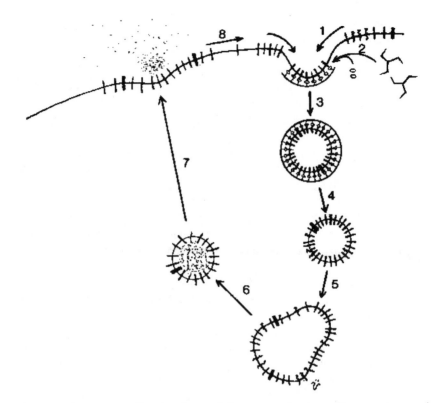

Figure 7. Receptor-mediated endocytosis in neuroendocrine cells serves to recycle synaptic vesicle proteins. Synaptic vesicle proteins at the presynaptic membrane, together with other receptors, are sequestered into clathrin-coated pits (1, 2). The coated vesicle forms (3), uncoats (4), and fuses with the endosome (5). The synaptic vesicle is reconstituted by the selective removal of its proteins from the endosome and is refilled with neurotransmitter (6). Following stimulation, the synaptic vesicle fuses with the presynaptic membrane and releases its contents (7). Synaptic vesicle proteins are once again incorporated into the presynaptic membrane and must again be recovered (8).

Kirchhausen et al., 1987b; Wong et al., 1990). These inserts appear to enhance the interaction of clathrin light-chains with calmodulin (Pley et al., 1995).

Taken together, it can be argued that these brain-specific modifications to the clathrin coated vesicle components could act together to increase the rate of synaptic vesicle recycling. Consistent with this view is the recent observation that neuronal synapses may not need to reform their synaptic vesicles through an endosomal intermediate but merely through a single budding step involving clathrin and dynamin (Takei et al., 1996).

In networks of neurons, the activity of individual synapses can be modified by repeated use. This is believed to be one of the earliest events underlying the formation and storage of learned behaviors. In the sea slug *Aplysia*, the defensive gill-withdrawal reflex can be enhanced by repeated stimulation. A number of biochemical events underlie these changes, including increased serotonin secretion. When the sensory neurons that control this reflex are exposed to serotonin, the internalization of cell-adhesion molecules from the presynaptic membrane is enhanced and the synthesis of clathrin light-chain is increased 10-fold (Bailey et al., 1992; Hu et al., 1993). This work suggests the exciting possibility that, by remodeling the surface of selected synapses, receptor-mediated endocytosis plays a crucial role in the early events leading to learning and memory formation.

B. Receptor-mediated Endocytosis in Plants

Until relatively recently, it was thought that endocytosis could not operate under the conditions of high turgor pressure that exist in plants. This was based on the assumption that the energy required to internalize solute and plasma membrane in a fully turgid cell would be prohibitive (Cram, 1980). However, there is a growing body of evidence that clathrin coated vesicles perform similar functions in plant cells to those described in other eukaryotes. At the ultrastructural level, clathrin coated pits and coated vesicles can be seen in association with the plant plasma membrane, Golgi apparatus, partially coated reticulum, and multivesicular bodies (Tanchak et al., 1984). The energetic requirements have since been recalculated to take into account the negligible impact that small coated vesicles have on solute uptake (Saxton and Breidenbach, 1988).

Plant coated vesicles are morphologically similar to their animal counterparts (Mersey et al., 1985). There are two populations: those forming at the plasma membrane, which are implicated in endocytosis, are typically 70–90 nm diameter, while those associated with the Golgi apparatus are smaller and direct the delivery of proteins to the vacuole (Robinson et al., 1989). The clathrin heavy-chain from soybean has a broadly similar sequence to other clathrins, although with some variation at the amino and extreme carboxy-termini (Blackbourn and Jackson, 1996).

The importance of clathrin light-chains for triskelion stability and uncoating mechanisms has encouraged the search for plant homologs. Unfortunately, the plant cell vacuole is a rich source of proteases capable of cleaving the coat associated proteins and there is evidence that coated vesicles have, on occasion, been charac-terized under conditions which have encouraged proteolysis of the light-chains (see Depta and Robinson, 1986; Robinson et al., 1991; Demmer et al., 1993). This factor and the attendant problems of isolating pure coated vesicles from plant tissues has led to some controversy as to the identity of putative light-chains. It now appears that two light-chain candidates of 57 kD and 60 kD isolated from carrot cells (Cole et al., 1987; Coleman et al., 1987) were the products of heavy-chain proteolysis

(Robinson and Depta, 1988). More recently, attempts have been made to identify light-chains associated with coated vesicles from developing cotyledons of pea. A total of four polypeptides of 31 kD, 40 kD, 46 kD, and 50 kD have been putatively identified as light-chains (Lin et al., 1992). These polypeptides were selected on the basis of mammalian light-chain characteristics, including the ability to bind to heavy-chains, sensitivity to elastase, and heat stability (Lin et al., 1992). Although contaminants such as ferritin may account for some of the bands (Hoh and Robinson, 1993), it seems probable that, as with mammalian cells, the plant light-chains are polymorphic. It remains to be seen whether there are noticeable differences between light-chains in different plant cells and tissues. It should also be born in mind that the coated vesicles isolated from peas will be predominantly involved in the transfer of storage proteins from the Golgi apparatus to the storage vacuoles (Harley and Beevers, 1989) and may not be involved in receptor mediated endocytosis.

If plant coated vesicles are highly purified under conditions that limit proteolysis, proteins in the 100 kD range resolve on SDS gels (Figure 8a) and one such band is recognized by a monoclonal antibody to β-adaptin (Figure 8b; Holstein et al., 1994). This result reinforces our impression that all the necessary molecular machinery for receptor-mediated endocytosis is present in plants.

The impermeable cellulosic cell wall has hindered direct analysis of either fluid phase or receptor-mediated endocytotic pathways. To overcome this problem, isolated protoplasts have been extensively studied using a variety of markers including colloidal gold conjugates (Villanueva et al., 1993), and membrane impermeant fluorescent probes (Hillmer et al., 1990; Robinson and Hedrich, 1991). Reservations have been expressed about the use of FITC-conjugates and Lucifer Yellow, as both free FITC and anionic Lucifer Yellow can enter plant cells and be sequestered into the vacuole by an anion carrier (O'Driscoll et al., 1991). In contrast, cationized ferritin (CF) has been used to great effect to demonstrate endocytotic membrane recovery (Tanchak et al., 1984; Fowke et al., 1991). As CF binds to the plasma membrane by electrostatic attraction, internalization is not receptor-mediated, but reflects nonspecific incorporation into the coated pit. Sections taken during the course of internalization reveal that the label appears successively in the partially coated reticulum, cisternae of the Golgi apparatus, the multivesicular body, and finally the vacuole (Fowke et al., 1991). The plant partially coated reticulum is analogous to the early endosomes of animals (Tanachak et al., 1988), and so early labeling of this structure is consistent with delivery to the first compartments of the endocytotic pathway. Labeling of the Golgi cisternae has also been reported for other protoplasts labeled with CF and lectin-gold conjugates (Joachim and Robinson, 1984; Hillmer et al., 1986). The multivesicular body is a similar structure in plants and animals and functions as a late endosome, processing endocytosed ligands (Tanchak and Fowke, 1987). Finally, the multivesicular body fuses with the vacuole, which, among other functions, serves as a plant lysosome (Nishimura and Beevers, 1979).

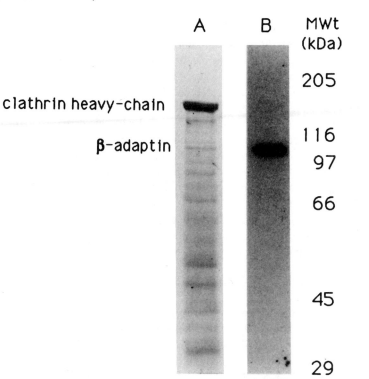

Figure 8. (a) Protein composition of soybean coated vesicles as determined by SDS-gel electrophoresis. (b) A β-adaptin homolog in soybean coated vesicles detected by western blotting using a monoclonal antibody to mammalian β-adaptin.

A consistent feature of endocytosis in plants is the apparent link between high rates of endocytic and exocytic vesicle trafficking (Steer, 1988). This is particularly clear in nongrowing, actively secreting cells, such as the maize root cap, and cells in suspension culture (Steer and O'Driscoll, 1991). In these examples, receptor-mediated endocytosis may serve to regulate the volume of the cytoplasm by the specific retrieval of membrane recently introduced into the plasma membrane (Mollenhauer et al., 1991). A number of extracellular macromolecules may also gain entry to the cell via a receptor. The case for internalization of oligosaccharide fungal elicitors is particularly compelling (Horn et al., 1989). Oligosaccharides are important plant signaling molecules and fungal elicitors induce a variety of plant defense responses (Ryan, 1987). The use of fluorescein- and ^{125}I-labeled elicitors revealed that this internalization was temperature- and energy-dependent, while unlabeled elicitors competitively inhibited uptake. The internalized elicitor finally accumulated in the vacuole (Horn et al., 1989). The polar vitamin, biotin, has also been shown to follow classical receptor-mediated conventions (Horn et al., 1990). Covalent attachment of biotin can facilitate internalization of a variety of exogenous macromolecules, including insulin, ribonuclease, and albumin. Excess free biotin inhibits this process and the exogenous proteins were entirely impermeable in the nonbiotinylated state (Horn et al., 1990). The characteristics of biotin and elicitor internalization indicate that different endocytotic routes exist (Horn et al., 1992; Low et al., 1993). While labeled fungal elicitors are delivered to the vacuole, biotin labeled macromolecules accumulate in the cytoplasm (Low et al., 1993). In neither case, however, is the nature of the receptor understood.

An alternative model system for studying endocytic recycling in plants is the growing pollen tube. These exhibit very high growth rates (up to 1 cm h^{-1}), and active recycling from the pollen tube tip (Steer, 1988). Antibodies to clathrin decorate the tip of growing pollen tubes suggesting a role for receptor-mediated, or at least clathrin-mediated endocytosis in remodeling the tip membrane (Blackbourn and Jackson, 1996).

C. Receptor-mediated Endocytosis in Single-celled Eukaryotes

Clathrin coated vesicles have been isolated from *Saccharomyces cerevisiae* (Mueller and Branton, 1984), and the yeast genes for the clathrin light-chains (Silveira et al., 1990) and heavy-chains (Lemmon et al., 1991) have been cloned and sequenced. Additionally, elements of the adaptor complexes have been sequenced and studied (Kirchhausen, 1990; Nakayama et al., 1991).

By exploiting the high rate of homologous recombination in yeast, the clathrin heavy-chain gene has been eliminated. Such cells grow slowly and are clearly abnormal, but the deletion is not usually lethal (Payne and Schekman, 1985; Payne et al., 1987). Yeast has only one clathrin light-chain and its deletion leads to a similar phenotype (Silveira et al., 1990). The severity of the heavy-chain deletion phenotype is affected by the genetic background. A number of suppressors of clathrin deficiency (*scd*) loci have been defined that modify the clathrin-deficiency phenotype (Lemmon and Jones, 1987). One such gene has been characterized as a polyubiquitin gene (Nelson and Lemmon, 1993). A plausible interpretation of these results is that the absence of clathrin leads to significant intracellular mistargeting, but this can be partially ameliorated by a more active proteolytic pathway.

Most of the phenotypic effects of clathrin deletion have been on intracellular pathways. Receptor-mediated endocytosis has been more difficult to study (Preston et al., 1987). One useful marker for receptor-mediated endocytosis is the α-mating factor. This pheromone is recognized by α-factor receptor which is linked to a heterotrimeric G-protein. It is known that the α-pheromone receptor is internalized in a ligand-induced, temperature, and energy dependent manner (Jenness and Spatrick, 1986; Chvatchko et al., 1986). Although the processes of endocytosis and signaling via the G-protein are not directly linked (Zanolari et al., 1992), endocytosis may regulate the number of receptors on the cell surface and thereby regulate the sensitivity of the cell response to the pheromone (Blumer and Thorner, 1991). Portions of the sequence from the cytoplasmic domain of the α-factor receptor have been identified as being essential for endocytosis to occur. The most important residue is a lysine in close proximity to a negatively charged amino acid. This is quite different from the classic tyrosine-turn sequence (Rohrer et al., 1993). Surprisingly, in strains deleted for clathrin heavy-chain, the rate of internalization of α-factor receptor was slowed by not more than 50% (Payne et al., 1988; Seeger and Payne, 1992). This suggests other nonclathrin pathways. Recently, for example, strains carrying mutations in actin genes have been shown to be defective in endocytosis (Kubler and Riezman, 1993).

The slime mold, *Dictyostelium discoideum*, is a species which is capable of both pinocytosis and phagocytosis. It shows a high degree of motility and undergoes a simple developmental program. Using antisense RNA, the effects of reducing the quantities of clathrin heavy-chain to an immunologically undetectable level has been examined (O'Halloran and Anderson, 1992b). Cells in which the expression of the clathrin heavy-chain have been disrupted are capable of growth, but at a much slower rate than in the wild-type. The antisense constructs also display certain physiological differences to the wild type that imply a defect in receptor-mediated endocytosis. In particular, they are unable to undertake pinocytosis (although phagocytosis is unimpaired), and osmoregulation is defective. In normal cells, starvation causes individual cells to secrete and respond to the chemo-attractant cAMP, as a result of which the cells aggregate to form a fruiting body. By contrast, the clathrin-deficient strains fail to respond to cAMP. This may be another example of receptor-mediated endocytosis regulating signal transduction (O'Halloran and Anderson, 1992b).

IV. CONCLUSIONS

Receptor-mediated endocytosis (like all biological mechanisms) displays a mixture of ancient fundamental features onto which have been layered increasingly complex innovations. Only by comparing the process in a range of diverse organisms can we begin to separate one from the other. Inevitably, most of our current understanding is based on systems for which well-characterized receptors and ligands exist. In this review, we have also attempted to high-light alternative cells and organisms in which the receptor–ligand complexes are less clearly defined, but in which receptor-mediated endocytosis is no less vital. In neuro-endocrine cells, for example, receptor-mediated endocytosis plays a critical role in synaptic vesicle biogenesis. Membrane recovery following secretion may also be an obligate function of endocytosis in higher plant cells. It is in this overall context that differences in coated vesicle components arouse great interest. For example, the clathrin light-chain appears to be a fundamental regulatory component, but cell-specific splicing (in neurons) and light-chain polymorphism (in animals and plants, but not yeast) seem to have been separately elaborated—presumably to solve particular problems that have arisen within multicellular organisms.

Many of the major protein components of coated vesicles are now well characterized. How several of them fit together is, at least in broad outline, also reasonably well understood. On the other hand, there are still many important gaps in our understanding of their functional significance. More attention will also surely be paid to the regulation of coated pit assembly. Receptor-mediated endocytosis is, above all, a dynamic process, but it is just these aspects that are so hard to study; how the coated pit invaginates and transforms into a coated vesicle is particularly difficult to imagine. However, the rapid development of cell free assays and the increasing application of targeted gene deletion should increasingly illuminate even these problems.

REFERENCES

Acton, S. L., & Brodsky, F. M. (1990). Predominance of clathrin light-chain LCb correlates with the presence of a regulated secretory pathway. J. Cell Biol. 111, 1419–1426.

Acton, S. L., Wong, D., Parham, P., Brodsky, F. M., & Jackson, A. P. (1993). Alteration of clathrin light-chain expression by transfection and gene disruption. Mol. Biol. Cell 4, 647–660.

Ahle, S., & Ungewickell, E. (1986). Purification and properties of a new clathrin assembly protein. EMBO J. 5, 3143–3149.

Ahle, S., & Ungewickell, E. (1989). Identification of a clathrin binding subunit in the HA2 adaptor protein complex. J. Biol. Chem. 264, 20089–20093.

Ahle, S., & Ungewickell, E. (1990). Auxilin, a newly identified clathrin-associated protein in coated vesicles from bovine brain. J. Cell Biol. 111, 19–29.

Ahle, S., Mann, A., Eichelsbacher, H., & Ungewickell, E. (1988). Structural relationships between clathrin assembly proteins from the Golgi and plasma membrane. EMBO J. 7, 919–929.

Anderson, R. G. W. (1993). Dissecting clathrin-coated pits. Trends Cell Biol. 3, 177–179.

Anderson, R. G. W., Brown, M. S., & Goldstein, J. L. (1977). Role of the coated endocytotic vesicle in the uptake of receptor-bound low density lipoprotein in human fibroblasts. Cell 10, 351–364.

Bailey, C. H., Chen, M., Keller, F., & Kendel, E. R. (1992). Serotonin-mediated endocytosis of apCAM: An early step of learning-related synaptic growth in Aplysia. Science 256, 645–649.

Bansal, A., & Gierasch, L. M. (1991). The NPXY internalization signal of the LDL receptor adopts a reverse-turn conformation. Cell 67, 1195–1201.

Bazinet, C., Katzen, A. L., Morgen, M., Mahowald, A. P., & Lemmon, S. K. (1993). The Drosophila clathrin heavy-chain gene—clathrin function is essential in a multicellular organism. Genetics 134, 1119–1134.

Benmerah, A., Bègue, B., Dautry-Varsat, A., & Cerf-Bensussan, N. (1996). The ear of α-adaptin interacts with the COOH-terminal domain of the Eps15 protein. J. Biol. Chem. 271, 12111–12116.

Bennett, M. K., Calakos, N., Kreiner, T., & Sheller, R. H. (1992). Synaptic vesicle membrane-proteins interact to form a multimeric complex. J. Cell Biol. 116, 761–775.

Blackbourn, H. D., & Jackson, A. P. (1996). Plant clathrin heavy chain: Sequence analysis and restricted localisation in growing pollen tubes. J. Cell Sci. 109, 777–787.

Blumer, K. J., & Thorner, J. (1991). Receptor-G protein signalling in yeast. Ann. Rev. Physiol. 53, 37–57.

Boll, W., Gallusser, A., & Kirchhausen, T. (1995). Role of the regulatory domain of the EGF-receptor cytoplasmic tail in selective binding of the clathrin-associated complex AP2. Curr. Biol. 5, 1168–1170.

Bretscher, M. S., Thompson, J. N., & Pearse, B. M. F. (1980). Coated pits as molecular filters. Proc. Natl. Acad. Sci. USA 77, 4156–4159.

Brodsky, F. M. (1988). Living with clathrin: Its role in intracellular membrane traffic. Science 242, 1396–1402.

Brodsky, F. M. (1992). Antigen processing and presentation: Close encounters in the endocytic pathway. Trends Cell Biol. 2, 109–115.

Brodsky, F. M., Galloway, C. J., Blank, G., Jackson, A. P., Seow, H. F., Drickamer, K., & Parham, P. (1987). Localization of clathrin light-chain sequences mediating heavy-chain binding and coated vesicle diversity. Nature 326, 203–205.

Bucci, C., Parton, R. G., Mather, L. H., Stannenburg, H., Simons, K., Hoflack, B., & Zerial, M. (1992). The small GTPase rab5 functions as a regulatory factor in the early endocytic pathway. Cell 70, 715–728.

Carter, L. L., Redelmeier, T. E., Woollenweber, L. A., & Schmid, S. L. (1993). Multiple GTP-binding proteins participate in clathrin-coated vesicle-mediated endocytosis. J. Cell Biol. 120, 37–45.

Chang, C. P., Kao, J. P. H., Lazar, C. S., Walsh, B. J., Wells, A., Wiley, H. S., Gill, G. N., & Rosenfeld, M. G. (1992). Ligand-induced internalization and increased cell calcium are mediated via distinct structural elements in the carboxyl terminus of the epidermal growth factor receptor. J. Biol. Chem. 266, 23467–23470.

Chang, C. P., Lazar, C. S., Walsh, B. J., Komuro, M., Collawn, J. F., Kuhn, L. A., Tainer, J. A., Trowbridge, I. S., Farquhar, M. G., Rosenfeld, M. G., Wiley, H. S., & Gill, G. N. (1993). Ligand-induced internalization of the epidermal growth factor receptor is mediated by multiple endocytic codes analogous to the tyrosine motif found in constitutively internalized receptors. J. Biol. Chem. 268, 19312–19320.

Chang, M. P., Mallet, W. G., Mostov, K. E., & Brodsky, F. M. (1993). Adaptor self-aggregation, adaptor-receptor recognition and binding of α-adaptin subunits to the plasma membrane contribute to recruitment of adapter (AP2) components of clathrin-coated pits. EMBO J. 12, 2169–2180.

Chen, M. S., Obar, R. A., Schraeder, C. C., Austin, T. W., Poodry, C. A., Wadsworth, S. C., & Vallee, R. B. (1991). Multiple forms of dynamin are encoded by *shibire*, a *Drosophila* gene involved in endocytosis. Nature 351, 583–586.

Chvatchko, Y., Howald, I., & Reizman, H. (1986). Two yeast mutants defective in endocytosis are defective in pheromone response. Cell 46, 355–364.

Cole, L., Coleman, J. O. D., Evans, D. E., Hawes, C. R., & Horsley, D. (1987). Antibodies to brain clathrin recognise plant coated vesicles. Plant Cell Rep. 6, 227–230.

Coleman, J., Evans, D., Hawes, C., Horsley, D., & Cole, L. (1987). Structure and molecular organization of higher plant coated vesicles. J. Cell. Sci. 88, 35–45.

Cram, W. J. (1980). Pinocytosis in plants. New Phytol. 84, 1–17.

Crowther, R. A., & Pearse, B. M. F. (1981). Assembly and packing of clathrin into coats. J. Cell Biol. 91, 790–797.

DeCamilli, P., & Jahn, R. (1990). Pathway to regulated endocytosis in neurones. Ann. Rev. Physiol. 52, 625–645.

DeLuca-Flaherty, C., McKay, D. B., Parham, P., & Hill, B. L. (1990). Uncoating protein Hsc70 binds a conformationally labile domain of clathrin light-chain LCa to stimulate ATP hydrolysis. Cell 62, 875–887.

Demmer, A., Holstein, S. E. H., Hinz, G., Schauermann, G., & Robinson, D. G. (1993). Improved coated vesicle isolation allows better characterization of clathrin polypeptides. J. Exp. Bot. 44, 23–33.

Depta, H., & Robinson, D. G. (1986). The isolation and enrichment of coated vesicles from suspension cultured carrot cells. Protoplasma 130, 162–170.

Donaldson, J. G., & Klausner, R. D. (1994). ARF: a key regulatory switch in membrane traffic and organelle structure. Cur. Op. in Cell Biol. 6, 527–532.

Dunn, K. W., McGraw, T. E., & Maxfield, F. R. (1989). Iterative fractionation of recycling receptors from lysosomally destined ligands in an early sorting lysosome. J. Cell Biol. 109, 3303–3314.

Emans, N., Gorvel, J. P., Walter, C., Gerke, V., Kellner, R., Griffiths, G., & Gruenberg, J. (1993). Annexin II is a major component of fusogenic endosomal vesicles. J. Cell Biol. 120, 1357–1369.

Felder, S., Miller, K., Moehren, G., Ullrich, A., Schlessinger, J., & Hopkins, C. R. (1990). Kinase-activity controls the sorting of the epidermal growth-factor receptor within the multivesicular body. Cell 61, 623–634.

Fowke, L. C., Tanchak, M. A., & Galway, M. E. (1991). Ultrastructural cytology of the endocytotic pathway in plants. In: *Endocytosis, Exocytosis and Vesicle Traffic in Plants* (Hawes, C. R., Coleman, J. O. D., & Evans, D. E., Eds.). Society for Experimental Biology, 45, Cambridge University Press, Cambridge, pp. 15–40.

Futter, C. E., Felder, J., Schlessinger, J., Ullrich, A., & Hopkins, C. R. (1993). Annexin I is phosphory-lated in the multivesicular body during the processing of the epidermal growth-factor receptor. J. Cell Biol. 120, 77–83.

Geuze, H. J., Slot, W., Strous, G. J. A. M., Lodish, H. F., & Schwartz, A. L. (1982). Immunocytochemical localization of the receptor for asialoglycoprotein in rat liver. J. Cell Biol. 92, 865–870.

Goldstein, J. L., Brown, M. S., Anderson, R. G. W., Russel, D. W., & Schneider, W. J. (1985). Receptor-mediated endocytosis: Concepts emerging from the LDL receptor system. Ann. Rev. Cell Biol. 1, 1–39.

Goud, B., & McCaffrey, M. (1991). Small GTP-binding proteins and their role in transport. Cur. Op. Cell Biol. 3, 626–633.

Goud, B., Huet, C., & Louvard, D. (1985). Assembled and unassembled pools of clathrin: A quantitative study using an enzyme immunoassay. J. Cell Biol. 100, 521–527.

Greene, L. E., & Eisenberg, E. (1990). Dissociation of clathrin from coated vesicles by uncoating ATPase. J. Biol. Chem. 265, 6682–6687.

Gruenberg, J., & Howell, K. E. (1989). Membrane traffic in endocytosis: Insights from cell-free assays. Ann. Rev. Cell Biol. 5, 453–482.

Griffiths, G., & Gruenberg, J. (1991). The argument for pre-existing early and late endosomes. Trends Cell Biol. 1, 59.

Guagliardi, L. E., Koppelman, B., Blum, J. S., Marks, M. S., Creswell, P., & Brodsky, F. M. (1990). Co-localization of molecules involved in antigen processing and presentation in an early endocytic compartment. Nature 343, 133–139.

Harley, S. M., & Beevers, L. (1989). Isolation and partial characterization of clathrin-coated vesicles from pea (*Pisum sativum L.*) cotyledons. Protoplasma 150, 103–109.

Harrison, S. C., & Kirchhausen, T. (1983). Clathrin, cages and coated vesicles. Cell 33, 650–652.

Heilker, R., Manning-Krieg, Zuber, J. F., & Spiess, M. (1996). *In vitro* binding of clathrin adaptors to sorting signals correlates with endocytosis and basolateral sorting. EMBO J. 15, 2893–2899.

Heuser, J. E. (1980). Three-dimensional visualization of coated vesicle formation in fibroblasts. J. Cell Biol. 84, 560–583.

Heuser, J. E. (1989a). Effects of cytoplasmic acidification on clathrin lattice morphology. J. Cell Biol. 108, 401–411.

Heuser, J. E. (1989b). The role of coated vesicles in the recycling of synaptic vesicle membrane. Cell Biol. Intl. Rep. 13, 1063–1076.

Heuser, J. E., & Keen, J. (1988). Deep-etch visualization of proteins involved in clathrin assembly. J. Cell Biol. 107, 877–886.

Heuser, J. E., & Steer, C. J. (1989). Trimeric binding of the 70KD uncoating ATPase to the vertices of clathrin triskelia: A candidate intermediate in the vesicle uncoating reaction. J. Cell Biol. 109, 1457–1466.

Hill, B. L., Drickamer, K., Brodsky, F. M., & Parham, P. (1988). Identification of the phosphorylation sites of clathrin light chain LCb. J. Biol. Chem. 263, 5499–5501.

Hillmer, S., Depta, H., & Robinson, D. G. (1986). Confirmation of endocytosis in higher plant protoplasts using lectin-gold conjugates. Eur. J. Cell Biol. 41, 142–149.

Hillmer, S., Hedrich, R., Robert-Nicoud, M., & Robinson, D. G. (1990). Uptake of lucifer yellow CH in leaves of *Commelina communis* is mediated by endocytosis. Protoplasma 158, 142–148.

Hinshaw, J. E., & Schmid, S. L. (1995). Dynamin self-assembles into rings suggesting a mechanism for coated vesicle budding. Nature 374, 190–192.

Hoh, B., & Robinson, D. G. (1993). The prominent 28kDa polypeptide in clathrin coated vesicle fractions from developing pea cotyledons is contaminating ferritin. Cell Biol. Intl. 17, 551–557.

Holstein, S. E. H., Drucker, M., & Robinson, D. G. (1994). Identification of a β-type adaptin in plant clathrin-coated vesicles. J. Cell Sci. 107, 945–953.

Hopkins, C. R., & Trowbridge, I. S. (1983). Internalization and processing of transferrin and transferrin receptor in human carcinoma A431 cells. J. Cell Biol. 97, 508–521.

Hopkins, C. R., Gibson, A., Shipman, M., & Miller, K. (1990). Movement of internalized ligand-receptor complexes along a continuous endosomal reticulum. Nature 346, 335–339.

Horn, M. A., Heinstein, P. F., & Low, P. S. (1989). Receptor-mediated endocytosis in plant cells. Plant Cell 1, 1003–1009.

Horn, M. A., Heinstein, P. F., & Low, P. S. (1990). Biotin-mediated delivery of exogenous macromolecules into soybean cells. Plant Physiol. 93, 1492–1496.

Horn, M. A., Heinstein, P. F., & Low, P. S. (1992). Characterization of parameters influencing receptor-mediated endocytosis in cultured soybean cells. Plant Physiol. 98, 673–679.

Hu, Y. H., Barzilai, A., Chen, M., Bailey, C. H., & Kendel, E. R. (1993). 5-HT and cAMP induce the formation of coated pits and vesicles and increase the expression of clathrin light-chain in sensory neurons of *Aplysia*. Neuron 10, 921–929.

Jackson, A. P., Seow, H-F., Holmes, N. J., Drickmer, K., & Parham, P. (1987). Clathrin light-chains contain brain-specific insertion sequences and a region of homology with intermediate filaments. Nature 326, 154–159.

Jackson, A. P., & Parham, P. (1988). Structure of human clathrin light-chains. J. Biol. Chem. 263, 16688–16695.

Jenness, D. D., & Spatrick, P. (1986). Down regulation of the α-factor pheromone in *S. cerevisiae*. Cell 46, 345–353.

Joachim, S., & Robinson, D. G. (1984). Endocytosis of cationic ferritin by bean leaf protoplasts. Eur. J. Cell Biol. 34, 212–216.

Kanaseki, T., & Kadota, K. (1969). The vesicle in a basket. J. Cell Biol. 42, 202–219.

Keen, J. H. (1990). Clathrin and associated assembly and disassembly proteins. Ann. Rev. Biochem. 59, 415–438.

Keen, J. H., & Black, M. M. (1986). The phosphorylation of coated membrane proteins in intact neurons. J. Cell. Biol. 102, 1325–1333.

Keen, J. H., Beck, K. A., Kirchhausen, T., & Jarrett, T. (1991). Clathrin domains involved in recognition by assembly protein AP2. J. Biol. Chem. 166, 7950–7956.

Keen, J. H., Willingham, M. C., & Pastan, I. H. (1979). Clathrin-coated vesicles: Isolation, dissociation and factor-dependent reassociation of clathrin baskets. Cell 16, 303–312.

Kirchhausen, T. (1990). Identification of a putative yeast homolog of the mammalian β-chains of the clathrin-associated protein complexes. Mol. Cell. Biol. 10, 6089–6090.

Kirchhausen, T. K., & Harrison, S. C. (1981). Protein organisation in clathrin trimers. Cell 23, 755–761.

Kirchhausen, T. K., & Toyoda, T. (1993). Immunoelectron microscopic evidence for the extended conformation of light-chains in clathrin trimers. J. Biol. Chem. 268, 10268–10273.

Kirchhausen, T., Davies, A. C., Frucht, S., O'Brine Greco, B., Payne, G. S., & Tubb, B. (1991). AP17 and AP19. The mammalian small chains of the clathrin-associated protein complexes show homology to Yap17p, their putative homolog in yeast. J. Biol. Chem. 266, 11153–11157.

Kirchhausen, T., Harrison, S. C., Parham, P., & Brodsky, F. M. (1983). Location and distribution of light-chains in clathrin trimers. Proc. Natl. Acad. Sci. USA 80, 2481–2485.

Kirchhausen, T., Harrison, S. C., & Heuser, J. (1986). Configuration of clathrin trimers: Evidence from electron microscopy. J. Ultrastruct. Mol. Struct. Res. 94, 199–208.

Kirchhausen, T. K., Harrison, S. C., Ping, E. P., Mattaliano, R. T., Ramachandran, K. L., Smart, J., & Brosius, J. (1987a). Clathrin heavy-chain: Molecular cloning and complete primary structure. Proc. Natl. Acad. Sci. USA 84, 8805–8809.

Kirchhausen, T., Nathanson, K. L., Matsui, W., Vaisberg, A., Chow, E. P., Burne, C., Keen, J. H., & Davies, A. E. (1989). Structural and functional division into two domains of the large (100–115kDa) chains of the clathrin-associated complex AP2. Proc. Natl. Acad. Sci. USA 86, 2612–2616.

Kirchhausen, T. K., Scarmento, P., Harrison, S. C., Monroe, J. J., Chow, E. P., Mattaliano, R. J., Ramachandra, K. L., Smart, J. E., Ahn, A. H., & Brosius, J. (1987b). Clathrin light-chains LCa and LCb are similar, polymorphic, and share repeated heptad motifs. Science 236, 320–324.

Kohtz, D. S., & Puszkin, S. (1988). A neuronal protein (NP185) associated with clathrin-coated vesicles. Characterization of NP185 with monoclonal antibodies. J. Biol. Chem. 263, 7418–7425.

Kosaka, T., & Ikeda, K. (1983a). Reversible blockage of membrane retrieval and endocytosis in the Garland cell of the temperature-sensitive mutant of Drosophila melanogaster, shibire. J. Cell Biol. 97, 499–507.

Kosaka, T., & Ikeda, K. (1983b). Possible temperature-dependent blockage of synaptic vesicle recycling induced by a single gene mutation in Drosophila. J. Neurobiol. 14, 207–225.

Kubler, E., & Riezman, H. (1993). Actin and fimbrin are required for the internalization step of endocytosis in yeast. EMBO J. 12, 2855–2862.

Lazarovits, J., & Roth, M. (1988). A single amino acid change in the cytoplasmic domain allows the influenza virus hemagglutinin to be endocytosed through coated pits. Cell 53, 743–752.

Lemmon, S. K., & Jones, E. W. (1987). Clathrin requirement for normal growth of yeast. Science 238, 504–509.

Lemmon, S. K., Pellicena-Palle, A., Conley, K., & Freund, C. L. (1991). Sequence of the clathrin heavy-chain from Saccharomyces cerevisiae and requirement of the COOH terminus for clathrin function. J. Cell Biol. 112, 65–80.

Letourneur, F., & Klausner, R. D. (1992). A novel di-leucine motif and a tyrosine-based motif independently mediate lysosomal targeting and endocytosis of CD3 chains. Cell 69, 1143–1157.

Lin, H.-B., Harley, S. M., Butler, J. M., & Beevers, L. (1992). Multiplicity of clathrin light-chain-like polypeptides from developing pea (Pisum sativum L.) cotyledons. J. Cell Sci. 103, 1127–1137.

Lin, S. H., Wong, M. L., Craik, C. S., & Brodsky, F. M. (1995). Regulation of clathrin assembly and trimerisation defined using recombinant triskelion hubs. Cell 83, 257–267.

Linden, C. D., Dedman, J. R., Chafouleas, J. G., Means, A. R., & Roth, T. F. (1981). Interactions of calmodulin with coated vesicles from brain. Proc. Natl. Acad. Sci. USA 78, 308–312.

Lindner, R., & Ungewickell, E. (1992). Clathrin-associated proteins of bovine brain coated vesicles. J. Biol. Chem. 267, 16567–16573.

Low, P. S., Legendre, L., Heinstein, P. F., & Horn, M. A. (1993). Comparison of elicitor and vitamin receptor-mediated endocytosis in cultured soybean cells. J. Exp. Bot. 44, 269–274.

Luzio, J. P., & Banting, G. (1993). Eukaryotic membrane traffic: Retrieval and retention mechanisms to achieve organelle residence. Trends Biochem. Sci. 18, 395–398.

Mahaffey, D. T., Moore, M. S., Brodsky, F. M., & Anderson, R. G. W. (1989). Coat proteins isolated from clathrin coated vesicles can assemble into coated pits. J. Cell Biol. 108, 1615–1624.

Maycox, P. R., Link, E., Reetz, A., Morris, S. A., & Jahn, R. (1992). Clathrin-coated vesicles in nervous tissue are involved primarily in synaptic vesicle recycling. J. Cell Biol. 118, 1379–1388.

Mayor, S., Presley, J. F., & Maxfield, F. R. (1993). Sorting of membrane-components from endosomes and subsequent recycling to the cell-surface occurs by a bulk flow process. J. Cell Biol. 121, 1257–1269.

Mayorga, L. S., Diaz, R., & Stahl, P. D. (1989). Regulatory role for GTP-binding proteins in endocytosis. Science 244, 1475–1477.

Mersey, B. G., Griffing, L. R., Rennie, P. J., & Fowke, L. C. (1985). The isolation of coated vesicles from protoplasts of soybean. Planta 163, 317–327.

Miettinen, H. M., Matter, K., Hunziker, W., Rose, J. K., & Mellman, I. (1992). Fc receptor endocytosis is controlled by a cytoplasmic domain determinant that actively prevents coated pit localization. J. Cell Biol. 116, 875–888.

Mollenhauer, H. H., Morre, D. J., & Griffing, L. R. (1991). Post Golgi apparatus structures and membrane removal in plants. Protoplasma 162, 55–60.

Morris, S. A., Schröder, S., Plessmann, U., Weber, K., & Ungewickell, E. (1993). Clathrin assembly protein AP180: Primary structure, domain organisation and identification of a clathrin binding site. EMBO J. 12, 667–675.

Mueller, S. C., & Branton, D. (1984). Identification of coated vesicles in *Saccharomyces cerevisiae*. J. Cell Biol. 98, 341–346.

Murphy, R. F. (1991). Maturation models for endosome and lysosome biogenesis. Trends Cell Biol. 1, 77–82.

Nakayama, Y., Goebl, M., O'BrineGreco, B., Lemmon, S., Pingchang Chow, E., & Kirchhausen, T. (1991). The medium chains of the mammalian clathrin-associated proteins have a homolog in yeast. Eur. J. Biochem. 202, 569–574.

Näthke, I., Hill, B. L., Parham, P., & Brodsky, F. M. (1990). The calcium-binding site of clathrin light chains. J. Biol. Chem. 265, 18621–18627.

Näthke, I. S., Heuser, J., Lupas, A., Stock, J., Turck, C. W., & Brodsky, F. M. (1992). Folding and trimerization of clathrin subunits at the triskelion hub. Cell 68, 899–910.

Nelson, K. K., & Lemmon, S. K. (1993). Suppressors of clathrin deficiency: Overexpression of ubiquitin rescues lethal strains of clathrin deficient Saccharomyces cerivisiae. Mol. Cell. Biol. 13, 521–532.

Nishimura, M., & Beevers, H. (1979). Hydrolysis of protein in vacuoles isolated from higher plant tissue. Nature 277, 412–413.

O'Driscoll, D., Wilson, G., & Steer, M. W. (1991). Lucifer Yellow and fluorescein isothiocyanate uptake by cells of *Morinda citrifolia* in suspension cultures is not confined to the endocytic pathway. J. Cell Sci. 100, 237–241.

O'Halloran, T. J., & Anderson, R. G. W. (1992a). Characterization of the clathrin heavy chain from Dictytostelium discoideum. DNA and Cell Biol. 11, 321–330.

O'Halloran, T. J., & Anderson, R. G. W. (1992b). Clathrin heavy chain is required for pinocytosis, the presence of large vacuoles and development in Dictytostelium. J. Cell Biol. 118, 1371–1377.

Ohno, H., Stewart, J., Fournier, M. C., Bosshart, H., Rhee, I., Miyatake, S., Saito, T., Gallusser, A., Kirchhausen, T., & Bonifacino, J. S. (1995). Interactions of tyrosine-based sorting signals with clathrin-associated proteins. Science 269, 1872–1877.

Page, L. J., & Robinson, M. S. (1995). Targeting signals and subunit interactions in coated vesicle adaptor complexes. J. Cell Biol. 131, 619–630.

Payne, G. S., & Schekman, R. (1985). A test of clathrin function in protein secretion and cell growth. Science 230, 1009–1014.

Payne, G. S., Hasson, T. B., Hasson, M. S., & Schekman, R. (1987). Genetic and biochemical characterization of clathrin-deficient *Saccharomyces cerevisiae*. Mol. Cell. Biol. 7, 3888–3898.

Payne, G. S., Baker, D., Van Tuinen, E., & Schekman, R. (1988). Protein transport to the vacuole and receptor mediated endocytosis by clathrin heavy chain-deficient yeast. J. Cell Biol. 106, 1453–1461.

Pearse, B. M. F. (1978). On the structural and functional components of coated vesicles. J. Mol. Biol. 126, 803–812.

Pearse, B. M. F. (1987). Clathrin and coated vesicles. EMBO J. 6, 2507–2512.

Pearse, B. M. F. (1988). Receptors compete for adaptors found in plasma membrane coated pits. EMBO J. 7, 3331–3336.

Pearse, B. M. F., & Robinson, M. S. (1984). Purification and properties of 100kDa proteins from coated vesicles and their reconstitution with clathrin. EMBO J. 3, 1951–1957.

Pearse, B. M. F., & Robinson, M. S. (1990). Clathrin, adaptors and sorting. Ann. Rev. Cell Biol. 6, 151–171.

Peeler, J. S., Donzell, W. C., & Anderson, R. G. W. (1993). The appendage domain of the AP2 subunit is not required for assembly or invagination of clathrin-coated pits. J. Cell. Biol. 120, 47–54.

Piper, R. C., Tai, C., Kulesza, P., Pang, S., Warnock, D., Baenziger, J., Slot, J. W., Geuze, H. J., Puri, C., & James, D. E. (1993). Glut-4 NH_2 terminus contains a phenylalanine-based targeting motif that regulates intracellular sequestration. J. Cell Biol. 121, 1221–1232.

Pley, U. M., Hill, B. L., Alibert, C., Brodsky, F. M., & Parham, P. (1995). The interaction of calmodulin with clathrin-coated vesicles, triskelions and light-chains. J. Biol. Chem. 270, 2395–2402.

Ponnambalam, S., Robinson, M. S., Jackson, A. P., Peiperl, L., & Parham, P. (1990). Conservation and diversity in families of coated vesicle adaptins. J. Biol. Chem. 265, 4814–4820.

Preston, R. A., Murphy, R. F., & Jones, E. W. (1987). Apparent endocytosis of fluorescein isothiocyanate conjugated dextran by Saccharomyces cerevisiae reflects uptake of low-molecular weight impurities, not dextran. J. Cell Biol. 105, 1981–1987.

Pryer, N. K., Wuestehube, L. J., & Schekman, R. (1992). Vesicle-mediated protein sorting. Ann. Rev. Biochem. 61, 471–516.

Regnier-Vigouroux, A., Tooze, S. A., & Huttner, W. B. (1991). Newly synthesized synaptophysin is transported to synaptic-like microvesicles via constitutive secretory vesicles and the plasma membrane. EMBO J. 10, 3589–3601.

Robinson, D. G., & Depta, H. (1988). Coated vesicles. Ann. Rev. Plant Physiol. Plant Mol. Biol. 39, 53–99.

Robinson, D. G., & Hedrich, R. (1991). Vacuolar lucifer yellow uptake in plants: Endocytosis or anion transport; a critical opinion. Botanica Acta 104, 257–264.

Robinson, D. G., Balusek, K., & Freundt, H. (1989). Legumin antibodies recognize polypeptides in coated vesicles isolated from developing pea cotyledons. Protoplasma 150, 79–82.

Robinson, D. G., Balusek, K., Depta, H., Hoh, B., & Holstein, S. E. H. (1991). Isolation and characterisation of plant coated vesicles. In: Endocytosis, Exocytosis and Vesicle Traffic in Plants (Hawes, C. R., Coleman, J. O. D., & Evans, D. E., Eds.). Society for Experimental Biology, 45, Cambridge University Press, Cambridge, pp. 65–79.

Robinson, M. S. (1989). Cloning of cDNAs encoding two related 100kD coated vesicle proteins (α-adaptins). J. Cell Biol. 108, 833–842.

Robinson, M. S. (1990). Cloning and expression of γ-adaptin. A component of clathrin-coated vesicles associated with the Golgi apparatus. J. Cell Biol. 111, 2319–2326.

Robinson, M. S. (1993). Assembly and targeting of adaptin chimeras in transfected cells. J. Cell Biol. 123, 67–77.

Robinson, M. S., & Kreis, T. E. (1992). Recruitment of coat proteins onto Golgi membranes in intact and permeabilized cells: Effects of Brefeldin A and G protein activators. Cell 69, 129–138.

Rohrer, J., Benedetti, H., Zanolari, B., & Riezman, H. (1993). Identification of a novel sequence mediating regulated endocytosis of the G-protein coupled α-pheremone receptor in yeast. Mol. Biol. Cell 4, 511–521.

Ryan, C. A. (1987). Oligosaccharide signalling in plants. Ann. Rev. Cell Biol. 3, 295–317.

Saxton, M. J., & Breidenbach, R. W. (1988). Receptor-mediated endocytosis in plants is energetically possible. Plant Physiol. 86, 993–995.

Schmid, S. L., Braell, W. A., Schlossman, D. M., & Rothman, J. E. (1984). A role for clathrin light-chains in the recognition of clathrin cages by uncoating ATPase. Nature 311, 228–231.

Schmid, S. L. (1992). The mechanism of receptor-mediated endocytosis: More questions that answers. Bioessays 14, 589–596.

Schmid, S. L. (1993). Coated-vesicle formation in vitro: Conflicting results using different assays. Trends Cell Biol. 3, 145–148.

Schröder, S., Morris, S., Knorr, R., Plessmann, U., Weber, K., Vinh, N. G., & Ungewickell, E. (1995). Primary structure of the neuronal clathrin-associated protein auxilin and its expression in bacteria. Eur. J. Biochem. 228, 297–304.

Seaman, M. N. J., Ball, C. J., & Robinson, M. S. (1993). Targeting of plasma membrane adaptors in vitro. J. Cell Biol. 123, 1093–1105.

Seeger, M., & Payne, G. S. (1992). Selective and immediate effects of clathrin heavy chain mutations on Golgi membrane-protein retention in *Saccharomyces cerevisiae*. J. Cell Biol. 118, 531–540.

Shih, W. G., Gallusser, A., & Kirchhausen, T. (1995). A clathrin binding site in the hinge of β2 chain of mammalian AP2 complexes. J. Biol. Chem. 270, 31083–31090.

Silveira, L. A., Wong, D. H., Masiarz, F. R., & Schekman, R. (1990). Yeast clathrin has a distinctive light chain that is important for cell growth. J. Cell Biol. 111, 1437–1449.

Smythe, E., Pypaert, M., Lucocq, J., & Warren, G. (1989). Formation of coated vesicles from coated pits in broken A431 cells. J. Cell Biol. 108, 843–853.

Smythe, E., Carter, L. L., & Schmid, S. (1992). Cytosol and clathrin-dependent stimulation of endocytosis *in vitro* by purified adaptors. J. Cell Biol. 119, 1163–1171.

Sorkin, A., & Waters, G. M. (1993). Endocytosis of growth factor receptors. Bioessays 15, 375–381.

Sorkin, A., & Carpenter, G. (1993). Interaction of activated EGF receptors with coated pit adaptins. Science 261, 612–615.

Steer, M. W. (1988). Plasma membrane turnover in plant cells. J. Exp. Bot. 39, 987–996.

Steer, M. W., & O'Driscoll, D. (1991). Vesicle dynamics and membrane turnover in plant cells. In: *Endocytosis, Exocytosis and Vesicle Traffic in Plants* (Hawes, C. R., Coleman, J. O. D., & Evans, D. E., Eds.). Society for Experimental Biology, 45, Cambridge University Press, Cambridge, pp. 129–142.

Takei, K., McPherson, P. S., Schmid, S. L., & De Camilli, P. (1995). Tubular membrane invaginations coated by dynamin rings are induced by GTPγS in nerve terminals. Nature 374, 186–190.

Takei, K., Mundigl, O., Daniell, L., & DeCamilli, P. (1996). The synaptic vesicle cycle: A single vesicle budding step involving clathrin and dynamin. J. Cell Biol. 133, 1237–1250.

Tanchak, M. A., Griffing, L. R., Mersey, B. G., & Fowke, L. C. (1984). Endocytosis of cationized ferritin by coated vesicles of soybean protoplasts. Planta 162, 481–486.

Tanchak, M. A., & Fowke, L. C. (1987). The morphology of multivesicular bodies in soybean protoplasts and their role in endocytosis. Protoplasma 138, 173–182.

Tanchak, M. A., Rennie, P. J., & Fowke, L. C. (1988). Ultrastructure of the partially coated reticulum and dictyosomes during endocytosis by soybean protoplasts. Planta 175, 433–441.

Terris, S., Hofman, A. C., & Steiner, D. F. (1979). Mode of uptake and degradation of ^{125}I-labelled insulin by isolated hepatocytes and H4 hepatoma cells. Can. J. Biochem. 57, 459–468.

Thurieau, C., Brodius, J., Burne, C., Jolles, P., Keen, J. H., Mattaliano, R. J., Chow, E. P., Ramachandran, K. L., & Kirchhausen, T. (1988). Molecular cloning and complete sequence of AP50, an assembly protein associated with clathrin-coated vesicles. DNA 7, 663–669.

Ungewickell, E. (1985). The 70kd mammalian heat shock proteins are structurally and functionally related to the uncoating protein that releases clathrin triskelia from coated vesicles. EMBO J. 4, 3385–3391.

Ungewickell, E., & Branton, D. (1981). Assembly units of clathrin coats. Nature 289, 420–422.

Ungewickell, E., & Ungewickell, H. (1991). Bovine brain clathrin light chains impede heavy chain assembly *in vitro*. J. Biol. Chem. 266, 12710–12714.

Ungewickell, E., Ungewickell, H., Holstein, S. E., Lindner, R., Prasad, K., Barouch, W., Martin, B., Greene, L. E., & Eisenberg, E. (1995). Role of auxilin in uncoating clathrin coated vesicles. Nature 378, 632–635.

van der Bliek, A. M., & Meyerowitz, E. M. (1991). Dynamin-like protein encoded by the *Drosophila shibire* gene, associated with vesicular traffic. Nature 351, 411–414.

van der Bliek, A. M., Redelmeier, T. E., Damke, H., Tisdale, E. J., Meyerowitz, E. M., & Schmid, S. L. (1993). Mutations in human dynamin block an intermediate stage in coated vesicle formation. J. Cell Biol. 122, 553–563.

van der Sluijs, P., Hull, M., Webster, P., Male, P., Gould, B., & Mellman, I. (1992). The small GTP-binding protein rab4 controls an early sorting event on the endocytotic pathway. Cell 70, 729–740.

Vaux, D. (1992). The structure of an endocytosis signal. Trends Cell Biol. 2, 189–192.

Vigers, G. P. A., Crowther, R. A., & Pearse, B. M. F. (1986). Three dimensional structure of clathrin cages in ice. EMBO J. 5, 529–534.

Villanueva, M. A., Taylor, J., Sui, X., & Griffing, L. R. (1993). Endocytosis in plant protoplasts: Visualization and quantitation of fluid-phase endocytosis using silver-enhanced bovine serum albumin-gold. J. Exp. Bot. 44, 275–281.

Watts, C., West, M. A., Reid, P. A., & Davidson, H. W. (1989). Processing of immunoglobulin-associated antigen in B lymphocytes. Cold Spring Harbor Symp. Quant. Biol. 54, 345–352.

Watts, C., & Marsh, M. (1992). Endocytosis: What goes in and how? J. Cell Sci. 103, 1–8.

Winkler, F. K., & Stanley, K. K. (1983). Clathrin heavy-chain light-chain interactions. EMBO J. 2, 1393–1400.

Wong, D. H., & Brodsky, F. M. (1992). 100 kD proteins of golgi and trans-Golgi network-associated coated vesicles have related but distinct membrane binding properties. J. Cell Biol. 117, 1171–1179.

Wong, D. H., Ignatius, M. J., Parosky, G., Parham, P., Trojanowski, J. Q., & Brodsky, F. M. (1990). Neuron-specific expression of high molecular weight clathrin light-chain. J. Neurosci. 10, 3025–3031.

Woodman, P. G., & Warren, G. (1991). Isolation of functional, coated, endocytic vesicles. J. Cell Biol. 112, 1133–1141.

Yamashiro, D. J., Tycko, B., Fluss, S. R., & Maxfield, F. R. (1984). Segregation of transferrin to a mildly acidic (pH 6.5) para-Golgi compartment in the recycling pathway. Cell 37, 789–800.

Zanolari, B., Raths, S., Singer-Kruger, B., & Riezman, H. (1992). Yeast pheromone receptor endocytosis and hyperphosphorylation are independent of G protein-mediated signal transduction. Cell 71, 755–763.

Zerial, M., & Stenmark, H. (1993). Rab GTPases in vesicular transport. Curr. Op. Cell Biol. 5, 613–620.

PHAGOCYTOSIS

Eric J. Brown and Thomas H. Steinberg

Biomembranes
Volume 4, pages 33–63.
Copyright © 1996 by JAI Press Inc.
All rights of reproduction in any form reserved.
ISBN: 1-55938-661-4.

I. INTRODUCTION

Phagocytosis is the process by which cells ingest other cells and cell fragments, bacteria, and particles >1 μm in diameter. This activity takes a variety of forms and achieves a number of goals. For example, free-living amoeba ingest bacteria and other organisms for nourishment and the roots of leguminous plants harbor nitrogen-fixing bacteria within compartments resembling phagocytic vacuoles and benefit from this nitrogen-fixing capacity. In mammalian cells, the most prominent roles for phagocytosis are degradative, either to remove tissue debris and dead cells (maintenance) or to destroy invading pathogens (host defense). The "professional phagocytes," the neutrophils, also known as the polymorphonuclear leukocytes (PMN) and the tissue macrophages (macrophage), are the best studied cells that fulfill these functions. However, many other cells are either highly phagocytic and have more specialized jobs (or have more fastidious taste), or display phagocytic activity under more limited circumstances. In the former instance, retinal pigment epithelial cells perform the critical and specialized role of eating packets of rod outer segment that are shed from photoreceptor cells, and defects in this degradative pathway result in retinal degeneration. In the latter, many endothelial and epithelial cells can be induced to ingest certain bacteria, and may unwittingly allow invading pathogens to breach mucosal and vascular barriers and to establish tissue infection.

Although the housekeeping roles of phagocytosis are arguably the most important for the well-being of the organism, most investigators have focused on phagocytosis by PMNs and macrophages as an essential component of host defense. These cells have different life histories which probably reflect their contributions to the inflammatory response. PMNs are circulating cells that emigrate from the vasculature and enter tissues in response to infection or other inflammatory stimuli. Here, they are essential for control of infection, but survive for only a few days. As illustrated by a variety of congenital and acquired diseases, the absence of PMNs is a major predisposition to infection for a wide range of bacteria. The macrophage is derived from circulating blood monocytes, and can survive in peripheral tissues for over a month. Fully differentiated tissue macrophages appear capable of only limited self replication. Macrophages are critical in host defense against organisms that evade destruction by PMNs in granuloma formation, in wound repair, and in normal tissue remodeling. Another important role for macrophages at sites of infection and inflammation is removal of the senescent and apoptotic PMNs which have migrated into the site. In each of these functions phagocytosis is important.

A major mechanism by which PMNs and macrophages recognize invasive microorganisms and devitalized tissue at sites of inflammation is through the process of opsonization. This involves interaction of host serum components, especially antibody and complement, with the abnormal or foreign tissue. In turn, these host components are recognized by specific receptors for immunoglobulin and for the complement component C3 on the phagocyte plasma membrane. These interactions trigger the membrane, cytoplasmic, and cytoskeletal events which result in internalization (phagocytosis) of the opsonized target. Finally, the act of ingestion and the receptors involved combine to trigger the intracellular events which lead to the destruction of the phagocytosed material. Thus, for modeling purposes, phagocytosis can be divided into three distinct components: (1) receptor-mediated interaction with the phagocytic target, most often through opsonin interaction with specific receptors on the phagocyte membrane; (2) the process of internalization of the phagocytic target into an intracellular compartment known as the phagosome; and (3) fusion of the phagosome with other intracellular compartments of the phagocyte and movement of the phagosome from the region near the plasma membrane toward the center of the cell.

Almost 20 years ago, Silverstein proposed a model for phagocytosis called "zippering" to explain the mechanism by which receptor–ligand interaction led to internalization. The "zipper hypothesis" stated that internalization of an opsonized particle requires sequential receptor–ligand interaction, with specific rearrangements of the cytoskeleton in the region of the newly ligated receptor as a consequence of receptor-mediated signal transduction. This model, which is still widely accepted, has two implications which have guided much subsequent research in the study of phagocytosis. First, it implies that signal transduction cascades activated during interaction of the phagocyte with the phagocytic target are a key component of the process of ingestion. Thus, it is important to understand which of the biochemical events that follow ligation of immunoglobulin or complement receptors are necessary for the phagocytic process and which are involved in other aspects of the inflammatory response. Second, the model suggests that signal transduction involved in phagocytosis is spatially, as well as temporally, confined since repeated receptor–ligand interactions are necessary for successful phagocytosis. This makes signal transduction for phagocytosis distinct from many of the signal transduction pathways commonly studied. Mitogenic signals or signals for induction of gene transcription must ultimately transmit information from the plasma membrane to the nucleus. Often the final outcome of receptor–ligand interaction does not occur until hours after the initiating event. In contrast, the cytoskeletal rearrangements that mediate particle ingestion require the movement of information only within a limited region around the ligated receptors, and often occur within a few minutes of contact between particle and cell. Therefore, there may be qualitative differences between signal transduction events involved in phagocytosis and those involved in information transfer to the nucleus.

II. PHAGOCYTIC RECEPTORS

Not all receptors on phagocytic cells are capable of ingestion. The molecular motifs which lead to phagocytic competence are not well understood, in part because there are likely to be several phagocyte-specific proteins involved in ingestion of particles, and in part because there may be multiple molecular mechanisms for phagocytosis. The best studied phagocytic receptors include receptors for the Fc domain of IgG (FcγR), for the third component of complement (CR), and for mannose/fucose carbohydrates (Table 1). A detailed description of the structure of each of these receptors is beyond the scope of this review. However, it is worth noting that FcγR are members of the immunoglobulin superfamily, and CR and mannose receptors are in other distinct protein families. There is little sequence homology among these

Table 1. Phagocytic Receptors

Receptor (CD)	Mr	Gene Family*	Other Names	Distribution	Ligand(s)
FcγRI (CD 64)	75 kDa	Ig		Monocytes, MΦ, IFN-γ-stimulated PMN	Monomeric IgG
FcγRII (CD32)	40 kDa	Ig		Leukocytes; Platelets	Immune complexes
FcγRIII (CD16)	50–70 kDA	Ig		PMN (PI-linked) MΦ, NK cells (transmembrane)	Immune complexes
CR1 (CD35)	160–220 kDa	RCA		Phagocytes, erythrocytes, B lymphocytes	C3b
CR3 (CD11b/CD18)	α 165 kDa β 95 kDa	Integrin	Mac-1; Mo-1 $\alpha_M\beta_2$	Phagocytes NK cells	iC3b, fibrinogen, others
CR4 (CD11c/CD18)	α 150 kDa β 95 kDa	Integrin	p150, 95 $\alpha_X\beta_2$	MΦ Dendritic Cells	Fibrinogen iC3b
Mannose Receptor	175 kDa	Lectin		MΦ	Mannose/fucose terminal residues

Note: *Ig = immunoglobulin superfamily; Integrin = integrin superfamily; RCA = regulators of complement activation; Lectin = carbohydrate binding family.

Schematic

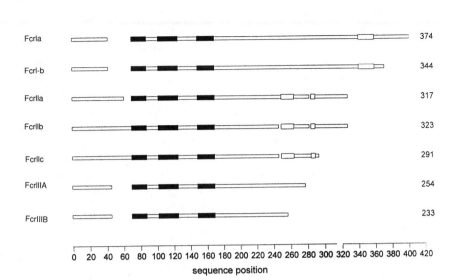

Figure 1. Alignment of human Fcγ receptors. Sequences of the seven human Fcγ receptors were compared using the multiple sequence alignment program, MACAW (Schuler et al., 1991). Regions of high homology among the seven receptors are in black boxes and are all within the extracellular domains of these receptors. Regions of homology within the cytoplasmic tails of the FcγRI family are in the light boxes; regions of homology within the cytoplasmic tails of the FcγRII family are designated by stippled boxes. For both families, these regions of homology begin in the membrane-proximal amino acids. The total size of the coding region for each gene is listed on the right.

receptors, and any structural motifs leading to their common phagocyte function are not obvious from inspection of the primary sequences. Indeed, even comparisons among the various FcγR reveals that the most divergent regions of the receptors are the intracytoplasmic domains, which would be expected to mediate interaction with signal transduction cascades and with cytoskeleton (Figure 1). Thus, the structure–function relationships which distinguish receptors involved in phagocytosis from those incompetent for ingestion are not yet understood.

A. IgG Fc Receptors

The phagocytic function of receptors for the Fc domain of IgG has been studied in more detail than for any other receptor. IgG Fc receptors are the products of three different gene families, which are called FcRI, FcRII, and FcRIII. The three gene families are homologous in their extracellular domains, particularly in the first approximately 100 amino acids (Figure 1). These receptors differ significantly in their cytoplasmic tails, presumably reflecting different effector functions in addition to ingestion. In man, expression of FcRI is limited to phagocytes, and FcRIII to phagocytes and NK cells, while FcRII is found on lymphocytes and platelets as well. FcRI is found predominantly on monocytes and macrophages, although it can also appear on IFN-γ stimulated PMN. FcRIII is found on PMNs in the circulation and appears to be expressed during the differentiation of monocytes to macrophages. All three receptor types have now been cloned (Ravetch et al., 1986; Kurosaki and Ravetch, 1989; Ravetch and Kinet, 1991). Cloning has revealed greater diversity than expected among IgG FcRs. For example, there are three different genes that encode FcRII, termed A, B, and C, which are apparently independently regulated. These three FcRII genes encode proteins that are 95% identical in their extracellular domains, but differ markedly in their intracytoplasmic tails. There are apparently mRNA splice variants in some FcRII genes, which also show tissue specific expression (Amigorena et al., 1992). Presumably, these different forms of FcRII, all of which are recognized by antibodies to the common extracellular domain, have different functions and activate different signal transduction pathways. While this has been studied to some extent, much more effort will be required before the functional consequences of this unexpected diversity are known. There are at least two forms of FcRIII, as well. PMNs express a phosphoinositolglycan-associated form of the receptor, FcγRIIIB, while macrophage and NK cells express a transmembrane form, FcγRIIIA. A single amino acid substitution in FcRIIIB (Ser^{203} for Phe^{203}), which is the product of a different gene from FcRIIIA, creates a recognition sequence for the phosphoinositolglycan transferase (Kurosaki and Ravetch, 1989). FcγRIIIA is coexpressed with an additional gene product, most often the γ chain of FcϵRI, but infrequently the ζ chain of the T cell antigen receptor complex. γ and ζ are homologous to each other, and are required both for efficient plasma membrane expression of FcγRIIIA and for signal transduction (Wirthmueller et al., 1992; Park et al., 1993).

    The roles of the various IgG Fc receptors in ligand binding have been thoroughly studied. FcRI has the highest affinity for monomeric IgG, with a $K_d \sim 10$ nM. There is subclass specificity of FcRI, which shows highest affinity for IgG1 and IgG3, and somewhat lower affinity for IgG2 or IgG4. The high affinity of FcRI for monomeric IgG implies that it will be saturated on circulating monocytes bathed in serum with an IgG concentration of >10 μM. This might suggest that there is some analogy in function between FcRI and the high-affinity IgE receptor of mast cells, which is also saturated with antibody at normal serum and extracellular fluid IgE levels, but as yet possible parallels have not been studied in detail. FcRII and FcRIII have at least 100-fold lower affinity for IgG monomer than FcRI. Both FcRII and FcRIII have a much higher affinity for IgG multimers than monomers, suggesting that these receptors preferentially recognize immune complexes. The increased affinity for complexes of IgG suggest that either (1) these receptors have more than a single Fc binding site or (2) the receptors tend to cluster in the membrane. These alternative hypotheses have not yet been distinguished.

Using monoclonal antibodies against different families of IgG Fc receptors, the functional roles of these receptors have been examined. Some of these studies have been difficult to interpret because intact blocking mAb have been used, which might lead to interaction of the Fab piece with one Fc receptor, and the Fc piece of the mAb with another class of FcR. With this caveat, it has been shown that on PMNs, FcRII can mediate the signal for degranulation and for activation of a respiratory burst, while FcRIII may be less competent at these functions. With respect to phagocytosis, FcRI, FcRII, and the transmembrane form of FcRIII all appear to be competent to mediate phagocytosis of appropriately opsonized targets. Interestingly, the PI-linked form of FcRIII appears to be unable to mediate ingestion (Buyon et al., 1990). These data have led to the hypothesis that on PMNs, which lack FcRI expression, FcRII is the primary signal transducing antibody and FcRIII may be present only to aid in recognition of immune complexes by PMN. However, recent data have shown that this hypothesis is too simple, since cross-linking of the PI-linked form of FcRIII can lead to increases in both $[Ca^{2+}]_i$ and polymerized actin in PMNs. These experiments make it likely that FcRIII has some as yet undefined role in signal transduction during activation of PMNs by immune complexes.

Recently, functions of various forms of FcRII have been studied. Neither human nor murine FcRIIA, when transfected into CHO cells, appear to mediate phagocytosis, endocytosis, or increases in $[Ca^{2+}]_i$, although there is some controversy about this (Joiner et al., 1990; Indik et al., 1991; Odin et al., 1991). On the other hand, when transfected into the murine macrophage cell line P388D1, the human FcRIIA is perfectly capable of mediating phagocytosis (Odin et al., 1991). The biology of two alternatively spliced forms of FcRIIb also has been studied. FcRIIb1 contains a 47-amino acid insertion in its cytoplasmic tail that is not present in FcRIIb2. The tissue distribution of this alternative splicing event is apparently quite specific: FcRIIb1 is found in lymphocytes; FcRIIb2 is found in macrophages (Ravetch et al., 1986). The 47-amino acid additional domain in FcRIIb1 mediates constitutive attachment of this receptor to the cytoskeleton. Interestingly, it also prevents the receptor from migrating to coated pits, and therefore prevents receptor-mediated internalization of immune complexes via the clathrin pathway. When the spliced domain is removed from its normal site within the cytoplasmic tail and moved to the carboxy-terminus of the tail, it retains its ability to mediate attachment to the actin cytoskeleton and to inhibit endocytosis. Perhaps this implies that a dynamic association of receptor with cytoskeleton is important for phagocytosis and that constitutive association is inhibitory. In lymphocytes, FcR ligation inhibits increases in $[Ca^{2+}]_i$, unlike the situation in macrophage (Choquet et al., 1993). This suggests that a dynamic association of the actin cytoskeleton with FcRs is important in Ca^{2+} signaling. This hypothesis is particularly appealing because FcR mediated increases in $[Ca^{2+}]_i$ are not associated with generation of IP_3, and may depend on the actin binding protein l-plastin (Rosales and Brown, 1992; Rosales et al., 1994).

The molecular dissection of structure/function relationships within the IgG Fc receptor family is in its infancy. We can expect to see much more work along the lines begun by these few studies in the near future.

B. Complement Receptors

The major complement opsonins are derived from the serum C3 component. C3b, the result of cleavage of C3 by either of two enzymes known as C3 convertases, binds covalently to complement-activating molecules and surfaces. It is the ligand for a receptor on PMNs and macrophages known as CR1 (CD35). CR1 consists of 16–24 repeats of a motif known as a short consensus repeat, which is a highly disulfide-bonded domain thought to confer a rod-like structure on CR1. Within CR1, these repeats are ~30% identical when compared to each other. The amino terminal-most two repeats and repeats 5–6 create two distinct binding sites for C3b.

The second family of complement receptors on phagocytes are the integrin receptors CR3 (CD11b/CD18) and CR4 (CD11c/CD18). CR3 binds the product of a further cleavage of C3b by complement factor I, known as iC3b. CR4 is thought to bind the same C3 fragment, but the data in favor of this are more controversial (Myones et al., 1988). CR3 is present on PMNs and monocytes; as monocytes migrate into tissue and become macrophages, CR4 expression increases. Differences in the function of CR3 and CR4 are not well studied, although a recent report suggests that CR4, unlike CR3, may be capable of activating the respiratory burst (Berton et al., 1992). These CD18 integrin receptors may have several ligands in addition to iC3b. There are reports that fibrinogen, clotting Factor X, *Bordetella* hemagglutinin, *E. coli* LPS, *Histoplasma capsulatum*, *Leishmania* lpg, gp65, and the endothelial adhesion molecule, ICAM-1 (CD54), all bind to these receptors (reviewed in Brown, 1991). There are two main concerns about the reports of other ligands for CR3. The first is that most of these studies were performed with intact macrophages or PMNs, leading to the possibility that the phagocytic cells will deposit C3b or iC3b onto the adhesion target. Certainly, macrophages make C3 and many other complement components. Recent studies have suggested that PMNs may also synthesize C3, and that mature circulating PMNs have endocytosed serum components during maturation and then secrete these components in response to activation (Borregaard et al., 1992; Botto et al., 1992). These data suggest that PMNs may also opsonize phagocytic targets with stored C3. The second concern about the apparently multiple ligands for these integrin receptors is that most studies have ignored the possibility that these integrins could be providing a signal necessary for firm adhesion, rather than being the ligand recognition receptor. Integrins are clearly able to enhance membrane–cytoskeleton interactions, and it is possible that, for low affinity receptors, cytoskeleton-mediated strengthening of a multipoint attachment is critical for stable interaction. Complement opsonization of sheep erythrocytes powerfully enhances the phagocytic function of IgG opsonin–receptor interaction, and certain anti-CD18 monoclonal antibodies can activate anti-inflammatory pathways in PMN (Gresham et al., 1991; Zhou et al., 1993). This alternative hypothesis has recently been discussed in detail (Brown, 1991).

In this context, studies examining the interaction of recombinant or purified CR3 or CR4 with putative ligands are essential for understanding the true spectrum of their binding abilities. These studies have recently been done to a limited extent with CR3, but none have yet been reported for CR4. It is absolutely clear that fibrinogen, ICAM-1, and iC3b are CR3 ligands, since both CR3 transfected into COS-1 cells and purified CR3 will bind these ligands (Diamond et al., 1990, 1991, 1993; Mosser et al., 1992; Van Strijp et al., 1993). However, the direct interaction of pathogens with CR3 is less certain; the few studies that have been performed with CR3 transfected into nonleukocytes suggest that adhesion without deliberate opsonization does not occur (Mosser et al., 1992). Recent studies with purified CR3 contradict this conclusion, and further study of purified and recombinant CR3 are necessary to determine the exact ligand specificities of this receptor. Studies with purified receptor have shown that, despite earlier studies on intact cells (Russell and Wright, 1988; Wright et al., 1989; Relman et al., 1990), CR3 does not bind the canonical integrin peptide ligand Arg–Gly–Asp (Van Strijp et al., 1993). This conclusion is reinforced by studies with mutant C3 in which the Arg–Gly–Asp sequence has been removed, but binding to CR3 has been retained (Taniguchi-Sidle and Isenman, 1992). The reason for the initial confusion seems to be that Arg–Gly–Asp binds to a distinct phagocyte integrin, called the Leukocyte Response Integrin (Gresham et al., 1989; Carreno et al., 1991; Gresham et al., 1992; Senior et al., 1992), which in turn activated CR3 to a high affinity state for its own ligands.

Both CR1 and the integrin complement receptors, CR3 and CR4, can exist in two different activation states. For the integrins, these two different activation states represent two strikingly different affinities for ligand. The increased affinity for ligand in the activated state is thought to reflect a conformational change in the receptors, which can be induced by Mn^{+2}, and a variety of leukocyte activators including fMetLeuPhe, C5a, platelet activating factor, and IL8. Recently, it has been proposed that CR3 conformation can be modulated by a membrane lipid (Herma-nowski-Vosatka et al., 1992), suggesting that cell activation may induce the high-affinity state of integrin receptors via effects on phospholipases. It is not clear whether this change in affinity with activation is necessary or sufficient to initiate ingestion. Certain neutrophil activators, such as fMetLeuPhe, will cause CR3 to achieve its high affinity ligand binding state without inducing phagocytosis. Bacteria and yeast opsonized with C3b or iC3b will be phagocytosed, even by unactivated phagocytes, but sheep erythrocytes similarly opsonized will not. This has created dissention about whether or not these receptors are truly phagocytic. The best interpretation of currently available data is that CR1, CR3, and CR4, unlike IgG FcRs, are not phagocytic in resting cells, but can be activated to mediate phagocytosis. Moreover, the presence of C3 opsonin on a phagocytic target mark-edly increases the phagocytic efficiency of even small amounts of IgG. No systematic study of signal transduction pathways activated by complement binding to complement receptors has yet been made.

C. Mannose Receptors

Certain phagocytic receptors can recognize invading microorganisms directly. The best studied example for this class of receptors is the mannose/fucose receptor. This is a 175 kD single-chain membrane protein (Stahl, 1990; Taylor et al., 1990) expressed exclusively on macrophage. The receptor, which has several lectin domains involved in ligand binding, was originally described as participating in receptor-mediated endocytosis of soluble ligands through clathrin-coated pits. It is now clear that the receptor also participates in phagocytosis of particles which express appropriate ligands. Unlike IgG Fc receptors, which do not ingest when transfected into nonphagocytic cells, mannose receptors apparently do. This may imply some differences in the molecular mechanisms regulating ingestion by Fc and mannose receptors, but this problem has not yet been investigated. Mannose receptors may be important in the non-opsonic recognition of potential pathogens, such as *Pneumocystis carinii* and *Mycobacterium tuberculosis*, which invade through the lung (Ezekowitz et al., 1991; Schlesinger, 1993). Macrophage mannose receptors may be important in HIV infection of macrophages (Lifson et al., 1986; Robinson et al., 1987). In the lung, there are many macrophages present in the alveoli, the site to which respiratory pathogens are often aspirated. These alveolar macrophages must recognize potential pathogens in the absence of serum opsonins, such as antibody and complement, often before a host immune response has initiated.

D. Cooperation between Phagocytic Receptors

An area which is probably of major importance in phagocytosis, but which has received little experimental attention, is the cooperation of membrane receptors in ingestion. This phenomenon was first described by Ehlenberger and Nussenzweig 15 years ago (Ehlenberger and Nussenzweig, 1977). They showed that addition of complement C3 to IgG-opsonized targets markedly enhanced their phagocytosis by both PMNs and macrophages. Complement opsonization alone did not lead to ingestion. Since those experiments, it has been shown that many pathogens will be ingested via complement opsonization alone, even though the standard phagocytic target, sheep erythrocytes, will not. This suggests there are other ligands on the microorganisms which interact with phagocytes, perhaps expressed at too low a level, or which interact with too low an affinity, to lead to phagocytosis on their own. This would be directly analagous to opsonization with low levels of IgG in the experiments of Ehlenberger and Nussenzweig, and suggests that cooperation between complement receptors and other membrane receptors for phagocytosis may be a general phenomenon. These results have two possible interpretations. The most straightforward explanation is that the additional complement ligands increase the efficiency of presentation of IgG to the phagocyte Fc receptors. The other possibility is that complement interaction with complement receptors modulates Fc receptor or cytoskeletal function in a way that makes the process of phagocytosis more efficient. The difference between the two hypotheses is significant, since the latter implies signal transduction through complement receptors, while the former does not. While the simpler explanation has been favored for many years, there has been little direct experimentation to distinguish between the two hypotheses. In fact, existing data suggest that the role for complement in these systems is more than simply to increase adhesion. First, for PMNs, phagocytosis of yeast opsonized with IgG or complement shows different dependence on intracytoplasmic Ca^{2+} ($[Ca^{2+}]_i$) (Lew et al., 1985). Second, data are accumulating that CR3 is involved in several FcR-dependent functions, including phagocytosis. This can be best demonstrated by showing that PMN genetically deficient in β_2 integrins, including CR3, are unable to achieve maximal rates of IgG-mediated phagocytosis (Gresham et al., 1991), even in the absence of exogenous complement. This is also true for IgG-dependent generation of LTB4 (Graham et al., 1993). This may be because IgG Fc receptors and CR3 can physically associate (Brown et al., 1988; Zhou et al., 1992, 1993). Which Fc receptor on phagocytes is responsible for interaction with CR3 is not known. Seghal et al. (1993) have suggested that CR3 associates specifically with the phosphoinositide-linked form of FcRIII, although their data do not rule out participation of FcRII. They also suggest that the interaction is mediated by direct interaction of the extracellular domains of these receptors in a carbohydrate-lectin interaction. We favor the hypothesis that FcRII and CR3 interact for the following reasons: (1) IgG receptor function is associated with CR3 on both monocytes and PMN (Graham et al., 1989) and the predominant Fc receptor expressed by both cell types is FcRII; (2) antibodies to FcRII, but not FcRIII, inhibit immune complex-stimulated LTB4 production (Graham et al., 1993), which is also dependent on FcR-CR3 cooperation; (3) we can demonstrate a direct physical association of CR3 with FcRII, but not FcRIII, on cells adherent to surfaces coated with monoclonal antibodies directed against CR3 (M.J. Zhou and E.J.B., unpublished); and (4) we can demonstrate direct cooperation between FcRII and CR3 for IgG-mediated phagocytosis in cells transfected with CR3 that express no FcRIII (I.L. Graham and E.J.B., unpublished). The summary of all the data published to date is that receptor–receptor cooperation in phagocytosis is an established phenomenon, but without an established molecular mechanism.

III. MODELS AND POTENTIAL MODELS FOR PHAGOCYTOSIS

A. Transfection of Phagocytic Receptors

A standard method for studying the function of a receptor is to transfect it into cells in which it is normally not expressed, and to examine the effect of specific mutations on the resultant phenotype. Transfection of wild type mannose receptor leads to yeast phagocytosis in Cos-1 cells (Ezekowitz et al., 1990), but no detailed structure–function analysis of the receptor has been made. Studies with several IgG Fc receptors have led to contradictory results (Joiner et al., 1990; Indik et al., 1991; Odin et al., 1991). Some groups have reported that Chinese Hamster Ovary (CHO) cells or murine fibroblast 3T3 cells transfected with FcRII do not ingest IgG-coated erythrocytes, while another group reported that Cos cells transfected with either FcRI or FcRII were competent for ingestion. A third group reported that CHO cells transfected with FcRII could not ingest IgG-coated erythrocytes, but could ingest IgG-coated *Toxoplasma gondii*. Superficially, there is little difference between the experiments with these contradictory results. The explanation of the differences is unclear. However, the fact that Fc receptors are competent for ingestion in at least some cases when expressed in fibroblasts suggests that there are no leukocyte-specific proteins other than the receptors which are required for phagocytosis. This is also suggested by reports of phagocytosis of microorganisms by epithelial cells and fibroblasts (Isberg and Leong, 1990; Kuroda et al., 1993). However, in all cases, phagocytosis by non-hemopoetic transfectants is much less efficient than macrophage- or PMN-phagocytosis. Thus, all cells can ingest particulate material; the unique features of "professional phagocytes" are their ability to recognize IgG and complement opsonins and mannose ligands, and their more rapid ingestion of appropriately opsonized particles. There may be leukocyte proteins which enable this more efficient ingestion, but this subject has not been studied.

B.   Unicellular Organisms

Phagocytosis as a host defense function clearly derives from the mechanisms used by *Dictyostelium*, amoebae and other unicellular organisms to obtain nutrition. In particular, these organisms are able to ingest bacteria in a manner which requires cytoskeletal reorganization and which can be inhibited by cytochalasin. This suggests that there are fundamental aspects of particulate ingestion which are conserved from these primitive free-living cells through the PMNs and macrophages of higher eukaryotes. Since *Dictyostelium* and amoebae may be experimentally more tractable than higher organisms, and are certainly more easily manipulated genetically, these organisms make appealing experimental models for an exploration of the fundamental mechanisms of ingestion. Reasonably, these studies have focused on cytoskeletal elements, which, as opposed to specific ligand receptors involved in ingestion, are well conserved through evolution. Studies of the role of non-muscle myosins in phagocytosis have been done almost exclusively in *Dictyostelium* and are reviewed in the section on cytoskeleton below.

IV.   PHAGOCYTOSIS AND SIGNAL TRANSDUCTION

Perhaps the most important recent advances in understanding phagocytosis have been in the area of signal transduction. Because the signals that mediate phagocytosis appear to operate locally, there are likely to be fundamental differences between signal transduction that generates the phagocytic response and the better understood pathways of membrane–nucleus signal transduction. Nonetheless, signal transduction pathways such as G-protein mediated activation of adenylate cyclases, phospholipases, or protein kinases, or tyrosine kinase activation of growth factor-dependent pathways, affect, and may be pivotal to, phagocytic signaling.

The first evidence that it might be fruitful to think about phagocytosis as a problem in localized signal transduction arose from the classic studies of Silverstein in the 1970s. His group demonstrated that phagocytosis required repeated receptor–ligand interactions (Griffin et al., 1975, 1976). This demonstrated that phagocytosis was fundamentally different from fluid-phase endocytosis, in which a single receptor–ligand interaction is sufficient to lead to internalization of ligand and movement through the endocytic pathway. In contrast to FcRs, complement receptors on resting peritoneal macrophage could bind opsonized particles, but could not activate the same pathway of ingestion (Bianco et al., 1975). These results implied specific signal transduction from IgG Fc receptors. This hypothesis of signal transduction in phagocytosis was reinforced by the discovery that complement receptors could, under certain circumstances, themselves become phagocytic. This demonstrated cellular control over the consequences of recognition of complement ligands. Finally, these studies demonstrated the local nature of signal transduction involved in phagocytosis. When macrophage bound two distinct particles coated with IgG and complement, only the IgG-coated particles were ingested (Griffin and Silverstein, 1974). The local nature of signals generated during phagocytosis has been repeatedly confirmed since those studies (Pryzwansky et al., 1981; Sawyer et al., 1985; Kim et al., 1992).

A. Protein Kinases

In recent years, these studies have been extended to examine particular signal transduction pathways. A pertussis toxin-sensitive GTP-binding protein has been implicated in macrophage phagocytosis (Brown et al., 1987), although PMN ingestion is insensitive to pertussis toxin (Rosales and Brown, 1991). IgG-mediated ingestion by human monocytes is blocked by protein kinase C inhibitors, although murine peritoneal macrophage phagocytosis is not, but is inhibited instead by blockade of tyrosine kinase activity (Zheleznyak and Brown, 1992; Greenberg et al., 1993). Interestingly, for monocytes, protein kinase C seems to be translocated to phagosomes and for murine macrophages, tyrosine kinases are concentrated in the phagosomes. Recently, one of the tyrosine-phosphorylated proteins which accumulate in phagosomes has been identified as paxillin, a component of the membrane cytoskeleton in many cells. These data suggest the possibility that different protein kinases are involved in regulation of ingestion in different cell types. The molecular consequences of these differences for the mechanism of phagocytosis, if any, are unknown.

B. Cytosolic Calcium

The role of increases in cytosolic free calcium ($[Ca^{2+}]_i$) in phagocytosis has been controversial. Originally, Lew et al. (1985) proposed that IgG Fc receptor-mediated phagocytosis by PMNs required an increase in $[Ca^{2+}]_i$, but complement-mediated ingestion did not. These experiments used quin-2 to chelate intracellular Ca^{2+} and studied the ingestion of opsonized yeast. As discussed above, because of the possibility that yeast interact with other receptors on phagocytes which may influence whether or not ingestion occurs, these experiments are difficult to interpret. Experiments from Silverstein's laboratory showed that IgG-mediated ingestion could occur in murine macrophages at very low $[Ca^{2+}]_i$ and without any rise in $[Ca^{2+}]_i$ during the phagocytic process (Di Virgilio et al., 1988). Our work on PMN phagocytosis has partially resolved these conflicting data. We showed that when PMNs ingest at optimal phagocytic rates, there are both Ca^{2+}-dependent and Ca^{2+}-independent pathways for ingestion. PMNs stimulated with the chemotactic peptide, fMet–Leu–Phe, require an increase in $[Ca^{2+}]_i$ for optimal ingestion; PMN stimulated with phorbol esters or with platelet activating factor do not (Rosales and Brown, 1991). Importantly, the Ca^{2+} required for the fMet–Leu–Phe-induced ingestion must arise from IgG Fc receptor ligation. This can be shown in a variety of ways including temporal separation of the fMet–Leu–Phe activation and the phagocytic event, and by using specific inhibitors of IgG-induced increase in $[Ca^{2+}]_i$ (Rosales et al., 1994). This requirement for Ca^{2+} release mediated by a particular receptor at the time of ingestion emphasizes the very local nature of signal transduction during phagocytosis. Monocyte phagocytosis is not stimulated by fMet–Leu–Phe (E.B., unpublished data). Thus, monocytes apparently lack some aspect of the Ca^{2+}-dependent pathway for ingestion. They do not lack fMet–Leu–Phe receptors, nor do they lack the ability to increase $[Ca^{2+}]_i$ upon Fc receptor stimulation. Thus, the proximal components of the signal transduction pathway are present; the defect in effector mechanism is not known. However, the apparent absence of this pathway could explain the discrepancy between PMN and macrophage experiments with respect to the Ca^{2+} dependence of ingestion. The increase in $[Ca^{2+}]_i$ which arises during ingestion may have another important role, regardless of whether it is involved in the phagocytic process. Phagosome–lysosome fusion, which is a necessary step in the destruction of ingested pathogens, may require an increase in $[Ca^{2+}]_i$ (Jaconi et al., 1990).

Various studies have examined the molecular mechanism of the release of Ca^{2+} in response to IgG Fc receptor ligation. Several groups have demonstrated that $[Ca^{2+}]_i$ increases in response to ligation of either FcRII or FcRIII (Kimberly et al., 1990; Rosales and Brown, 1991; Odin et al., 1991). This rise in $[Ca^{2+}]_i$ is independent of generation of IP_3, although the Ca^{+2} is apparently released from the IP_3-sensitive intracellular pool (Rosales and Brown, 1992). Little is known about the mechanism of release, although our own recent data implicate the actin cytoskeleton as directly involved in the release, since inhibition of actin microfilament assembly or inhibition of actin association with a bundling protein known as l-plastin (Lin et al., 1988) prevent Ca^{2+} release in response to Fc receptor ligation, but have no effect on IP_3-dependent mechanisms of Ca^{2+} release from intracellular stores (Rosales et al., 1994). The implications of this association of cytoskeletal alterations with signal transduction during phagocytosis are discussed below.

C. Phospholipases

Recent studies have addressed the possibility that activation of phospholipases C and D (PLC and PLD) are important for ingestion. This work has been done using PMN ingestion of opsonized yeast (Fallman et al., 1989, 1992) and is subject to the criticisms detailed above, specifically that yeast may interact with unidentified PMN receptors and affect interpretation of results. However, these studies have suggested that both PLC and PLD are activated during phagocytosis. The kinetics of activation of both are slow, but not inconsistent with the kinetics of ingestion, since phagocytosis is itself a process which requires 10–30 minutes to complete. While the kinetics and extent of PLD activation especially correlate with ingestion, there have been to date no direct tests of whether PLD activation is necessary for phagocytosis or merely a consequence of the multiple membrane perturbations which necessarily accompany the ingestion of large particles.

D. Cytokines

In some situations, signal transduction for phagocytosis may arise, in part, indirectly from the interaction of phagocytic receptor with its ligand. This appears to be the case for TNF-stimulated phagocytosis, which requires the respiratory burst (Gresham et al., 1988, 1990). TNF-α primes PMNs for increased respiratory burst and phagocytosis in response to IgG Fc receptors. Inhibition of generation of reactive oxygen intermediates with multiple pharmacologic inhibitors, and study of cells from patients with chronic granulomatous disease, demonstrate that TNF-activated phagocytosis requires respiratory burst activity. This pathway appears to require hydroxyl-radical generated through the Haber–Weiss reaction and the effect on phagocytosis may be mediated by direct phospholipid hydrolysis by hydroxyl-radical (Smiley et al., 1991). More recent studies have shown that TNF-α may also prime for the activation of other pathways involved in stimulation of IgG-mediated ingestion (Della Bianca et al., 1993).

V. PHAGOCYTOSIS AND CYTOSKELETON

In the classic model of phagocytosis, envelopment of the opsonized particle requires remodeling of the actin cytoskeleton. This has been demonstrated by inhibition with cytochalasins, and by immunofluorescent- and electron-microscopic observation of accumulation of actin in the phagocytic cup. Presumably, this cytoskeleton organization involves interaction not only of soluble cytoskeletal proteins with each other, but of the assembling skeleton with the plasma membrane as well. Unlike the erythrocyte membrane, for which detailed understanding of the interactions now exists, little is known about the proteins involved in membrane–cytoskeleton interaction in phagocytes. In particular, which plasma membrane proteins mediate membrane–cytoskeleton interaction is not known for phagocytes. Two models from other cells may be instructive. First is the example of integrin-mediated adhesion in fibroblasts. In these cells, transmembrane receptors of the integrin superfamily recognize extracellular matrix proteins and mediate a transmembrane interaction with the actin cytoskeleton. The actual sites of interaction between integrins and the actin microfilaments may be quite complex and are known to involve several interacting proteins including talin, vinculin, and α-actinin in a formation termed a focal contact (Burridge and Fath, 1989; Burridge et al., 1990). Importantly, focal contacts also represent the sites of concentration of many enzymes involved in signal transduction including tyrosine kinases, serine/threonine kinases, and phosphatases (Burridge and Fath, 1989). Teleologically, this makes sense, since the cytoskeletal assembly at adhesion points is a platform on which relevant receptors, activatable enzymes, and their potential substrates can be brought into close proximity. It is likely, but not certain, that equivalent structures exist in leukocytes. Because leukocytes do not contain the actin microfilament bundles termed stress fibers and because focal contacts were initially defined as the sites at which stress fibers anchored in the plasma membrane, phagocytes do not exhibit focal contacts by this definition. However, adherent phagocytes have multiple small punctate accumulations of actin on their ventral surface known as podosomes. These contain talin and vinculin, and it is likely that they are the focal contact equivalent in these cells. It has been shown that the leukocyte-specific CD18 (β_2) integrins can mediate a regulatable interaction with the cytoskeleton through α-actinin in PMN (Pavalko and Laroche, 1993). Whether these integrin–cytoskeleton contacts represent sites of accumulation of enzymes and substrates involved in signal transduction in phagocytes is not yet known. The extent to which substrate adhesion and phagocytosis are similar processes is also not known. Adhesion to target particles is an essential step in phagocytosis, and there are morphologic similarities between substrate adhesion and phagocytosis. There are biochemical events common to the two processes, such as phospholipase A_2 activation and PKC activation, as well (Lennartz et al., 1990; Lefkowith et al., 1991, 1992; Chun and Jacobson, 1992). Thus, it is reasonable that many of the events coordinating cytoskeleton and membrane movement during phagocytosis and adhesion are similar. Phagocytosis, unlike adhesion, requires membrane fusion to complete internalization of an opsonized particle. This aspect of the process has no obvious parallel in cell adhesion, and little is known about the molecular mechanisms involved in this step of phagocytosis.

A.   Cytoskeleton and Signal Transduction

Based, in part, on analogy with substrate adhesion, several recent discoveries have a bearing on the role of the cytoskeleton in phagocytosis. The first is the discovery of a tyrosine kinase that localizes to focal contacts and becomes both tyrosine phosphorylated and activated in response to adhesion (Guan and Shalloway, 1992; Burridge et al., 1992; Kornberg et al., 1992; Romer et al., 1992). This kinase of ~125 kD, called pp125^{FAK}, was cloned as a protein constitutively phosphorylated during *v-src* mediated transformation of fibroblasts. It has a typical tyrosine kinase domain, but has none of the usual motifs for membrane localization of either receptor tyrosine kinases or *src* family members. It appears to be phosphorylated by *src* or an analogue, also present in focal contacts, during adhesion. Constitutive phosphorylation of pp125^{FAK} is associated with loss of anchorage-dependence for proliferation during fibroblast transformation. These data place pp125^{FAK} on the pathway for mitogenesis. Its role in adhesive or phagocytic events is less clear. While it is present in platelets and is phosphorylated by platelet aggregation (Lipfert et al., 1992), it is not even known whether it is present in phagocytes. Nonetheless, it is intriguing to note that protein(s) of this M_r are phosphorylated in adherent PMNs in response to TNF-α (Fuortes et al., 1993), and to speculate that pp125^{FAK} or an analagous protein plays a key role in tyrosine phosphorylation-dependent events during adhesion and phagocytosis by these cells.

B.   Paxillin

Another focal contact protein which may play a role in these events is paxillin (Turner et al., 1990; Burridge et al., 1992). Paxillin was originally identified as a vinculin-binding protein in focal contacts. Paxillin is phosphorylated on tyrosine during substrate adhesion (Turner et al., 1990; Burridge et al., 1992). Recently, it has been reported as well that paxillin is phosphorylated on tyrosine during IgG-mediated phagocytosis (Greenberg et al., 1993). This result demonstrates that paxillin is present at sites of phagocytic activity and suggests that its function may be regulated during ingestion. Our own data suggest that paxillin phosphorylation during PMN Fcγ receptor-mediated phagocytosis requires CR3 (Graham and Brown, unpublished). If tyrosine phosphorylation of paxillin is required for ingestion, this observation could provide an explanation for the role of CR3 in IgG-mediated ingestion (Graham et al., 1989; Gresham et al., 1991).

C.   MARCKS

A third protein which may be important in phagocytosis is a 68–80 kD M_r protein known by the acronym MARCKS, for Myristolylated Alanine Rich C-Kinase Substrate. MARCKS was originally identified as a prominent target for protein kinase C. The cDNA for MARCKS encodes only a ~30 kD core protein; the

anomalous migration of MARCKS on SDS-PAGE is attributable to its hyper-phos-
phorylated state. Recently, MARCKS has been shown to be an actin cross-linking
protein whose membrane association is regulated by phosphorylation. Not only
PKC, but also calcium–calmodulin dependent protein kinases can regulate
MARCKS function (Rosen et al., 1990; Hartwig et al., 1992). These data suggest
that MARCKS may play an important role in the actin cytoskeleton rearrangements
that accompany adhesion and phagocytosis. Disregulation of MARCKS interaction
with actin and with membranes may explain the inhibition of ingestion by both
PKC inhibitors and calmodulin-dependent kinase inhibitors. Interestingly, expression of a close homologue of MARCKS, termed F52 or MacMARCKS (Aderem,
1992; Blackshear et al., 1992) can be induced in macrophages by exposure to
lipopolysaccharide. To date, possible functional differences between the two family
members have not been thoroughly investigated, but the existence of two related,
but distinct proteins with similar functions may allow a further level of complexity
in the regulation of the function of the actin cytoskeleton during phagocytosis.

D. GTP Binding Proteins

Proteins of the rac/rho class of ras-like GTPases are important in regulating
cytoskeletal organization and are reasonable candidates for molecules that may be
involved in the membrane remodeling of phagocytosis. Racs appear to be preferentially expressed in myeloid cells (Didsbury et al., 1989). Rac is required for the
formation of membrane ruffling in response to growth factors (Ridley et al., 1992),
and rac and rho appear to organize stress fibers (Ridley and Hall, 1992). Rac
proteins participate in the generation of respiratory burst activity via the NADPH
oxidase in cell free systems (Knaus et al., 1991).

E. Myosins

Finally, there has been a recent interest in the role of nonmuscle myosins in
movement and phagocytosis. The existence of myosin in nonmuscle cells has been
known for many years. Nonmuscle myosin has been considered a likely candidate
for the molecular motor driving the membrane remodeling that occurs during the
phagocytic process, because of the importance of actin in phagocytosis. However,
to date, little progress has been made in unravelling the role of myosin in particle
engulfment, due largely to the overwhelming diversity in myosins that has recently
emerged. Thus, nine distinct classes of myosin are recognized. Myosin II, the
two-headed filamentous myosin, is present in leukocytes, but the available evidence
suggests it is unlikely to be involved in phagocytosis. For example, in *Dictyostelium*, myosin II accumulates in the uropod during movement, suggesting that
it may be involved in a contractile event rather than the initial extension of the
lamella forward (Fukui et al., 1989). The hypothesis that myosin II is primarily
involved in lamellar contraction rather than extension has been supported by
experiments with electro-permeabilized macrophages. In addition, mutant *Dic-*

tyostelium lacking myosin II or with a truncated protein were capable of chemotaxis and phagocytosis (De Lozanne and Spudich, 1987; Knecht and Loomis, 1987). These data all suggested that myosin II was not essential for ingestion or adhesion.

For these reasons, attention has turned to "unconventional" myosins as potentially important motors in phagocytosis. The largest and best understood class of unconventional myosins are in the myosin I family. Even within this class, there is great diversity, and currently nine myosin I genes have been identified in *Dictyostelium* alone. All have a single globular head domain, with actin and ATP-binding sites homologous to myosin II, associated with a carboxy-terminal tail which has no homology to myosin II. Myosin I molecules have an additional actin binding site in this tail region. Unlike myosin II, myosin I does not form filamentous structures. The myosin I light-chain, at least in higher eukaryotes, is the calcium-binding protein, calmodulin. Unlike myosin II, myosin I is found in lamellapodia during migration and in the phagocytic cup during ingestion by *Dictyostelium*. This finding has led to the hypothesis that myosin I may play a key role in these functions. However, as of this writing, no myosin I protein has been definitively identified in a leukocyte, leaving open the question of its role(s), in movement or phagocytosis in these specialized cells.

VI. REGULATION OF PHAGOCYTOSIS

It is a hallmark of phagocytic cells from vertebrates that they require activation to exhibit maximal ingestion. Presumably, this is because phagocytosis is associated with the induction and/or secretion of many pro-inflammatory mediators: arachidonate metabolites, lysosomal enzymes, toxic oxygen metabolites, and so forth, which are generally deleterious to the host unless confined to the site of inflammation or infection. Among activators of the full phagocytic potential of these cells are many molecules expected to be found at sites of inflammation: complement-derived and bacterial chemotactic peptides, platelet activating factor, tumor necrosis factor, and GM-CSF. In addition, extracellular matrix proteins can provide a significant stimulus to phagocytosis. This suggests that interaction with extracellular matrix proteins or with other molecules concentrated at sites of inflammation or infection is an important mechanism for signaling to the host defense cell that it is outside the vasculature, in a site where maximal phagocytic function is required. This, in turn, implies that these stimulants to phagocytosis transduce signals across the phagocyte plasma membrane which alter cell behavior, presumably mediated by activation of specific biochemical pathways (after receptor–ligand interaction) that effect rearrangement of the membrane-associated cytoskeleton.

A. Cytokine Activation of Phagocytosis

The first example of modulation of phagocytic function came from the studies of Bianco and Silverstein and then of Griffin on complement receptor-mediated phagocytosis in macrophages (Bianco et al., 1975; Griffin and Griffin, 1979, 1980; Griffin and Mullinax, 1981). These workers showed that inflammatory macrophages would ingest particles via their C3 receptors, while resident peritoneal macrophages would not. Furthermore, the ability to ingest via complement receptors could be conferred on resident phagocytes in as little as 1 hour by a cytokine generated from T cell–macrophage collaboration. These were the first data which suggested that macrophage phagocytic phenotype could be regulated by products found at sites of inflammation. Recently, GM-CSF and TNF-α have been identified as the cytokines responsible for complement receptor activation (G. Bancroft, personal communication), although there may be other cytokines which can cause this phenotypic change as well (Sampson et al., 1991).

B. Extracellular Matrix Activation of Phagocytosis

We and others have shown that extracellular matrix proteins can activate complement receptor-mediated phagocytosis in monocytes and macrophages as well (Brown, 1986; Brown and Lindberg, 1993). Furthermore, these proteins can enhance IgG Fc receptor-mediated phagocytosis by both neutrophils and macrophages (Brown et al., 1988; Gresham et al., 1989) as well. Details of the biochemical pathways for activation are not fully understood. However, certain facts are known. First, there is more than one receptor responsible for extracellular matrix enhancement of phagocytosis, although all seem to be integrins. The integrin $\alpha_3\beta_1$ is apparently necessary for stimulation by the basement membrane protein entactin, and $\alpha_6\beta_1$ is necessary for laminin-mediated enhancement (Bohnsack, 1992). In addition, there is an integrin receptor which is present on PMNs and monocytes which recognizes multiple Arg–Gly–Asp-containing proteins and appears to be necessary for phagocytosis enhancement by these extracellular matrix proteins (Gresham et al., 1989). This receptor is related to the β_3 integrin family, but seems to be distinct from either $\alpha_v\beta_3$ or $\alpha_{IIb}\beta_3$, the two known members of this integrin family, both immunologically and in peptide specificity (Gresham et al., 1992; Carreno et al., 1993). For these reasons, this phagocyte receptor, which can activate the NADPH oxidase as well as phagocytosis (Zhou et al., 1993), has been called the Leukocyte Response Integrin (LRI). Unfortunately, LRI has not been cloned, so its molecular characterization and relation to other members of the integrin family remain unclear. While the mechanism of signal transduction through LRI also is not completely understood, it is known to be inhibited by pertussis toxin, protein kinase C inhibitors, intracytoplasmic Ca^{2+} chelators, and cAMP and not affected by tyrosine kinase inhibitors (Zhou and Brown, 1993). These data are consistent with ligand-dependent activation of phospholipase C via a pertussis toxin-sensitive G-protein. Signal transduction is also inhibited by cytochalasin B, suggesting a role for the cytoskeleton, so whether the phospholipase and protein kinase activities are required for cytoskeletal assembly or directly for activation of the phagocytic receptors is not known. Since phagocytosis is intimately associated with cytoskeleton–membrane interactions and many forms of signal transduction depend on these connections as well, there may be no fundamental difference between these possibilities.

An important feature of signal transduction through LRI is that activation by LRI ligands can be blocked by antibodies to a second membrane protein (Brown et al., 1990). This protein appears to co-immunoprecipitate with β_3 integrins and to co-localize with LRI in a cytoskeleton-dependent manner. Because of this close functional and physical association, we have called this protein Integrin-Associated Protein (IAP). IAP has now been cloned and shown to be a member of the immunoglobulin superfamily, with multiple transmembrane domains (Lindberg et al., 1993). Thus, we hypothesize that physical interaction of an integrin and an immunoglobulin family member on the same cell is necessary for leukocyte activation by extracellular matrix proteins. Further investigations will reveal how the interaction of these two plasma membrane proteins leads to activation of the enzymes necessary for NADPH oxidase activity and stimulated phagocytosis.

C. Two Pathways for Activation of PMN Phagocytosis

Recently, we have also begun investigations of the mechanisms for activating Fc receptor-mediated phagocytosis in PMNs by soluble ligands, such as the bacterial peptide fMet–Leu–Phe, the complement-derived chemotaxin C5a, and the product of arachidonate metabolism platelet-activating factor (PAF). We have shown that PMNs possess two distinct mechanisms for activation of Fc receptor-mediated phagocytosis, based on whether or not an increase in $[Ca^{2+}]_i$ is required (Rosales and Brown, 1991). PAF is an example of a PMN activator which can stimulate phagocytosis even when $[Ca^{2+}]_i$ is clamped at a very low level. Since phorbol esters can activate phagocytosis by this pathway as well, it is likely to involve protein kinase C activation. fMet–Leu–Phe is an example of an activator which requires an increase in $[Ca^{2+}]_i$ to activate phagocytosis. Although fMet–Leu–Phe can itself increase $[Ca^{2+}]_i$, this increase is neither necessary nor sufficient for enhanced phagocytosis. The $[Ca^{2+}]_i$ increase needed for phagocytosis must arise from Fc receptor engagement (Rosales and Brown, 1991; Rosales et al., 1994). This startling result implies that there must be spatial heterogeneity of Ca^{2+} concentration within the PMN cytoplasm in response to various agonists and that engagement of Fc receptors releases Ca^{2+} from intracellular stores into a region of the cell important for phagocytosis. Presumably, this is a site which is in proximity to the Fc receptor itself and is important for signaling some of the cytoskeletal rearrangements which are essential for ingestion. The actual ingestive force generated by the cytoskeleton must be activatable in the absence of increases in $[Ca^{2+}]_i$, as demonstrated by the fact that PAF and phorbol esters can stimulate ingestion even at very low $[Ca^{2+}]_i$, and that macrophages can phagocytose maximally with $[Ca^{2+}]_i$ clamped at a very low level as well (DiVirgilio et al., 1988).

Because of the importance of Fc receptor-mediated increase in $[Ca^{2+}]_i$ for phagocytosis, we have gone on to investigate the pathway by which this occurs. Fc-receptor mediated increase in $[Ca^{2+}]_i$ in PMNs comes entirely from intracellular stores, whether FcRII or FcRIII is activated to signal release (Kimberly et al., 1990; Rosales and Brown, 1992). Although inositol trisphosphate (IP_3) is the best known intracellular messenger for release of Ca^{2+} from intracellular stores, Fc receptor-mediated release of Ca^{2+} is completely IP_3 independent. We have now shown that Fc receptor-mediated release of Ca^{2+} from intracellular stores is instead dependent on the actin cytoskeleton (Rosales et al., 1994). This suggests the existence of a novel mechanism for Ca^{2+} release and emphasizes the intimate association between cytoskeleton, Fc receptors, effector mechanisms such as phagocytosis, and more conventional signal transduction pathways. Currently, we envision that the actin cytoskeleton may modulate signal transduction cascades through the multiple actin-binding proteins whose association with actin is regulated by Ca^{2+} or phosphorylation. Actin association with these proteins may provide a scaffolding on which specific enzymes and substrates can associate.

VII. INTRACELLULAR EVENTS FOLLOWING PHAGOCYTOSIS

So far, this review has concentrated on the biochemistry of the process of ingestion and its regulation. After a target particle or organism is internalized, the phagosome undergoes characteristic events which culminate in transport from the lamella to the cell body and fusion with intracellular acidic compartments, including lysosomes. While a detailed review of this topic is not possible here, certain recent observations need to be highlighted, because of their significance for host defense and intracellular infection.

A. Fusion Events Following Phagocytosis

Phagosomes fuse with intracellular vesicular compartments. This fusion begins rapidly after initiation of ingestion, sometimes even before the phagosome is fully formed. Presumably, this is the reason that lysosomal contents are found in the extracellular *milieu* after phagocytosis has occurred. In PMNs, phagosome fusion with lysosomes is dependent on an increase in $[Ca^{2+}]_i$ (Jaconi et al., 1990). This implies that the Fc receptor-mediated increase in $[Ca^{2+}]_i$ discussed above is critical not only for ingestion and respiratory burst, but also for subsequent normal intracellular processing of phagocytosed material. However, phagosomes can undergo Ca^{2+}-independent fusion events as well (Mayorga et al., 1991; Pitt et al., 1992). So far, Ca^{2+}-independent fusion has been studied only in macrophages and only for phagosome–endosome fusion. Thus, it is unclear whether dependence on $[Ca^{2+}]_i$ is a difference between phagosome–lysosome and phagosome–endosome fusion, or a difference between macrophages and PMNs. The study of phagosome fusion with intracellular acidic compartments is receiving increasing attention because of its significance for infectious diseases such as tuberculosis, which is becoming a major public health problem once again.

B. Intracellular Infection

Many pathogens have subverted the process of phagocytosis, either by professional phagocytes or other cells, to establish intracellular infection protected from antibodies, complement, and other agents of host defense. Study of classic examples of invasion into epithelial cells by *Yersinia*, *Salmonella*, and *Listeria* make it clear that all enter these cells through a process morphologically indistinguishable from phagocytosis. The receptor(s) through which this occurs has been extensively studied for *Yersinia*. These organisms express a protein, called invasin, which binds to β_1 integrins on epithelial cells (Isberg and Leong, 1990). Invasin binds to these receptors with higher affinity than their extracellular matrix ligands, which in some way induces the phagocytic process (Tran Van Nhieu and Isberg, 1993).

Some organisms, for example mycobacteria, *Leigonella*, and *Leishmania*, have found macrophages to be a convenient host. It is thought that these organisms live inside phagosomes, although for mycobacteria there is some evidence that they may escape into the cytosol (McDonough et al., 1993). They can survive either by resisting the low pH and acid hydrolases of the phagosome which follows phagosome–lysosome fusion, like *Leishmania*; they can live in abnormal phagosomes which are incapable of fusing with lysosomes, like *Legionella*; or they can inhibit phagosome–lysosome fusion by some other means. Although mycobacteria exist in a morphologically normal phagocytic vacuole, the pH of this vacuole is abnormally high. This prevents activity of many of the proteolytic enzymes present in lysosomes. The mechanism by which mycobacteria do this is not completely understood. Mycobacteria produce NH_3, which can raise the pH of their environment. They may inhibit fusion of phagosomes with lysosomes entirely, or they may inhibit the fusion of the Na^+/H^+ ATPase-containing vesicles with the phagosome. Since the Na^+/H^+ ATPase is responsible for lowering intravesicular pH, this would account for the higher pH of the mycobacterial phagosome. In any case, the inhibition of acidification requires living organisms. An intriguing possibility for failure of phagosome–lysosome fusion is raised by *Histoplasma*, a fungus which grows as a yeast inside macrophages. *Histoplasma* do not trigger an increase in $[Ca^{2+}]_i$ when they are ingested, perhaps suggesting that some organisms simply suppress a necessary signal for phagosome–lysosome fusion (Eissenberg and Goldman, 1991).

VIII. BEYOND THE ZIPPER HYPOTHESIS

The picture of the phagocytic process that emerges from the work detailed above is one of considerable complexity and diversity. Nevertheless, most of these data can be incorporated within the classic zippering model of phagocytosis, wherein specific receptor–ligand interactions must proceed circumferentially around the opsonized particle and lead to the generation of localized (and probably also global) signals that, in turn, regulate membrane and cytoskeletal remodeling and membrane fusion events. In this model, diversity in the phagocytic process occurs because ingestible particles express different ligands that initiate or modulate phagocytosis, because of differences in phagocytic receptors expressed by different phagocytic cells, or because of differences in the state of activation of these cells in response to secreted molecules or interactions with extracellular matrix proteins. However, several recent studies have documented phagocytic processes that appear to differ in a fundamental way from those described above. The uptake of *Salmonella* by epithelial cells appears to violate the rules laid out above (Bliska et al., 1993). Invasive *Salmonella* induce plasma membrane ruffling in these cells which leads to engulfment of the bacteria after they are trapped within a membrane ruffle in a fashion that appears to be distinct from the sequential membrane interactions postulated for the zippering mechanism, and which has the morphologic appearance of macro-pinocytosis (Bliska et al., 1993; Francis et al., 1993; Pace et al., 1993). These bacteria induce increases in cytoplasmic calcium, which appear to be required for invasion. Non-invasive bacteria can also be ingested within these membrane ruffles if they present along with the invasive *Salmonella*, demonstrating that this process lacks the selectivity of classic phagocytosis (Pace et al., 1993). Finally, global membrane ruffling induced by epidermal growth factor also induced ingestion of bacteria by Hep-2 cells (Francis et al., 1993). Although uptake of bacteria in this system involves the actin cytoskeleton, these experiments show that the requirement for local and sequential receptor–ligand engagement is circumvented, and that the cytoskeletal rearrangement that results in particle ingestion can be induced by certain global signals.

The mode of invasion employed by intracellular parasites such as *Trypanosoma cruzi* represents an even more striking departure from the prevalent model of ingestion (Tardieux et al., 1992). These parasites invade NRK cells apparently by recruiting lysosomes to the plasma membrane and inducing fusion of these lysosomes to form the parisitophorous vacuole. The actin cytoskeleton is not involved in this process, and depolymerization of actin with cytochalasin D actually facilitates invasion. In contrast, movement of lysosomes along microtubules affected invasion, and maneuvers that induced lysosomes to move to the cell periphery along microtubules enhanced the entry of trypanosomes into cells. For *Toxoplasma*, parasite-derived membrane stored in intracellular organelles known as rhoptries may be crucial for invasion of mammalian cells. The concept has emerged that *Toxoplasma* inject themselves into the host cell, since cytochalasin treatment of the parasite, but not of the host, inhibits infection *in vitro*. Moreover, the parasitophorous vacuole appears not to contain host plasma membrane components. Thus these parasites, and perhaps others as well, have devised strategies to enter cells that appear to bypass normal phagocytic mechanisms entirely.

IX. CONCLUSIONS

In this review, we have attempted to summarize the current understanding of the phagocytic process. Much work has been done in this area, and many of the receptors that initiate the phagocytic process have been characterized. The challenges of the next few years will be to understand the details of the signaling process and to identify the molecules involved in the engulfment process and subsequent steps, and to further explore the variations and alternatives to the basic phagocytic process that are utilized by different microbial pathogens.

ACKNOWLEDGMENTS

Work in the authors' labs reported in this review is supported by grants AI24674, GM38330, AI33348, GM45815, and DK46686 from the National Institutes of Health. T.H. Steinberg is an Established Investigator of the American Heart Association.

REFERENCES

Aderem, A. (1992). The role of myristoylated protein kinase C substrates in intracellular signaling pathways in macrophages. Curr. Top. Microbiol. Immunol. 181, 189–207.

Amigorena, S., Bonnerot, C., Drake, J. R., Choquet, D., Hunziker, W., Guillet, J. G., Webster, P., Sautes, C., Mellman, I., & Fridman, W. H. (1992). Cytoplasmic domain heterogeneity and functions of IgG Fc receptors in B lymphocytes. Science 256, 1808–1812.

Berton, G., Laudanna, C., Sorio, C., & Rossi, F. (1992). Generation of signals activating neutrophil functions by leukocyte integrins: LFA-1 and gp150/95, but not CR3, are able to stimulate the respiratory burst of human neutrophils. J. Cell Biol. 116, 1007–1017.

Bianco, C., Griffin, F. M., Jr., & Silverstein, S. C. (1975). Studies of the macrophage complement receptor. Alteration of receptor function upon macrophage activation. J. Exp. Med. 141, 1278–1291.

Blackshear, P. J., Verghese, G. M., Johnson, J. D., Haupt, D. M., & Stumpo, D. J. (1992). Characteristics of the F52 protein, a MARCKS homologue. J. Biol. Chem. 267, 13540–13546.

Bliska, J. B., Galan, J. E., & Falkow, S. (1993). Signal transduction in the mammalian cell during bacterial attachment and entry. Cell 73, 903–920.

Bohnsack, J. F. (1992). CD11/CD18-independent neutrophil adherence to laminin is mediated by the integrin VLA-6. Blood 79, 1545–1552.

Borregaard, N., Kjeldsen, L., Rygaard, K., Bastholm, L., Nielsen, M. H., Sengelov, H., Bjerrum, O. W., & Johnsen, A. H. (1992). Stimulus-dependent secretion of plasma proteins from human neutrophils. J. Clin. Invest. 90, 86–96.

Botto, M., Lissandrini, D., Sorio, C., & Walport, M. J. (1992). Biosynthesis and secretion of complement component (C3) by activated human polymorphonuclear leukocytes. J. Immunol. 149, 1348–1355.

Brown, E. J. (1986). The interaction of connective tissue proteins with phagocytic cells. J. Leuk. Biol. 39, 579–591.

Brown, E. J. (1991). Complement receptors and phagocytosis. Curr. Opin. Immunol. 3, 76–82.

Brown, E. J., Bohnsack, J. F., & Gresham, H. D. (1988). Mechanism of inhibition of immunoglobulin G-mediated phagocytosis by monoclonal antibodies that recognize the Mac-1 antigen. J. Clin. Invest. 81, 365–375.

Brown, E. J., & Goodwin, J. L. (1988). Fibronectin receptors of phagocytes: Characterization of the Arg-Gly-Asp binding proteins of human monocytes and polymorphonuclear leukocytes. J. Exp. Med. 167, 777–793.

Brown, E. J., Hooper, L., Ho, T., & Gresham, H. D. (1990). Integrin-associated protein: A 50-kD plasma membrane antigen physically and functionally associated with integrins. J. Cell Biol. 111, 2785–2794.

Brown, E. J., & Graham, I.L. (1991). Macrophage and inflammatory cell matrix receptors: LFA-1, Mac-1, p150,95 family. In: *Receptors for Extracellular Matrix* (McDonald, J. A. et al., Eds.). Academic Press, Inc., San Diego, pp. 39–79.

Brown, E. J., & Lindberg, F. P. (1993). Matrix receptors of myeloid cells. In: *Blood Cell Biochemistry*, Vol. 5, *Macrophages and Related Cells* (Horton, M. A., Ed.). Plenum Press, New York, pp. 279–306.

Brown, E. J., Newell, A. M., & Gresham, H. D. (1987). Molecular regulation of phagocyte function: Evidence for involvement of a GTP binding protein in opsonin mediated phagocytosis by monocytes. J. Immunol. 139, 3777–3782.

Burridge, K., & Fath, K. (1989). Focal contacts: Transmembrane links between the extracellular matrix and the cytoskeleton. Bioessays 10, 104–108.

Burridge, K., Nuckolls, G., Otey, C., Pavalko, F., Simon, K., & Turner, C. (1990). Actin-membrane interaction in focal adhesions. Cell Differ. Dev. 32, 337–342.

Burridge, K., Turner, C. E., & Romer, L. H. (1992). Tyrosine phosphorylation of paxillin and pp125[FAK] accompanies cell adhesion to extracellular matrix: A role i cytoskeletal assembly. J. Cell Biol. 119, 893–903.

Buyon, J. P., Slade, S. G., Reibman, J., Abramson, S. B., Philips, M. R., Weissmann, G., & Winchester, R. (1990). Constitutive and induced phosphorylation of the alpha- and beta-chains of the CD11/CD18 leukocyte integrin family. Relationship to adhesion-dependent functions. J. Immunol. 144, 191–197.

Carreno, M. P., Gresham, H. D., & Brown, E. J. (1991). Characterization of a novel integrin involved in regulation of phagocytosis. FASEB J. 5, A549.

Carreno, M. P., Gresham, H. D., & Brown, E. J. (1993). Isolation of the leukocyte response integrin (LRI): A novel RGD-binding protein involved in regulation of phagocyte function. Clin. Immunol. Immunopathol. 69, 43–51.

Choquet, D., Partiseti, M., Amigorena, S., Bonnerot, C., & Fridman, W. H. (1993). Cross-linking of IgG receptors inhibits membrane immunoglobulin-stimulated calcium influx in B lymphocytes. J. Cell Biol. 121, 355–363.

Chun, J. -S., & Jacobson, B. S. (1992). Spreading of HeLa cells on a collagen substratum requires a second messenger formed by the lipoxygenase metabolism of arachidonic acid released by collagen receptor clustering. Mol. Biol. Cell 3, 481–492.

De Lozanne, A., & Spudich, J. A. (1987). Disruption of the Dictyostelium myosin heavy chain gene by homologous recombination. Science 236, 1086–1091.

Della Bianca, V., Grzeskowiak, M., Renzi, E., & Rossi, F. (1993). The potentiation by TNF-α and PMA of Fc receptor-mediated phagocytosis in neutrophils is independent of reactive oxygen metabolites produced by NADPH oxidase and of protein kinase C. Biochem. Biophys. Res. Commun. 193, 919–926.

Diamond, M. S., Garcia-Aguilar, J., Bickford, J. K., Corbi, A. L., & Springer, T. A. (1993). The I domain is a major recognition site on the leukocyte integrin Mac-1 (CD11b/CD18) for four distinct adhesion ligands. J. Cell Biol. 120, 1031–1043.

Diamond, M. S., Staunton, D. E., de Fougerolles, A. R., Stacker, S. A., Garcia-Aguilar, J., Hibbs, M. L., & Springer, T. A. (1990). ICAM-1 (CD54): A counter-receptor for Mac-1 (CD11b/CD18). J. Cell Biol. 111, 3129–3139.

Diamond, M. S., Staunton, D. E., Marlin, S. D., & Springer, T. A. (1991). Binding of the integrin Mac-1 (CD11b/CD18) to the third immunoglobulin-like domain of ICAM-1 (CD54) and its regulation by glycosylation. Cell 65, 961–971.

Didsbury, J., Weber, R. F., Bokoch, G. M., Evans, T., & Snyderman, R. (1989). rac, a novel ras-related family of proteins that are botulinum toxin substrates. J. Biol. Chem. 264, 16378–16382.

DiVirgilio, F., Meyer, B. C., Greenberg, S., & Silverstein, S. C. (1988). Fc receptor-mediated phagocytosis occurs in macrophages at exceedingly low cytosolic Ca^{2+} levels. J. Cell Biol. 106, 657–666.

Ehlenberger, A. G., & Nussenzweig, V. (1977). The role of membrane receptors for C3b and C3d in phagocytosis. J. Exp. Med. 145, 357–371.

Eissenberg, L. G., & Goldman, W. E. (1991). Histoplasma variation and adaptive strategies for parasitism: New perspectives on histoplasmosis. [Review]. Clin. Microbiol. Rev. 4, 411–421.

Ezekowitz, R. A., Sastry, K., Bailly, P., & Warner, A. (1990). Molecular characterization of the human macrophage mannose receptor: Demonstration of multiple carbohydrate recognition-like domains and phagocytosis of yeasts in Cos-1 cells. J. Exp. Med. 172, 1785–1794.

Ezekowitz, R. A. B., Williams, D. J., Koziel, H., Armstrong, M. Y. K., Warner, A., Richards, F. F., & Rose, R. M. (1991). Uptake of *Pneumocystis carinii* mediated by the macrophage mannose receptor. Nature 351, 155–158.

Fallman, M., Gullberg, M., Hellberg, C., & Andersson, T. (1992). Complement receptor-mediated phagocytosis is associated with accumulation of phosphatidylcholine-derived diglyceride in human neutrophils. Involvement of phospholipase D and direct evidence for a positive feedback signal of protein kinase C. J. Biol. Chem. 267, 2656–2663.

Fallman, M., Lew, D. P., Stendahl, O., & Andersson, T. (1989). Receptor-mediated phagocytosis in human neutrophils is associated with increased formation of inositol phosphates and diacylglycerol: Elevation in cytosolic free calcium and formation of inositol phosphates can be dissociated from accumulation of diacylglycerol. J. Clin. Invest. 84, 886–891.

Francis, C. L., Ryan, T. A., Jones, B. D., Smith, S. J., & Falkow, S. (1993). Ruffles induced by Salmonella and other stimuli direct macropinocytosis of bacteria. Nature 364, 639–642.

Fukui, Y., Lynch, T. J., Brzeska, H., & Korn, E. D. (1989). Myosin I is located at the leading edges of locomoting Dictyostelium amoebae. Nature 341, 328–331.

Fuortes, M., Jin, W.-W., & Nathan, C. (1993). Adhesion-dependent protein tyrosine phosphorylation in neutrophils treated with tumor necrosis factor. J. Cell Biol. 120, 777–784.

Graham, I. L., Gresham, H. D., & Brown, E. J. (1989). An immobile subset of plasma membrane CD11b/CD18 (Mac-1) is involved in phagocytosis of targets recognized by multiple receptors. J. Immunol. 142, 2352–2358.

Graham, I. L., Lefkowith, J. B., Anderson, D. C., & Brown, E. J. (1993). Immune complex-stimulated neutrophil LTB4 production is dependent on beta2 integrins. J. Cell Biol. 120, 1509–1517.

Greenberg, S., Chang, P., & Silverstein, S. C. (1993a). Tyrosine phosphorylation is required for Fc receptor-mediated phagocytosis in mouse macrophages. J. Exp. Med. 177, 529–534.

Greenberg, S., Chang, P., & Silverstein, S. C. (1993b). Multiple proteins, including pzxillin, become phosphorylated on tyrosine residues during Fc receptor-mediated phagocytosis in mouse macrophages. Clin. Res. 41, 136a.(Abstract)

Gresham, H. D., McGarr, J. A., Shackelford, P. G., & Brown, E. J. (1988). Studies on the molecular mechanisms of receptor-mediated phagocytosis: Amplification of ingestion is dependent on the generation of reactive oxygen metabolites and is deficient in PMN from patients with chronic granulomatous disease. J. Clin. Invest. 82, 1192–1201.

Gresham, H. D., Goodwin, J. L., Anderson, D. C., & Brown, E. J. (1989). A novel member of the integrin receptor family mediates Arg-Gly-Asp-stimulated neutrophil phagocytosis. J. Cell Biol. 108, 1935–1943.

Gresham, H. D., Graham, I. L., Anderson, D. C., & Brown, E. J. (1991). Leukocyte adhesion deficient (LAD) neutrophils fail to amplify phagocytic function in response to stimulation: Evidence for

CD11b/CD18-dependent and -independent mechanisms of phagocytosis. J. Clin. Invest. 88, 588–597.

Gresham, H. D., Adams, S. P., & Brown, E. J. (1992). Ligand binding specificity of the leukocyte response integrin expressed by human neutrophils. J. Biol. Chem. 267, 13895–13902.

Gresham, H. D., Zheleznyak, A., Mormol, J. S., & Brown, E. J. (1990). Studies on the molecular mechanisms of human neutrophil Fc receptor-mediated phagocytosis: Evidence that a distinct pathway for activation of the respiratory burst results in reactive oxygen metabolite-dependent amplification of ingestion. J. Biol. Chem. 265, 7819–7826.

Griffin, F. M., Jr., Griffin, J. A., Leider, J. E., & Silverstein, S. C. (1975). Studies on the mechanism of phagocytosis: I. Requirements for circumferential attachment of particle-bound ligands to specific receptors on the macrophage plasma membrane. J. Exp. Med. 142, 1263–1282.

Griffin, F. M., Jr., Griffin, J. A., & Silverstein, S. C. (1976). Studies on the mechanism of phagocytosis: II. The interaction of macrophages with anti-immunoglobulin IgG-coated bone marrow-derived-lymphocytes. J. Exp. Med. 144, 788–809.

Griffin, F. M., Jr., & Griffin, J. A. (1980). Augmentation of macrophage complement receptor function in vitro. II. Characterization of the effects of a unique lymphokine upon the phagocytic capabilities of macrophages. J. Immunol. 125, 884.

Griffin, F. M., Jr., & Mullinax, P. J. (1981). Augmentation of macrophage complement receptor function in vitro. III. C3b receptors that promote phagocytosis migrate within the plane of the macrophage plasma membrane. J. Exp. Med. 154, 291.

Griffin, F. M., Jr., & Silverstein, S. C. (1974). Segmental response of the macrophage plasma membrane to a phagocytic stimulus. J. Exp. Med. 139, 323–336.

Griffin, J. A., & Griffin, F. M., Jr. (1979). Augmentation of macrophage complement receptor function in vitro. I. Characterization of the cellular interactions required for the generation of a T-lymphocyte product that enhances macrophage complement receptor function. J. Exp. Med. 150, 653.

Guan, J. L., & Shalloway, D. (1992). Regulation of focal adhesion-associated protein tyrosine kinase by both cellular adhesion and oncogenic transformation. Nature 358, 690–692.

Hartwig, J. H., Thelen, M., Rosen, A., Janmey, P. A., Nairn, A. C., & Aderem, A. (1992). MARCKS is an actin filament crosslinking protein regulated by protein kinase C and calcium-calmodulin. Nature 356, 618–622.

Hermanowski-Vosatka, A., Van Strijp, J. A. G., Swiggard, W. J., & Wright, S. D. (1992). Integrin modulating factor-1: A lipid that alters the function of leukocyte integrins. Cell 68, 341–352.

Indik, Z., Kelly, C., Chien, P., Levinson, A. I., & Schreiber, A. D. (1991). Human $Fc_{gamma}RII$, in the absence of other Fc_{gamma} receptors, mediates a phagocytic signal. J. Clin. Invest. 88, 1766–1771.

Isberg, R. R., & Leong, J. M. (1990). Multiple β1 chain integrins are receptors for invasin, a protein that promotes bacterial penetration into mammalian cells. Cell 60, 861–871.

Jaconi, M. E. E., Lew, D. P., Carpentier, J.-L., Magnusson, K. E., Sjogren, M., & Stendahl, O. (1990). Cytosolic free calcium elevation mediates the phagosome-lysosome fusion during phagocytosis in human neutrophils. J. Cell Biol. 110, 1555–1564.

Joiner, K. A., Fuhrman, S. A., Miettinen, H. M., Kasper, L. H., & Mellman, I. (1990). Toxoplasma gondii: Fusion competence of parasitophorous vacuoles in Fc receptor-transfected fibroblasts. Science 249, 641–646.

Kim, E., Enelow, R. I., Sullivan, G. W., & Mandell, G. L. (1992). Regional and generalized changes in cytosolic free calcium in monocytes during phagocytosis. Infect. Immun. 60, 1244–1248.

Kimberly, R. P., Ahlstrom, J. W., Click, M. E., & Edberg, J. C. (1990). The glycosyl phosphatidylinositol-linked Fc gamma RIIIon PMN mediates transmembrane signaling events distinct from Fc gamma RII. J. Exp. Med. 171, 1239–1255.

Knaus, U. G., Heyworth, P. G., Evans, T., Curnutte, J. T., & Bokoch, G. M. (1991). Regulation of phagocyte oxygen radical production by the GTP-binding protein Rac 2. Science 254, 1512–1515.

Knecht, D. A., & Loomis, W. F. (1987). Antisense RNA inactivation of myosin heavy chain gene expression in Dictyostelium discoideum. Science 236, 1081–1086.

Kornberg, L., Earp, H. S., Parsons, J. T., Schaller, M., & Juliano, R. L. (1992). Cell adhesion or integrin clustering increases phosphorylation of a focal adhesion-associated tyrosine kinase. J. Biol. Chem. 267, 23439–23442.

Kuroda, K., Brown, E. J., Telle, W. B., Russell, D. G., & Ratliff, T. L. (1993). Characterization of the internalization of Bacillus Calmette-Guerin (BCG) by human bladder tumor cells. J. Clin. Invest. 91, 69–76.

Kurosaki, T., & Ravetch, J. V. (1989). A single amino acid in the glycosyl phosphatidylinositol attachment domain determines the membrane topology of Fc gamma RIII. Nature 342, 805–807. (Published erratum appears in Nature 1990 Jan. 25, 343(6256), 390.)

Lefkowith, J. B., Lennartz, M. R., Rogers, M., Morrison, A. R., & Brown, E. J. (1992). Phospholipase activation during monocyte adherence and spreading. J. Immunol. 149, 1729–1735.

Lefkowith, J. B., Rogers, M., Lennartz, M. R., & Brown, E. J. (1991). Essential fatty acid deficiency impairs macrophage spreading and adherence: Role of arachidonate in cell adhesion. J. Biol. Chem. 266, 1071–1076.

Lennartz, M. R., Lefkowith, J. B., & Brown, E. J. (1990). Signal transduction during IgG-mediated phagocytosis in human monocytes- role of arachidonic acid. J. Cell Biol. 111, 214a. (Abstract)

Lew, D. P., Andersson, T., Hed, J., Di Virgilio, F., Pozzan, T., & Stendahl, O. (1985). Ca^{2+}-dependent and Ca^{2+}-independent phagocytosis in human neutrophils. Nature 315, 509–511.

Lifson, J., Coutrë, S., Huang, E., & Engleman, E. (1986). Role of envelope glycoprotein carbohydrate in human immunodeficiency virus (HIV) infectivity and virus-induced cell fusion. J. Exp. Med. 164, 2101–2106.

Lin, C. S., Aebersold, R. H., Kent, S. B., Varma, M., & Leavitt, J. (1988). Molecular cloning and characterization of plastin, a human leukocyte protein expressed in transformed human fibroblasts. Mol. Cell Biol. 8, 4659–4668.

Lindberg, F. P., Gresham, H. D., Schwarz, E., & Brown, E. J. (1993). Molecular cloning of Integrin-Associated Protein: An immunoglobulin family member with multiple membrane spanning domains implicated in alpha-v, beta-3-dependent ligand binding. J. Cell Biol. 123, 485–496.

Lipfert, L., Haimovich, B., Schaller, M. D., Cobb, B. S., Parsons, J. T., & Brugge, J. S. (1992). Integrin-dependent phosphorylation and activation of the protein tyrosine kinase pp125[FAK] in platelets. J. Cell Biol. 119, 905–912.

Mayorga, L. S., Bertini, F., & Stahl, P. D. (1991). Fusion of newly formed phagosomes with endosomes in intact cells and in a cell-free system. J. Biol. Chem. 266, 6511–6517.

McDonough, K. A., Kress, Y., & Bloom, B. R. (1993). Pathogenesis of tuberculosis: Interaction of Mycobacterium tuberculosis with macrophages. Infect. Immun. 61, 2763–2773. (Published erratum appears in Infect. Immun. 1993 Sep., 61(9), 4021–4024.)

Mosser, D. M., Springer, T. A., & Diamond, M. S. (1992). Leishmania promastigotes require opsonic complement to bind to the human leukocyte integrin Mac-1 (CD11b/CD18). J. Cell Biol. 116, 511–520.

Myones, B. L., Dalzell, J. G., Hogg, N., & Ross, G. D. (1988). Neutrophil and monocyte cell surface p150,95 has iC3b receptor (CR4) activity. J. Clin. Invest. 82, 640–651.

Odin, J. A., Edberg, J. C., Painter, C. J., Kimberly, R. P., & Unkeless, J. C. (1991). Regulation of phagocytosis and $[Ca^{2+}]_i$ flux by distinct regions of an Fc receptor. Science 254, 1785–1788.

Pace, J., Hayman, M. J., & Galan, J. E. (1993). Signal transduction and invasion of epithelial cells by S. typhimurium. Cell 72, 505–514.

Park, J.-G., Isaacs, R. E., Chien, P., & Schreiber, A. D. (1993). In the absence of other Fc receptors, FcgammaRIIIA transmits a phagocytic signal that requires the cytoplasmic domain of its gamma subunit. J. Clin. Invest. 92, 1967–1973.

Pavalko, F. M., & Laroche, S. M. (1993). Activation of human neutrophils induces an interaction between the integrin β_2-subunit (CD18) and the actin binding protein α-actinin. J. Immunol. 151, 3795–3807.

Pitt, A., Mayorga, L. S., Stahl, P. D., & Schwartz, A. L. (1992). Alterations in the protein composition of maturing phagosomes. J. Clin. Invest. 90, 1978–1983.

Pryzwansky, K. B., Steiner, A. L., Spitznagel, J. K., & Kapoor, C. L. (1981). Compartmentalization of cyclic AMP during phagocytosis by human neutrophilic granulocytes. Science 211, 407–410.

Ravetch, J. V., Luster, A. D., Weinshank, R., Kochan, J., Pavlovec, A., Portnoy, D. A., Hulmes, J., Pan, Y. C., & Unkeless, J. C. (1986). Structural heterogeneity and functional domains of murine immunoglobulin G Fc receptors. Science 234, 718–725.

Ravetch, J. V., & Kinet, J. P. (1991). Fc receptors. Annu. Rev. Immunol. 9, 457–492.

Relman, D., Tuomanen, E., Falkow, S., Golenbock, D. T., Saukkonen, K., & Wright, S. D. (1990). Recognition of a bacterial adhesin by an integrin: Macrophage CR3 (αMβ2, CD11b/CD18) binds filamentous hemagglutinin of bordetella pertussis. Cell 61, 1375–1382.

Ridley, A. J., & Hall, A. (1992). The small GTP-binding protein rho regulates the assembly of focal adhesions and actin stress fibers in response to growth factors. Cell 70, 389–399.

Ridley, A. J., Paterson, H. F., Johnston, C. L., Diekmann, D., & Hall, A. (1992). The small GTP-binding protein rac regulates growth factor-induced membrane ruffling. Cell 70, 401–410.

Robinson, W. E., Jr., Montefiori, D. C., & Mitchell, W. M. (1987). Evidence that mannosyl residues are involved in human immunodeficiency virus type 1 (HIV-1) pathogenesis. AIDS Res. Hum. Retroviruses 3, 265–282.

Romer, L. H., Burridge, K., & Turner, C. E. (1992). Signaling between the extracellular matrix and the cytoskeleton: Tyrosine phosphorylation and focal adhesion assembly. Cold Spring Harbor Symp. Quant. Biol. 57, 193–202.

Rosales, C., & Brown, E. J. (1991). Two mechanisms for IgG Fc-receptor-mediated phagocytosis by human neutrophils. J. Immunol. 146, 3937–3944.

Rosales, C., & Brown, E. J. (1992). Signal transduction by neutrophil immunoglobulin G Fc receptors. Dissociation of [Ca^{+2}] rise from IP$_3$. J. Biol. Chem. 267, 5265–5271.

Rosales, C., Jones, S. L., McCourt, D., & Brown, E. J. (1994). Bromophenacyl bromide binding to the actin bundling protein l-plastin inhibits IP$_3$-independent [Ca^{+2}]$_i$ rise in human neutrophils. Proc. Natl. Acad. Sci. USA (in press).

Rosen, A., Keenan, K. F., Thelen, M., Nairn, A. C., & Aderem, A. (1990). Activation of protein kinase C results in the displacement of its myrisoylated, alanine-rich substrate from punctate structures in macrophage filopodia. J. Exp. Med. 172, 1211–1215.

Russell, D. G., & Wright, S. D. (1988). Complement receptor type 3 (CR3) binds to an Arg-Gly-Asp-containing region of the major surface glycoprotein, gp63, of Leishmania promastigotes. J. Exp. Med. 168, 279–292.

Sampson, L. L., Heuser, J., & Brown, E. J. (1991). Cytokine regulation of complement receptor-mediated ingestion by mouse peritoneal macrophages: M-CSF and IL-4 activate phagocytosis by a common mechanism requiring autostimulation by β interferon. J. Immunol. 146, 1005–1013.

Sawyer, D. W., Sullivan, J. A., & Mandell, G. L. (1985). Intracellular free calcium localization in neutrophils during phagocytosis. Science 230, 663–665.

Schlesinger, L. S. (1993). Macrophage phagocytosis of virulent but not attenuated strains of *Mycobacterium tuberculosis* is mediated by mannose receptors in addition to complement receptors. J. Immunol. 150, 2920–2930.

Schuler, G. D., Altschul, S. F., & Lipman, D. J. (1991). A workbench for multiple alignment construction and analysis. Proteins 9, 180–190.

Sehgal, G., Zhang, K., Todd, R. F., Boxer, L. A., & Petty, H. R. (1993). Lectin-like inhibition of immune-complex receptor-mediated stimulation of neutrophils—effects on cytosolic calcium release and superoxide production. J. Immunol. 150, 4571–4580.

Senior, R. M., Gresham, H. D., Griffin, G. L., Brown, E. J., & Chung, A. E. (1992). Entactin stimulates neutrophil adhesion and chemotaxis through interactions between its Arg-Gly-Asp (RGD) domain and the leukocyte response integrin (LRI). J. Clin. Invest. 90, 2251–2257.

Smiley, P. L., Stremler, K. E., Prescott, S. M., Zimmerman, G. A., & McIntyre, T. M. (1991). Oxidatively fragmented phosphatidylcholines activate human neutrophils through the receptor for platelet-activating factor. J. Biol. Chem. 266, 11104–11110.

Stahl, P. D. (1990). The macrophage mannose receptor: Current status. Am. J. Respir. Cell Mol. Biol. 2, 317–318.

Taniguchi-Sidle, A., & Isenman, D. E. (1992). Mutagenesis of the Arg-Gly-Asp triplet in human complement component C3 does not abolish binding of iC3b to the leukocyte integrin complement receptor type III (CR3, CD11b/CD18). J. Biol. Chem. 267, 635–643.

Tardieux, I., Webster, P., Ravesloot, J., Boron, W., Lunn, J. A., Heuser, J. E., & Andrews, N. W. (1992). Lysosome recruitment and fusion are early events required for Trypanosome invasion of mammalian cells. Cell 71, 1117–1130.

Taylor, M. E., Conary, J. T., Lennartz, M. R., Stahl, P. D., & Drickamer, K. (1990). Primary structure of the mannose receptor contains multiple XX motifs resembling carbohydrate-recognition domains. J. Biol. Chem. 265, 12156–12162.

Tran Van Nhieu, G., & Isberg, R. R. (1993). Bacterial internalization mediated by β_1 chain integrins is determined by ligand affinity and receptor density. EMBO J. 12, 1887–1895.

Turner, C. E., Glenney, J., Jr., & Burridge, K. (1990). Paxillin: A new vinculin-binding protein present in focal adhesions. J. Cell Biol. 111, 1059–1068.

Van Strijp, J. A. G., Russell, D. G., Tuomanen, E., Brown, E. J., & Wright, S. D. (1993). Ligand specificity of purified complement receptor type 3 (CD11b/CD18, Mac-1, alphaM beta2): Indirect effects of an Arg-Gly-Asp sequence. J. Immunol. 151, 3324–3336.

Wirthmueller, U., Kurosaki, T., Murakami, M. S., & Ravetch, J. V. (1992). Signal transduction by Fc[gamma]RIII (CD16) is mediated through the [gamma] chain. J. Exp. Med. 175, 1381–1390.

Wright, S. D., Levin, S. M., Jong, M. T. C., Chad, Z., & Kabbash, L. G. (1989). CR3 (CD11b/CD18) expresses one binding site for Arg-Gly-Asp-containing peptides and a second site for bacterial lipopolysaccharide. J. Exp. Med. 169, 175–183.

Zheleznyak, A., & Brown, E. J. (1992). IgG-mediated phagocytosis by human monocytes requires protein kinase C activation: Evidence for protein kinase C translocation to phagosomes. J. Biol. Chem. 267, 12042–12048.

Zhou, M., Todd, R. F., III, Van de Winkel, J. G. J., & Petty, H. R. (1993). Cocapping of the leukoadhesin molecules complement receptor type 3 and lymphocyte function-associated antigen-1 with Fc_{gamma} receptor III on human neutrophils: Possible role of lectin-like interactions. J. Immunol. 150, 3030–3041.

Zhou, M.-J., Poo, H., Todd, R. F., III, & Petty, H. R. (1992). Surface-bound immune complexes trigger transmembrane proximity between complement receptor type 3 and the neutrophil's cortical microfilaments. J. Immunol. 148, 3550–3553.

Zhou, M.-J., & Brown, E. J. (1993). Leukocyte response integrin and integrin associated protein act as a signal transduction unit in generation of a phagocyte respiratory burst. J. Exp. Med. 178, 1165–1174.

ANNEXIN HYPOTHESIS FOR EXOCYTOSIS

A. Lee Burns and Harvey B. Pollard

I. INTRODUCTION

Membrane fusion processes in cells underlie the structural integrity of the cytoplasm and are crucial to directing synthesis and export of enzymes, hormones, and transmitters. Export is accomplished by a process of exocytosis, in which the membranes of secretory vesicles fuse with the plasma membrane. The fusion mechanism may involve a class of calcium-binding proteins, termed annexins, and the data described below forms the basis of the annexin hypothesis for exocytosis.

Biomembranes
Volume 4, pages 65–74.
Copyright © 1996 by JAI Press Inc.
All rights of reproduction in any form reserved.
ISBN: 1-55938-661-4.

The first annexin to be discovered was synexin (annexin VII) and we shall focus our attention on its function and molecular biology in secretory cells.

II. EXOCYTOSIS AND THE DISCOVERY OF SYNEXIN

Calcium is important for this process in many cellular systems, as first recognized in the chromaffin cell. This requirement for calcium led to the question of whether a mediating protein might be involved, and calcium-binding proteins such as calmodulin, myosin, and tubulin historically received a lot of attention. Synexin was discovered in chromaffin cells, and it exhibited properties which rendered it a prime candidate for driving the calcium-dependent portion of exocytotic fusion (Creutz et al., 1978, 1979). An advantage of the chromaffin cell system is the abundance of secretory granules (chromaffin granules) containing epinepherine, ATP, enkephalins, and other bioactive agents. A convenient assay for synexin action was the ability of synexin to aggregate and fuse these granules in a calcium-dependent manner. A number of properties of this reaction have proved remarkably similar to those observed in secretory cells. Electron micrographs of the above granules fused with synexin actually appeared similar to pictures of granules within cells that had been stimulated to secrete with veratridine (Pollard et al., 1982) or nicotine (Ornberg et al., 1986). It is interesting to note that both *in vitro* aggregation of granules and exocytosis from isolated chromaffin cells are inhibited by phenothiazine drugs (trifluoperazine and promethazine; Creutz, 1981; Pollard et al., 1983). Additional support for the involvement of synexin in exocytosis was the observation that it preferentially bound to the inside of the cellular membrane, but not to the outside of the cell (Scott et al., 1985). Indeed, synexin has high affinity for acidic phospholipids which are preferentially located on the inner aspect of the plasma membrane and on the outer aspect of granule membranes. By contrast, synexin has very low affinity for phospholipids such as phosphatidyl choline (PC) or sphingomyelin, which are typically found on the outside of cell membranes and inside the granule membrane (Hong et al. 1982). More recently, synexin was detected in close proximity to granule membranes, in the cytoplasm, and in the nucleus of chromaffin cells by immunolocalization with different anti-synexin antisera (Kuijpers et al., 1992). Surprisingly, after these cells were stimulated with nicotine, there was a specific, rapid decrease in gold labeling in all compartments, and a 51% change in synexin associated with granules. Taken together, the above results strongly support the concept that synexin plays a key role in the process of exocytotic secretion.

In addition to synexin, lipocortin I (annexin I) and calpactin I (annexin II) can aggregate phospholipid vesicles in a solution containing calcium (Blackwood and Ernst, 1990). Furthermore, aggregated chromaffin granules can be fused by hetero-tetrameric calpactin I in the presence of micromolar calcium and arachidonic acid (Drust and Creutz, 1988). In as much as many mammalian cells contain more than one annexin, it can not be excluded that membrane fusion may involve several different annexins acting in concert or individually in different circumstances.

Conditions for involvement of different proteins might include: (1) differential sensitivity to changes in $[Ca^{2+}]$; (2) dependence on different divalent cations and metabolites; (3) fusion of membranes within cells; or (4) secretion in a manner specific to various tissues or developmental stage. For example, it might be argued that calpactin I, with a high calcium sensitivity, might function at low calcium concentrations, while synexin might be recruited at higher calcium concentrations. Indeed, it has been claimed that during exocytosis, the free $[Ca^{2+}]$ just under the plasma membrane may reach 100s of μM (Simon and Llinas, 1985). Another example supporting the concept of concerted action is the observation that calcium and barium can stimulate different and coincident processes in chromaffin cells (Heldman et al., 1989). Our search to discover a possible mediator of barium activity in chromaffin cells resulted in the isolation of lipocortin I from bovine lung (Lee et al., 1991). Possibly coincidentally, Ba^{2+} driven secretion is a much slower process than is that driven by Ca^{2+}, while lipocortin I-driven membrane fusion supported either by Ca^{2+} or Ba^{2+}, is also much slower that the same reactions driven by synexin and Ca^{2+}. Thus, while synexin might act rapidly and selectively with Ca^{2+}, lipocortin I could drive either Ca^{2+} or Ba^{2+} dependent events in a slower manner. Incidentally, slower does not necessarily refer to the speed of fusion, but perhaps to the probability that any one interaction may lead to fusion. Finally, most of these proteins differ either qualitatively or quantitatively in subcellular localization, in tissue distribution, and in calcium dependence of binding to various phospholipids and these variations may permit cells to respond by secreting differently in subtle ways to a wide variety of stimuli.

III. RECOGNIZING THE ANNEXIN FAMILY

In the mid-1980s, separate groups interested in important and distinct conditions (inflammation, blood coagulation, secretion, and others) identified several proteins that bound to membranes in the presence of calcium. At the time, it was not known how or whether these proteins were related to other, more well-defined calcium-binding proteins (calmodulin, parvalbumin, tropomyosin, etc.). However, the annexin family only became established after it became apparent that the group shared a conserved sequence of sixteen amino acids, termed the endonexin fold (Geisow and Walker, 1986). This domain was predicted from sequences derived from cDNAs of lipocortin I and calpactin I (annexins I and II) (Huang et al., 1986; Wallner et al., 1986) and from peptide sequences of endonexin I (annexin IV) and calelectrin (annexin VI; Geisow and Walker, 1986). As more annexins were discovered, the family was defined as a group of calcium-dependent membrane binding proteins with unique N-termini and a conserved C-terminal tetrad (or octad for annexin VI) repeat. During the ensuing years, annexins have been identified in a variety of other metazoans, including slime mold (Doring et al., 1991), hydra

(Schlaepfer et al., 1992), *Drosophila* (Johnston et al., 1990) and a plant (Smallwood et al., 1990).

   Cloning of annexin VII from human (Burns et al., 1989), mouse (Zhang-Keck et al., 1993), and slime mold (Doring et al., 1991) has permitted the identification of conserved regions as well as two polymorphisms. First, sequences of annexin VII cDNAs revealed that annexin VII mRNA in brain, cardiac, and skeletal muscles is alternatively spliced and results in the insertion of twenty two additional amino acids into the hinge region of the N-terminus (Magendzo et al., 1991). This addition actually does not change the aggregation activity when compared with the shorter form, but may alter the muscle annexin VII in other ways. The second polymorphism is in the utilization of alternate polyadenylation signals resulting in two sizes of mRNAs. Although the sequence of human and mouse annexin VIIs are very highly conserved (92%), like other mammalian annexins, the *Dictyostelium* annexin VII is quite divergent (42% conserved) in the C-terminus and has an expanded, repetitive GYP motif (GYPPQQ) occurring fourteen times in the N-terminus. When the annexin VII sequences are compared among themselves and with other aggregating annexins, certain conserved amino acid residues become apparent and currently are being altered by site-directed mutagenesis.

IV. MODEL FOR FUSION

Several years ago, we developed the Hydrophobic Bridge Hypothesis of membrane fusion as a way to help visualize the process of exocytosis and also to assist in the design of experiments with annexin VII (Figure 1) (Pollard et al., 1991). When cells are stimulated to secrete, the concentration of calcium in the cell increases mostly from extracellular sources. Higher levels of calcium permit annexin VII to dimerize or multimerize into cigar-shaped structures ($100 \text{ Å} \times 50 \text{ Å}$) in electron micrographs (Creutz et al., 1979). End views of these annexin VII structures in the same EM field appear to be dimeric cylinders joined to other dimers. A fascinating aspect of this polymerization is that it can be reversed in a low dielectric medium. Thus, annexin VII changes from a monomeric, cytoplasmic protein at low calcium levels to a polymeric, membrane-bound form at higher calcium concentrations (Figure 1A). The initial binding likely occurs by direct interactions of polymeric annexin VII with the phospholipids on two juxtaposed membranes, although it is also possible that calcium mediates the initial binding between the two types of molecules. Next, the constant movement of the hydrocarbon chains of the acidic phospholipids in the *cis* leaflets permits penetration by the annexin VII multimer. At this stage, the two *cis* leaflets become continuous (Figure 1B), and that may correspond to a previously determined rapid membrane-mixing rate constant (Nir et al., 1986; Pollard et al., 1991). Annexin VII polymers dissociate in the absence of a hydrophobic driving force, which allows both *trans* leaflets to migrate over the annexin VII molecules and to join. This would be sufficient to accomplish fusion

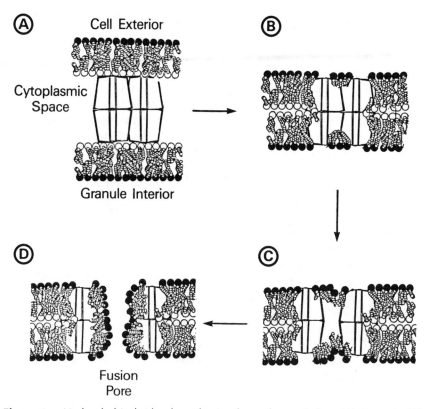

Figure 1. Hydrophobic bridge hypothesis of membrane fusion. (A) Annexin VII multimers simultaneously attach to granule and cytoplasmic membranes. (B) Acidic phospholipids (light-colored) migrate across the annexin VII bridge using conformational changes appearing in the sub-nanosecond time frame. (C) Annexin VII polymer dissociates in the low dielectric medium. (D) Outer-leaflet phospholipids (dark colored) move over the newly exposed hydrophobic annexin VII surfaces to complete fusion.

between the granule and the cytoplasmic membranes. Alternatively, the sequence of events could be that the calcium-dependent binding of annexin VII to membranes occurs prior to multimerization of annexin VII and granule aggregation. In this case, annexin VII first would bind to membranes at lower (40 μM) calcium concentration, followed by the self-association of annexin VII and membrane aggregation at higher (200 μM) amounts of calcium (Creutz and Sterner, 1983). In any case, the mixing of the *cis* and *trans* leaflets of the membranes could proceed as described above.

V. MOLECULAR DISSECTION OF ANNEXINS

Recently, two annexin chimeras were designed to determine which annexin domains are responsible for aggregation of liposomes *in vitro*. It has been known for several years, that annexin V cannot aggregate liposomes, whereas annexins I and VII can aggregate and actually fuse liposomes under proper conditions. As a first step, domains of annexins VII and I were replaced with corresponding domains of annexin V. In one case, a chimera (annexin V\VII) with an N-terminus of annexin V and a C-terminus of annexin VII retained the annexin VII ability to aggregate phosphatidyl-serine (PS) liposomes, but lost the fusion activity (Figure 1A). Therefore, the long, hydrophobic N-terminus of annexin VII is not required for aggregation, but may be important for fusion of PS liposomes (Burns et al., unpublished observations). Another chimera of annexin I\V was constructed with a very short N-terminus (8 amino acids) plus the first repeat (amino acids 41–118) of annexin I joined to the last three repeats (amino acids 92–320) of annexin V (Ernst et al., 1991). The annexin I\V chimera and annexin I supported aggregation of PS liposomes in the presence of 160 μM Ca^{2+}, whereas in 28 μM Ca^{2+} only unmodified annexin I retained the activity. Furthermore, an anti-annexin I monoclonal antibody inhibits aggregation by both the chimera and annexin I, and was shown by expression of cDNA deletion mutants to bind to an epitope within repeat I of annexin I (Ernst et al., 1991). Interestingly, this antibody–antigen reaction occurs, even when the annexin I is bound to phospholipids. Since both chimeras are active (Figure 2), one might conclude that the first repeat contains information for the promotion of liposome aggregation (Figure 2). Comparison of this repeat among the above annexins reveals only six amino acids which are identical in annexins I and VII, but different in annexin V (Figure 2). Unfortunately, none of these differences remain specific to the aggregating group, when additional annexins (annexins II and IV) are included in the comparison (Barton et al., 1991). Even though this inconsistency exists, these six residues may be useful targets for modification to define specific residues responsible for aggregation of phospholipids by annexins. Alternatively, the differences in activity between annexins V and I or VII may result from differences in secondary structure or hydrophobicity.

Another series of experiments have been performed in another and our laboratories to characterize the calcium-binding sites in several annexins. We decided to modify acidic residues in the endonexin fold of each repeat (i.e., D182N, E183Q in repeat I) and in a region 38 amino acids downstream (i.e., D220L in repeat I) in repeats 1, 2, and 4 of annexin VII based on amino acid sequence conservation and modeling. Subsequent information from the annexin V crystal structure indicated that the former sites are very close to carbonyl-oxygens coordinating calcium in the (M,L)KG(A,L)GT motif (Huber et al., 1990). In the latter case, the carboxylic bonds of aspartic- and glutamic-residues actually coordinate with calcium ions, if it can be assumed that the structure of different annexins are similar. Preliminary data indicate that some of the mutated proteins expressed in *E. coli* still were able to bind phospholipids in a calcium-dependent manner (Srivastava et al., unpub-

(A)

PROTEIN	N-TERMINI	+	C-TERMINAL TETRAD REPEATS				AGGREGATES
			(1)	(2)	(3)	(4)	
ANNEXIN I	I	+	I	I	I	I	YES
ANNEXIN I\V	(I)	+	I	V	V	V	YES
ANNEXIN V	V	+	V	V	V	V	NO
ANNEXIN V\VII	V	+	VII	VII	VII	VII	YES
ANNEXIN VII	VII	+	VII	VII	VII	VII	YES

(B)

```
            *                      *        *            *   *       *
  I:  DVAALHKAIMVKGVDEATIIDILTKRNNAQRQQIKAAYLQETGKPLDETLKKALTGHLEEVVLALLKTPAQF
  V:      AET R  MKGL T  ES LTLLTS S A   E S   FKTLF RD LDD   SE T KF KLIV  MKPSRLY
VII:      AEI R  MKGF T  QA VDVVAN S D   K K   FKTSY KD IKD   SE S NM ELIL  FMPPTYY
```

Figure 2. Activity of annexin chimera and sequence comparison of repeat I. (A) Representation of annexins I, V, and VII plus two chimera, where different domains were switched between annexins. The annexin I\V chimera contains a short N-terminus (amino acids 1–2 and 41–46) and the first repeat (amino acids 47–118) of annexin I added to the last three C-terminal repeats (amino acids 92–320) of annexin V (Ernst et al., 1991). The annexin V\VII construct was made by replacing the 167 amino acids in the N-terminus of annexin VII with the first 19 amino acids of annexin V (Burns et al., 1989; unpublished observations). Note that all proteins except annexin V can aggregate. (B) Protein sequences of repeat I from the three annexins. Numbers for residues in the first repeat of annexins I, V, and VII are 47–118, 20–91, and 168–239, respectively. Amino acids which are the same in annexins I and VII, but different in annexin V are starred.

lished observations). Similar mutations (D161A, E245A, and D321A) in carboxylic coordinating positions in annexin II have been studied by another group of investigators (Jost et al., 1992). Calcium concentration at half-maximal binding to PS was most affected in the mutant protein with changes in three repeats (33-fold decrease) and was intermediate for proteins containing less changes. Even though these modified annexin II molecules were still able to bind to membranes in a calcium dependent manner, the carboxyl-oxygens of the acidic amino acids are important for calcium binding (Jost et al., 1992). The retention of activity in some of the endonexin fold mutants could indicate that these mutations were not severe enough to perturb the proper folding of the calcium-binding site. The data also suggest that other weaker calcium-binding sites (i.e., perhaps lanthanium sites identified in annexin V or the presumed calcium-binding site in the channel, Huber et al., 1990) as predicted by the crystal structure could provide additional sites critical for binding. Additional experiments are required to explain how the calcium binding and annexin structure influences aggregation and fusion.

VI. CONCLUSION

Since the discovery of annexin VII fifteen years ago, results from several laboratories have shown that annexins I, II, and VII aggregate chromaffin granules and liposomes *in vitro*, and that annexin VII also drives fusion of granule ghosts and liposomes. These data lend support to the hypothesis that these annexins may be secretory factors *in vivo*. The above mutagenesis experiments revealed the importance in annexin I of repeat one for aggregation activity and of repeats 2, 3, and 4 for calcium-binding sites. Furthermore, since all annexins bind to phospholipids in a calcium-dependent manner, membranes continue to be a focus for understanding annexin activity. Recently, annexin II was detected in postpartum mammary epithelial cells, when calcium-dependent casein secretion starts (Burgoyne, 1992) and immunogold labeling of annexin VII in chromaffin cells was shown to decrease following treatment with a secretagogue (Kuijpers et al., 1992). In addition, annexin VII (plus annexins I and V) has been reported to insert into artificial membranes in the presence of calcium and to produce voltage-dependent channels (Pollard et al., 1991). Although the relationship between the aggregation and channel activities is not clear at this time (Pollard et al., 1991), the data indicate that the annexins can penetrate membranes as well as adhere to membrane surfaces. Thus, while the actual functions of annexins *in vivo* have yet to be defined, the available data considerably strengthens the concept of their likely involvement in the membrane fusion event occurring during exocytosis.

REFERENCES

Barton, G. J., Newman, R. H., Freemont, P. S., & Crumpton, M. J. (1991). Amino acid sequence analysis of the annexin super-gene family of proteins. Eur. J. Biochem. 198, 749–760.

Blackwood, R. A., & Ernst, J. D. (1990). Characterization of Ca2(+)-dependent phospholipid binding, vesicle aggregation and membrane fusion. Biochem. J. 266, 195–200.

Burgoyne, R. D. (1992). Annexin II: Involvement in exocytosis. In: The Annexins (Moss, S. E., ed.), pp. 69–76. Portland Press Ltd., London.

Burns, A. L., Magendzo, K., Shirvan, A., Alijani, M., Rojas, E., & Pollard, H. B. (1989). Calcium Channel Activity of purified human synexin and structure of the human synexin gene. Proc. Natl. Acad. Sci. USA 86, 3798–3802.

Creutz, C. E., Pazoles, C. J., & Pollard, H. B. (1978). Identification and purification of an adrenal medullary protein (synexin) that causes calcium-dependent aggregation of chromaffin granules. J. Biol. Chem. 253, 2858–2866.

Creutz, C. E., Pazoles, C. J., & Pollard, H. B. (1979). Self-association of synexin in the presence of calcium: Correlation with synexin-induced membrane fusion and examination of the structure of synexin aggregates. J. Biol. Chem. 254, 553–558.

Creutz, C. E. (1981). Cis-unsaturated fatty acids induce the fusion of chromaffin granules aggregated by synexin. J. Cell Biol. 91, 247–256.

Creutz, C. E., & Sterner, D. C. (1983). Calcium dependence of the binding of synexin to isolated chromaffin granules. Biochem. Biophys. Res. Commun. 114, 355–364.

Drust, D. S., & Creutz, C. E. (1988). Aggregation of chromaffin granules by calpactin at micromolar levels of calcium. Nature (London) 331, 88–91.

Doring, V., Schleicher, M., & Noegel, A. A., (1991). *Dictyostelium* annexin VII (synexin): cDNA sequence and isolation of a gene disruption mutant. J. Biol. Chem. 266, 17509–17515.

Ernst, J. D., Hoye, E., Blackwood, R. A., & Mok, T. L. (1991). Identification of a domain that mediates vesicle aggregation reveals functional diversity of annexin repeats. J. Biol. Chem. 266, 6670–6673.

Geisow, M. J., & Walker, J. H. (1986). New proteins involved in cell regulation by Ca^{2+} and phospholipids. Trends Biochem. Sci. 11, 120–123.

Heldman, E., Levine, M., Raveh, L., & Pollard, H. B., (1989). Barium ions enter chromaffin cells via voltage-dependent calcium channels and induce secretion by a mechanism independent of calcium. J. Biol. Chem. 264, 7914–7920.

Hong, K., Duzgunes, N., Ekerdt, R., & Papahadjopoulos, D. (1982). Synexin facilitates fusion of specific phospholipid membranes at divalent cation concentrations found intracellularly. Proc. Natl. Acad. Sci. USA 79, 4642–4644.

Huang, K-S., Wallner, B. P., Mattaliano, R. J., Tizard, R., Burne, C., Frey, A., Hession, C., McGray, P., Sinclair, L. K., Chow, E. P., Browning, J. L., Ramachandran, K. L., Tang, J., Smart, J. E., & Pepinsky, R. B. (1986). Two human 35kd inhibitors of phospholipase A2 are related to substrates of pp60 v-src and of the epidermal growth factor receptor/kinase. Cell 46, 191–199.

Huber, R., Romisch, J., & Paques, E-P. (1990). The crystal and molecular structure of human annexin V, an anticoagulant protein that binds to calcium and membranes. EMBO J. 9, 3867–3874.

Jost, M., Thiel, C., Weber, K., Gerke, V. (1992). Mapping of three unique Ca^{2+}-binding sites in human annexin II. Eur. J. Biochem. 207, 923–930.

Johnston, P. A., Perin, M. S., Reynolds, G. A., Wasserman, W. S. A., & Sudhof, T. C. (1990). Two novel annexins from *Drosophila melanogaster*. J. Biol. Chem. 265, 11382–11388.

Kuijpers, G. A. J., Lee, G., & Pollard, H. B. (1992). Immunolocalization of synexin (annexin VII) in adrenal chromaffin granules and chromaffin cells: Evidence for a dynamic role in the secretory process. Cell Tissue Res. 269, 323–330.

Lee, G., de la Fuente, M., & Pollard, H. B. (1991). A barium-dependent chromaffin granule aggregating protein from bovine adrenal medulla and other tissues. Ann. N. Y. Acad. Sci. 635, 477–479.

Magendzo, K., Shirvan, A., Cultraro, C., Srivastava, M., Pollard, H. B., & Burns, A. L. (1991). Alternative splicing of human synexin mRNA in brain, cardiac, and skeletal muscle alters the unique N-terminal domain. J. Biol. Chem. 266, 3228–3232.

Nir, S., Stutzin, A., & Pollard, H. B. (1987). Effect of synexin on aggregation and fusion of chromaffin granule ghosts at pH 6. Biochim. Biophys. Acta 903, 309–318.

Ornberg, R. L., Duong, L. T., & Pollard, H. B. (1986). Intergranular vesicles: New organelles in the secretory granules of adrenal chromaffin cells. Cell Tissue Res. 245, 547–553.

Pollard, H. B., Creutz, C. E., Fowler, V. M., Scott, J. H., & Pazoles, C. J. (1982). Calcium-dependent regulation of chromaffin granule movement, membrane contact, and fusion during exocytosis. Cold Spring Harbor Symp. Quant. Biol. 46, 819–834.

Pollard, H. B., Scott, J. H., & Creutz, C. E. (1983). Inhibition of synexin activity and exocytosis from chromaffin cells by phenothiazine drugs. Biochem. Biophys. Res. Comm. 113, 908–915.

Pollard, H. B., Rojas, E., Pastor, R. W., Rojas, E. M., Guy, H. R., & Burns, A. L. (1991). Synexin: Molecular mechanism of calcium-dependent membrane fusion and voltage-dependent calcium-channel activity. Ann. N. Y. Acad. Sci. 635, 328–351.

Schlaepfer, D. D., Fisher, D. A., Brandt, M. E., Bode, H. R., Jones, J. M., & Haigler, H. T. (1992). Identification of a novel annexin in *Hydra vulgaris*. J. Biol. Chem. 267, 9529–9539.

Scott, J. H., Creutz, C. E., Pollard, H. B., & Ornberg, R. L. (1985). Synexin binds in a calcium dependent fashion to oriented chromaffin cell plasma membranes. FEBS Letts. 180, 17–23.

Simon, S. M., & Llinas, R. R. (1985). Compartmentalization of the submembrane calcium activity during calcium influx and its significance in transmitter release. Biophys. J. 48, 485–498.

Smallwood, M., Keen, J. N., & Bowles, D. J. (1990). Purification and partial sequence analysis of plant annexins. Biochem. J. 270, 157–161.

Wallner, B. P., Mattaliano, R. J., Hession, C., Cate, R. L., Tizard, R., Sinclair, L. K., Foeller, C., Chow, E. P., Browning, J. L., Ramachandran, K. L., & Pepinsky, R. B. (1986). Cloning and expression of human lipocortin, a phospholipase A2 inhibitor with potential anti-inflammatory activity. Nature (London) 320, 77–81.

Zhang-Keck, Z. -Y., Burns, A. L., & Pollard, H. B. (1993). Mouse synexin (annexin VII) polymorphisms and a phylogenetic comparison with other annexins. Biochem. J. 289, 735–741.

SMALL SYNAPTIC VESICLES

Nandini V. L. Hayes and Anthony J. Baines

Biomembranes
Volume 4, pages 75–122.
Copyright © 1996 by JAI Press Inc.
All rights of reproduction in any form reserved.
ISBN: 1-55938-661-4.

I. INTRODUCTION

The secretion of neurotransmitters and neuropeptides is mediated by regulated exocytosis at the nerve terminal. At least two different regulated secretory pathways exist in nerve terminals: they are specialized for secretion from two distinct classes of regulated secretory vesicles—small synaptic vesicles (SSVs) and large dense core vesicles (LDCVs). The existence of two pathways enables a single neuron to secrete a cocktail of different neurotransmitters.

Electrophysiological studies of nerve endings in muscle provided the first evidence that chemical neurotransmitters were released in quanta or packets of molecules (Katz, 1969). It is clear that each quantum of neurotransmitter (approximately 1,000–5,000 molecules) is packaged into a single synaptic vesicle. The contents of SSVs are released by exocytosis in response to an action potential at the active zone, a region of the presynaptic membrane that appears somewhat dense and thickened in thin section electron microscopy. It has also become apparent that there must be endocytosis at nerve terminals to retrieve membrane components inserted into the presynaptic membrane by exocytosis, and to maintain the correct area of the plasma membrane: The nerve terminal is active in both endocytotic and exocytotic membrane traffic.

To understand how neurotransmitters are selectively incorporated into synaptic vesicles, and are eventually secreted by fusion of exocytotic vesicles with the cell surface, it is necessary to understand the structure of synaptic vesicles, and how that structure relates to their function. Ultimately, this means understanding the nature, activities, and interactions of the synaptic vesicle proteins, and the proteins of the presynaptic membrane. Molecular genetics and protein biochemical approaches have been combined in the last ten years with electron microscopy and electrophysiology to characterize the proteins of synaptic vesicles. It now seems likely that the majority of the proteins associated with SSVs have been identified. The result of these studies has been the beginning of a coherent molecular model of how a nerve ending responds to the arrival of an action potential. The models now emerging also benefit from the study of naturally occurring neurotoxins, some of which act directly upon proteins involved in neuronal exocytosis.

The dynamic membrane traffic at nerve endings may also serve as a model for membrane trafficking in non-neuronal cells. It is becoming clear that many synaptic vesicle-associated proteins are, in fact, members of larger families of proteins which often have a more widespread distribution, especially in neuro-endocrine tissue but also in many other cell types. An emerging concept is that proteins previously considered unique to nervous systems have been adapted from non-neuronal ancestral proteins to perform specialized functions at the nerve terminal.

In this chapter, we will introduce the classes of synaptic vesicle, and then discuss the structure and components of SSVs in some detail. We will review the relationship of SSV structure to their dynamics in the nerve terminal, especially the mechanisms that organize SSV within the synaptic cytoplasm, and at the active zone where exocytosis takes place. Finally, we will comment on the recycling of SSV from components retrieved from synaptic plasma membranes by endocytosis.

A. Small Synaptic Vesicles

SSVs store and release the so-called classical neurotransmitters such as acetylcholine, γ-aminobutyric acid (GABA), glutamate, and glycine. These vesicles are remarkably homogenous both in size (40–60 nm in diameter; Peters et al., 1991) and density, and they account for approximately 6% of total brain protein (Goelz et al., 1981; Huttner et al., 1983). Electron microscopic studies reveal that they have a clear core, except for catecholamine containing SSVs, which appear to have an electron-opaque core under certain fixation conditions (De Camilli and Jahn, 1990); these particular vesicles are sometimes referred to as small dense core vesicles. SSVs undergo a local exo/endocytotic recycling at the nerve terminal and can be reloaded with cytoplasmic neurotransmitter after each cycle as the machinery that is required to synthesize the classical neurotransmitters is present in the presynaptic nerve terminal and the neurotransmitter transporters are present in the SSV membrane.

SSV exocytosis is initiated by the Ca^{2+} influx that is mediated by voltage-sensitive Ca^{2+} channels upon depolarization of the presynaptic membrane. Within 200 μsec of calcium influx, SSV exocytosis has occurred. A second essential requirement for exocytosis is GTP. The mechanism of exocytosis is discussed in Section VII.

B. Large Dense Core Vesicles

LDCVs are involved in the storage and secretion of neuropeptides and peptides. They are larger (80–120 nm in diameter) than SSVs and are more heterogeneous in size and appearance. LDCVs characteristically have an electron-dense appearance in electron microscopy, due to the very high concentrations of neuropeptides that these vesicles contain (Klevin et al., 1982). LDCVs have been considered to be the neuronal equivalent of secretory granules in endocrine cells as they have a similar morphology, content, and mechanism of assembly (Navone et al., 1984). LDCVs are recycled via the Golgi apparatus. The neuropeptides are dependent on a constant supply of pre-assembled LDCVs from the perikarya as neuropeptide and pro-peptide precursors can only be synthesized and loaded in the perikarya.

LDCVs do not cluster at active zones and this suggests that a docking mechanism distinct from that controlling the docking of SSVs must be present (De Camilli and Jahn, 1990). Neurotransmitter release from this class of vesicle occurs approximately 50 msec after the calcium influx has occurred. This time lag again indicates a distinct release mechanism when compared to SSV. For a more detailed review on LDCV, see De Camilli and Jahn (1990).

C. Coated Vesicles

SSVs and LDCVs are exocytotic vesicles. A third class of vesicle is common in synaptic termini—coated endocytotic vesicles. Their function seems to be associated with recycling exocytotic vesicle components from presynaptic plasma membranes. The major coat protein of synaptic coated vesicles is the hetero-hexamer, clathrin. Individual clathrin molecules (triskelions) have a three-legged structure: they interact with each other to form a basket-like lattice (the name clathrin indicates "lattice protein"; Pearse and Robinson, 1990). Each clathrin triskelion contains three heavy-chains and three light-chains (the light-chain polypeptides being derived from two genes—LCa or LCb). Heavy- and light-chains are the products of two separate genes (for a review see Brodsky et al., 1991). Two neuron-specific isoforms of LCa and one additional isoform of LCb exist (Jackson, 1992; Stamm et al., 1992).

Clathrin molecules in the coat interact with transmembrane "cargo" proteins through so-called adaptor proteins which adapt the (constant) clathrin coat to the (variable) membrane protein cargos of coated vesicles (reviewed in Pearse and Robinson, 1990 and Keen, 1990). Two adaptor protein complexes, AP-1 and AP-2, are found in all cells: AP-1 is specific for coated vesicles associated with the *trans*-Golgi network and AP-2 is specific for coated vesicles involved in the endocytosis of plasma membrane components. Their subunits are referred to as adaptins and two neuron-specific α and β adaptins have been identified (Robinson, 1989; Ponnambalam et al., 1990). Other adaptors, AP180/AP3/np-185 (Ahle and Ungewickell, 1986; Keen, 1990; Puszkin et al., 1992) and auxilin (Ahle and Ungewickell, 1990), are neuron-specific. A recent study indicates that coated vesicles which have been isolated from nerve terminals are derived from SSVs (Maycox et al., 1992). These coated vesicles contain AP-2 adaptor components, as well as auxilin and AP180.

It may be supposed that the specialized adaptor components of coated synaptic vesicles relate to the specific recycling of synaptic vesicle components—this is discussed more fully in Section VIII.

II. PROPERTIES OF SMALL SYNAPTIC VESICLES

A. Function of SSVs

In chemical synapses, neurotransmitters are stored in and released from SSVs in the nerve terminal of presynaptic cells. Therefore, SSVs have two distinct functions: the first is uptake and storage of neurotransmitters; the second is membrane fusion and exocytosis. These functions are carried out by different proteins; SSV proteins can be divided into two classes corresponding to these functions. Fusion and exocytosis are likely to involve additional proteins located close to the exocytotic active zone, and probably some cytoplasmic proteins too.

B. Distribution of SSVs

SSVs are located in all presynaptic nerve terminals of the CNS and PNS. Small vesicles which are morphologically and biochemically similar to SSVs are found in other types of nerve endings, for example, in sensory nerve endings and in nerve endings thought to be specialized for the secretion of neurotransmitters into the blood rather than to adjacent neurons (the so-called neurosecretory cells; Scharrer and Scharrer, 1954; Navone et al., 1989).

C. Purification of SSVs

As SSVs are abundant at the presynaptic terminal and very homogenous in size and density, they can be purified to a high degree and high yield. It must be remembered though, that the purified vesicles contain a mixed population of synaptic vesicles with regard to neurotransmitter content, unless they are taken from a specialized source such as *Torpedo* electroplax organ where all SSV serve the same communication function.

Synaptic vesicles were originally purified by Whittaker et al. (1964). They were isolated from a crude guinea pig synaptosomal preparation which was subjected to hypo-osmotic lysis (to release the synaptic vesicles) and differential ultracentrifugation. Further purification was achieved by centrifugation through a sucrose gradient; later workers added a gel filtration step on glass bead columns (e.g., Nagy et al., 1977). This method can be directly scaled-up for use with bovine or ovine brain tissue. Hell et al. (1988) have modified this method by pulverizing frozen tissue instead of osmotically lysing synaptosomes.

Synaptic vesicles containing a single neurotransmitter (acetylcholine) were obtained from *Torpedo* electric organ (Whittaker, 1984). Synaptic vesicles from this source were isolated mainly by sucrose density gradient centrifugation in zonal rotors (Tashiro and Stadler, 1978) or by sucrose density gradient flotation centrifugation (Carlson et al., 1978) and subjected to a final round of purification by chromatography on controlled-pore glass bead columns. Another source of cholinergic SSVs was found to be the myenteric plexus longitudinal muscle preparation from guinea pig ileum (Huttner et al., 1983; Agoston et al., 1985). This preparation was also used for isolating SSVs containing neuropeptides (Tashiro and Stadler, 1978).

It is possible that these preparations contain extravesicular soluble material and membrane components which may cause problems in evaluating studies which try to identify components of SSVs.

D. Neurotransmitters of SSVs and Their Uptake

Although many chemicals function as neurotransmitters, SSVs only contain the classical neurotransmitters. These can be classified into two groups: the amino acid or type 1 neurotransmitters are usually fast acting and include glutamate, γ-aminobutyric acid (GABA), and glycine. The type 2 class of neurotransmitters include acetylcholine, dopamine, noradrenaline, adrenalin, purines, and 5-hydroxytyramine (5-HT; for a review see McMahon and Nicholls, 1991).

After neuronal exocytosis SSVs reform and refill with neurotransmitter from cytoplasmic pools. The precise mechanism for this uptake and storage of neurotransmitters by SSVs is still unclear, but it seems certain that the transporters that take up neurotransmitters into SSVs are different from those that mediate uptake at the presynaptic membrane (McMahon and Nicholls, 1991). Neurotransmitter uptake is driven by a proton electrochemical gradient which is generated across the membrane of SSVs by a proton pump (Maycox et al., 1990). All SSVs have a proton pump which belongs to a family of vacuolar H^+-ATPases (or V-type pumps) which is found in endosomes, Golgi membranes, and clathrin coated vesicles. These proton pumps consist of a large hetero-oligomeric complex with 8–9 different subunits (for reviews see Stone et al., 1990 and Sudhof and Jahn, 1991). Recently, the vacuolar H^+-ATPase has been partially purified from rat brain SSVs (Floor et al., 1990). The results from this study suggest that the vacuolar H^+-ATPase is essential for catecholamine uptake.

SSVs contain four specific classes of neurotransmitter transporter which are coupled to the electrochemical proton gradient. One class is associated with acetylcholine uptake (Anderson et al., 1982); a second is specific for the biogenic amines and is probably identical to that of chromaffin granules (Sternbach et al., 1990; Isambert et al., 1992); a third is specific for glutamate (Maycox et al., 1988); and the fourth for GABA (Hell et al., 1988; Burger et al., 1991) and glycine (Christensen et al., 1990). The uptake of different neurotransmitters is differentially coupled to the proton electrochemical gradient (Maycox et al., 1988). As the proton pump is electrogenic, the generation of a pH gradient depends on negative counterions provided by chloride influx via chloride channels in the vesicle membrane. A chloride channel has been purified from bovine brain coated vesicles (Xie et al., 1989), but has not to date been identified in synaptic vesicles. The uptake of glutamate anions is largely dependent on the vesicular membrane potential (McMahon and Nicholls, 1991) although at low chloride concentrations, both a change in the membrane potential and the vesicular pH gradient appear to drive glutamate uptake (Tabb et al., 1992). The uptake of positively charged amines is driven mainly by the pH gradient while acetylcholine, GABA, and glycine uptake by both (McMahon and Nicholls, 1991).

Several of the vesicular transporters including the glutamate (Maycox et al., 1988) and GABA (Hell et al., 1988) transporters have been solubilized and functionally reconstituted into liposomes. Recently, the amino acid sequence of two vesicular amine transporters from brain and adrenal medulla has been published (Lui et al., 1992). The sequence predicts proteins which have twelve transmembrane domains; these proteins and the synaptic vesicle protein, SV2 (see forthcoming), belong to the ubiquitous "duodecimal" family of transmembrane transporters that are characterized by having twelve putative transmembrane helices (Henderson, 1993).

III. TRANSMEMBRANE PROTEINS ASSOCIATED WITH SMALL SYNAPTIC VESICLES

The transmembrane proteins of small synaptic vesicles are likely to mediate channel and pump functions (including neurotransmitter uptake and proton pumping) and interactive functions (include aspects of the docking and fusion of SSV with presynaptic plasma membranes). In this section, we describe, in turn, proteins with channel functions (synaptophysin, synaptoporin, and SV2) and some proteins possibly involved in docking and exocytosis (synaptotagmins and synaptobrevins) together with one whose functions are not so well established—p29.

A. Synaptophysin

Synaptophysin (or p38) is an abundant integral membrane glycoprotein associated with SSVs in both the CNS and PNS (Wiedenmann and Franke, 1985; Jahn et al., 1985). Immunocytochemical studies showed that synaptophysin was found in all SSVs independent of their neurotransmitter content (Navone et al., 1986). It is also found in neuroendocrine cells (Navone et al., 1986) in clear small synaptic-like microvesicles vesicles (SLMVs) distinct from the peptide-containing granule cells in endocrine tissue (for a review, see De Camilli and Jahn, 1990). The primary sequences for rat (Buckley et al., 1987; Sudhof et al., 1987), human (Sudhof et al., 1987), and bovine synaptophysin (Johnson et al., 1989) have been elucidated (Leube et al., 1987; Sudhof et al., 1987) and show that this protein is highly conserved between species. It has cytoplasmic amino- and carboxy-terminals which sandwich four transmembrane regions. The carboxy-tail region contains ten copies of an imperfect tyrosine and glycine-rich repeat sequence (YPG(Q)QG) which is not α-helical, but nevertheless is thought to be rigid. The transmembrane and cytoplasmic regions exhibit a remarkable degree of homology (there were only 3% amino acid substitutions) while there was 22% amino acid substitution in the intravesicular regions. There was a single N-linked glycosylation site located on the first intravesicular loop; a second potential glycosylation site was present in bovine synaptophysin, but this was not thought to be used. Synaptophysin contained a cytoplasmic calcium-binding region which may be physiologically important during neuronal exocytosis (Rhem et al., 1986), although whether synaptophysin is a true Ca^{2+} binding protein has been challenged (Brose et al., 1992).

Cross-linking and sedimentation studies indicated that synaptophysin could form hexameric oligomers (Thomas et al., 1988) although a tetrameric structure has also been proposed (Rhem et al., 1986; Johnston and Sudhof, 1990). Synaptophysin hexamers have been reported to form voltage-sensitive channels when reconstituted into lipid bilayers (Thomas et al., 1988). This finding and limited structural similarities between synaptophysin and the gap junction proteins, the connexins (Leube et al., 1987; Sudhof et al., 1987), has led to the suggestion that synaptophysin contributed to the formation of exocytotic fusion pores (see Section VII for discussion). There is evidence though to suggest that synaptophysin is required for calcium-dependent neuronal exocytosis. When calcium-dependent glutamate secretion was reconstituted into *Xenopus* oocytes by microinjection of rat cerebellar mRNA, the introduction of anti-synaptophysin antibodies or antisense oligonucleotides to synaptophysin inhibited calcium-dependent secretion of glutamate (Alder et al., 1992). However, these experiments did not distinguish between synaptophysin having a regulatory or a structural role.

Synaptophysin is a substrate for endogenous protein kinases in intact synaptosomes and in purified SSVs (Pang et al., 1988; Rubenstein et al., 1993). It was phosphorylated on serine by calcium/calmodulin protein kinase II (Rubenstein et al., 1993) and by the proto-oncogene product pp60[c-src] on tyrosine residues (Barnekow et al., 1990). Both kinases have been located on the membrane of SSVs (Benfenati et al., 1992b).

B. Synaptoporin

Synaptoporin is a 37 kD SSV-associated protein which is homologous (58% identity) to synaptophysin (Knaus et al., 1990). In fact, synaptoporin was identified by the use of low-stringency hybridization with synaptophysin cDNA probes (Knaus et al., 1990). Synaptoporin co-purified with synaptic vesicles and appeared to be neuron-specific, with the exception that very small amounts have been detected in neuro-endocrine tissue (Knaus et al., 1990). The primary structure of synaptoporin revealed that it could be assigned to a proposed synaptophysin/connexin hexameric channel superfamily; like synaptophysin, synaptoporin had four potential transmembrane regions (Knaus et al., 1990). The transmembrane segments were highly conserved between the two proteins. There were two putative glycosylation sites on the first intracellular loop and a single site on the second intracellular loop. The serine-rich cytoplasmic carboxy-tail of synaptoporin contained five copies of an imperfect pentapeptide repeat (YXQ/SXX). Two serine residues located in the tail region conformed to the consensus sequence for calcium/calmodulin-dependent protein kinase II.

Synaptoporin may be involved in the regulation of intravesicular ion and solute composition or it may function as the vesicular part of a transient bilayer channel at the presynaptic membrane; the latter of these suggestions is discussed below.

C. SV2

Synaptic vesicle protein 2 (SV2) was originally identified by monoclonal antibodies raised against a preparation of highly purified *Torpedo* cholinergic synaptic vesicles (Stadler and Dowe, 1982; Carlson and Kelly, 1983). This proteoglycan had the characteristics of an integral membrane protein (Carlson and Kelly, 1983; Buckley and Kelly, 1985). Cross-reactive proteins have been detected in neuro-endocrine tissue, but were absent from some secretory cells including exocrine cells (Buckley and Kelly, 1985).

cDNA encoding this protein was recently analyzed (Bajjalieh et al., 1992; Feany et al., 1992). The amino acid sequence of the protein predicted from these analyses indicated that the SV2 protein is composed of 12 hydrophobic stretches which are long enough to be transmembrane regions; this is consistent with epitope-mapping data which indicated that SV2 was a transmembrane protein (Stadler and Dowe, 1982; Carlson and Kelly, 1983; Buckley and Kelly, 1985). There were three consensus sequences for N-glycosylation in a long hydrophilic region situated between the putative transmembrane regions 7 and 8 (Bajjalieh et al., 1992).

Two forms of SV2 were detectable in *Torpedo* cholinergic synaptic vesicles, the H (250 kD) and L (100 kD) forms (Scranton et al., 1993). Both forms of this protein were keratan sulphate proteoglycans (SV2pg). At least one epitope was unique to the *Torpedo* H form. In addition, two forms of SV2 have been detected in rat brain (Bahr and Parsons, 1992; Bajjalieh et al., 1992). One form had a higher level of expression in the cortex and hippocampus, the other in subcortical regions such as the basal ganglia and thalamus (Bajjalieh et al., 1992).

SV2 appears to belong to the "duodecimal" superfamily of transporters (reviewed in Henderson, 1993); members of this family carry out diverse transport functions in most living organisms. Its structure is most closely related to one of the human glucose transporters (GLUT1) and plasma membrane transporters for neurotransmitters (Bajjalieh et al., 1992; Feany et al., 1992). Recently, SV2 was identified as the receptor for vesamicol, an inhibitor of acetylcholine transport in *Torpedo* synaptic vesicles (Bahr and Parsons, 1992). This raised the possibility that this protein may mediate the uptake of neurotransmitters into synaptic vesicles.

D. Synaptotagmins

The synaptotagmins are a family of putative calcium-binding synaptic vesicle-associated proteins which have a differential distribution in the brain (Geppert et al., 1991; Wendland et al., 1991). Synaptotagmin I, or p65, was the first integral SSV-associated protein to be identified. This abundant 65kD synaptic vesicle protein was also found in neuro-endocrine tissue on synaptic-like microvesicles (Matthew et al., 1981). Synaptotagmins I and II contain a small glycosylated intravesicular amino-terminus, a single transmembrane domain, and a large cytoplasmic tail region. Most of the tail region consists of two copies of a repeat sequence that is homologous to the phospholipid- and calcium-binding regulatory C2 domain of protein kinase C (Perin et al., 1991). The C2 domain confers calcium sensitivity to protein kinase C activity and allows it to translocate to phospholipid membranes as a function of calcium concentration (Bell and Burns, 1991). This suggests that synaptotagmins I and II are calcium- and phospholipid-binding proteins and may play a role in the fusion and/or docking of SSVs. Both recombinant and native synaptotagmin I do, indeed, bind physiological concentrations of calcium in the presence of negatively charged phospholipids (Brose et al., 1992).

In SSVs, native synaptotagmin I is a homo-oligomer, probably a homotetramer (Perin et al., 1991; Brose et al., 1992). Proteolytic cleavage at a single site adjacent to the transmembrane region produced a large fragment which contained the two C2 domains (Perin et al., 1990). This cleavage abolished the binding of synaptotagmin I to calcium and to phospholipids. The amino-terminal fragment remained as a high molecular weight complex in the membrane, but the carboxy-terminal fragment was released as a soluble monomer. This suggested that the tetramerization was mediated by the amino-terminus and was required for the formation of the calcium/phospholipid complex (Brose et al., 1992). Therefore, the multiple calcium-binding sites which are present on a single complex may conceivably satisfy the nonlinear dependence for calcium during exocytosis (Davletov et al., 1993) leading to the proposal that synaptotagmin may be the calcium sensor that triggers exocytosis of SSV (Brose et al., 1992).

Synaptotagmin I interacts with the α-latrotoxin receptor (see Section VII). This interaction was found to inhibit the *in vitro* phosphorylation of synaptotagmin I in the presence of either ATP or GTP (Petrenko et al., 1991), raising the possibility that synaptotagmin I might be phosphorylated by casein kinase II. Of the more abundant serine/threonine kinases, casein kinase II uses GTP as efficiently as ATP (Tuazon and Trough, 1991). Recently, Davletov et al. (1993) reported that synaptotagmin I was one of the major substrates for casein kinase II in brain. It was phosphorylated on a single threonine residue within the region separating the two C2 domains from the transmembrane region. An equivalent threonine is also present in synaptotagmin II; it is very likely that this isotype will also be phosphorylated by casein kinase II. There is also evidence to suggest that synaptotagmin I is phosphorylated on the cytoplasmic domain in intact synaptic vesicles by calcium/calmodulin protein kinase II (Popoli, 1993).

E. Synaptobrevins

The 18kD integral protein, synaptobrevin (or VAMP, vesicle associated membrane protein), was identified by two independent groups. A cDNA encoding an 18kD vesicle-associated membrane protein was cloned from *Torpedo* electric organ (Trimble et al., 1988). Its rat brain counterpart synaptobrevin was identified as an SSV antigen; the synaptobrevin antigen was present in all regions of rat brain on SSVs independent of their neurotransmitter content (Baumert et al., 1989). Immunoreactivity was also observed in some neuro-endocrine tissue. Synaptobrevin cDNA has been cloned from bovine brain and *Drosophila melanogaster*; the protein displayed a high degree of interspecies homology (Sudhof et al., 1989). The protein consists of four distinct domains: a large cytoplasmic amino-terminal domain, a highly conserved "core" region, the transmembrane domain, and a small carboxy-terminal segment (possibly only one amino acid residue) that is probably lumenal (Sudhof et al., 1989). Two highly homologous (77% identity) isotypes of synaptobrevins have been identified in rat and human brain (Elferink et al., 1989; Archer et al., 1990). In the rat, synaptobrevin I consists of 118 amino acids and synaptobrevin II of 116 amino acids (Elferink et al., 1989).

F. p29

The integral membrane phosphoprotein p29 copurifies with SSVs and the small clear synaptic-like microvesicles that are located near to the *trans* side of the Golgi apparatus in neuro-endocrine cells and cell lines (Baumert et al., 1990). It has been detected in practically all nerve terminals investigated. Immunologically, p29 is related to synaptophysin, but at the time of this writing the sequence of p29 had not been published so it is not clear whether they share transmembrane topology. Like synaptophysin, p29 can be phosphorylated by an endogenous tyrosine kinase in SSV fractions.

IV. PERIPHERAL PROTEINS OF SMALL SYNAPTIC VESICLE CYTOPLASMIC MEMBRANES

Cytoplasmically disposed peripheral membrane proteins have the potential to display regulated interactions with vesicle membranes. While they cannot fulfill functions such as channel activities, they might be expected to play parts in regulating, for example, the disposition of SSV within the cytoplasm and relative to the active zone. In this section we deal with four types of peripheral membrane proteins—the synapsins, proteins that organize SSV relative to the cytoskeleton; calmodulin-dependent protein kinase II, which acts both as a membrane binding site for the synapsins and regulates their interactions with SSV membranes; the low M_r GTP binding protein, Rab3a, which seems to have a role in making vesicles competent to undergo exocytosis; and the proteins SVAPP120 and amphiphysin, the functions of which are not yet clear.

A. The Synapsins

The SSV population that comprises the reserve pool (as discussed in Section VA) and is not immediately available for exocytosis is restrained to a region of the presynaptic terminal classified as the synaptic vesicle domain (Gotow et al., 1991). There appears to be a network of filaments which cross-link synaptic vesicles to each other and to the cytoskeleton in this domain (Landis, 1988; Landis et al., 1988; Hirokawa et al., 1989; Gotow et al., 1991). The molecular shape of these filaments suggested that they are members of the synapsin family (Hirokawa et al., 1984; Landis et al., 1988). This family of closely related phospho-proteins are highly concentrated on SSVs (De Camilli et al., 1990) in the presynaptic terminal both in the central and peripheral nervous systems. The four homologous proteins, synapsin Ia and Ib (synapsin I, previously known as protein I) and synapsin IIa and IIb (synapsin II, previously known as protein II) are thought to be essential for the structural organization of the presynaptic terminal and to play a major part in the regulation of neurotransmitter release.

Primary Structure of the Proteins and Gene Analysis

The amino acid sequences of rat and bovine synapsin I and II (Sudhof et al., 1989) and human synapsin I (Sudhof, 1990) have been predicted from analysis of cDNA. They do not appear to be part of any known superfamily of proteins. Synapsins I and II are encoded by two different genes and each primary transcript can be differentially spliced to give the a and b isoforms of each protein. The mRNAs for synapsins Ia and Ib encode proteins of 704- and 608-amino acids, respectively (rat) and for synapsins IIa and IIa, 586- and 479-amino acids, respectively (rat). The amino-terminal region of all of the isoforms share a remarkable degree of homology. The amino-terminal region is also highly conserved between different mammalian species (Sudhof, 1990). It has been reported that a protein homologous to synapsin I is present in *Drosophila* (Mitschulat, 1989). The a and b isoforms only differ at the C-terminus—in synapsin I this is due to the alternative use of two different splice acceptor sites of the primary RNA transcript at the last intron/exon boundary (Sudhof et al., 1989).

A domain model based on the primary amino acid sequence has been proposed by Sudhof et al. (1989). The large homologous amino-terminal region which is common to all synapsins (residues 1–420) can be divided into domains A, B, and C. At the amino-terminus of domain A, there is a single phosphorylation site (site 1) for cAMP-dependent protein kinase and calcium/calmodulin dependent protein kinase I (ser[9] in synapsin I and ser[10] on synapsin II; Czernic et al., 1987). Domain B, the least homologous of these three domains, is rich in small-chain amino acids, such as alanine and serine. The central large homologous region, domain C, is rich in both hydrophilic (39%) and hydrophobic residues (27%; Sudhof et al., 1989). Secondary structure predictions suggest that domain C has multiple stretches of sequence which have the potential to form amphipathic α-helices and β-sheets (Sudhof et al., 1989). On the carboxy-side of domain C, the structures of the four synapsins are very different. Domain D is only found in synapsin I. This domain contains three phosphorylation sites, two for calcium/calmodulin-dependent protein kinase II (site 2 on ser[566]) and (site 3 on ser[603]; Czernic et al., 1987), and one for a proline directed kinase on ser[551] (Hall et al., 1990). Domain D is rich in basic residues and is sensitive to cleavage by the proteolytic enzyme, collagenase, due to the large number of glycine and proline residues which are also present. Therefore, the digestion of synapsin I by collagenase produces a 46kD fragment (Ueda and Greengard, 1977) which approximately corresponds to domains A–C. This region of the molecule is commonly known as the head region, while the collagenase-sensitive portion of the molecule is referred to as the tail.

Synapsin II has a short proline-rich sequence (domain G) which is adjacent to domain C. Domains E, H, and I are all variably spliced regions. Domain E is shared by synapsins Ia and IIa, while domain G is shared by synapsins IIa and IIb. Domain H is unique to synapsin IIa and domain I is unique to synapsin IIb. The functional significance of these variably spliced regions has yet to be investigated, but it is known that there is an isotype-specific distribution of the synapsins in some neuronal tissues, which hints at functional differences for the different isotypes (Sudhof et al., 1989).

As synapsin I is a neuron-specific protein, it has been a good candidate in which to look for a neuron-specific promoter. This led to the analysis of the 5′ flanking region of synapsin I, which revealed that a constitutive promoter was present. In addition, both positive and negative *cis*-acting sequence promoter elements control neuron-specific expression of the rat synapsin I gene (Howland et al., 1991; Thiel et al., 1991). The rat synapsin I gene promoter lacks TATA- and CAAT-box elements (Howland et al., 1990). A functional silencer element has been found in the human synapsin I gene which contributes to the repression of transcription in non-neuronal cells (Li et al., 1993).

Physical Properties

The physical properties of the synapsins have been well documented (for a review see Valtorta et al., 1992a). Quick-freeze deep-etch rotary shadowing of purified bovine synapsin I revealed a tadpole-shaped molecule which had a spherical head of approximately 14 nm in diameter and a 37 nm tail (Hirokawa et al., 1989). Fluorescence anisotropic studies and hydrodynamic data indicate that synapsin I is an asymmetric molecule (Ueda and Greengard, 1977; Benfenati et al., 1990). It would be predicted that the basic- and proline-rich tail region of synapsin I which is soluble in aqueous solutions would largely contribute to the elongation of the molecule. In contrast, synapsins II have much shorter tail regions and, therefore, have a less elongated structure (Huang et al., 1982). However, Landis et al. (1988) found that rotary shadowing of freshly prepared synapsin I had a fibrous appearance and Gotow et al. (1991) also found little evidence for distinct tadpole-shaped synapsin I molecules.

Synapsin I has an isoelectric point near 10.7. This basicity is mainly contributed by the tail region. On the other hand, synapsin II, which has a shorter and much less basic tail region, has a pI near to neutral (from data in Sudhof et al., 1989).

Even though hydrophobicity plots show that the head regions of both synapsin I and II are hydrophobic, there are no obvious transmembrane regions (Sudhof et al., 1989). Physico-chemical data suggests that in dilute solutions (< 0.2 mg/ml), the head region of synapsin I has a tertiary structure which serves to protect the hydrophobic amino acids from an aqueous environment (Benfenati et al., 1990). At high concentrations, the head region of the synapsins have a tendency to self-associate, although synapsin I is stable in solution up to 32 mg/ml (Ueda and Greengard, 1977; Font and Aubert-Foucher, 1989; Benfenati et al., 1990). The synapsins also have a very high surface activity. Film balance studies of synapsin fragments generated by cysteine-specific cleavage showed that this surface activity is mainly contributed by the head region (Ho et al., 1991).

Binding Activities of Synapsin I

Synapsin I has been shown to interact *in vitro* with both protein and lipid components on SSVs (Benfenati et al., 1989a, 1989b), as well as several cytoskeletal components—microfilaments (Bahler and Greengard, 1987), microtubules (Baines and Bennett, 1986), spectrin (Baines and Bennett, 1985; Okabe and Sobue, 1987), and neurofilaments (Goldenring et al., 1986; Steiner et al., 1987a). These findings suggest that synapsin I might mediate the linkages of SSVs to the cytoskeleton and control their ability to move within the presynaptic terminal. This section will describe the interactions of synapsin I with SSVs and the cytoskeleton.

Synapsin I and SSV membranes.   The binding characteristics of exogenous synapsin I to SSVs that have been depleted of endogenous synapsin I have been carefully characterized (Schiebler et al., 1986). SSVs were found to bind to the dephosphorylated form of synapsin I with high affinity ($K_d \sim 10$ nM). The maximum amount of synapsin I which could be bound corresponded to the amount of synapsin I which copurified with SSVs (synapsin I comprised 6–7% of total vesicle protein (Huttner et al., 1983). When synapsin I was phosphorylated on sites 2 and 3 (by calcium/calmodulin protein kinase II), a fivefold reduction in the affinity for synaptic vesicles was observed, but the phosphorylation of synapsin I on site 1 (by cAMP-dependent kinase) resulted in only a small decrease in the binding affinity for synaptic vesicles. Numerous studies have indicated that a large proportion of the SSV surface is surrounded by synapsin I molecules (for example, Schiebler et al., 1986 and Ho et al., 1991).

The SSV components which bind synapsin I have been partially characterized. Synapsin I binds to both lipid- and protein-components of the SSV membrane (Benfenati et al., 1989a, 1989b). The affinities for synapsin I binding to artificial membranes, which were composed of phospholipids, were similar to those observed for the binding to native SSVs. The binding appeared to be specific for acidic phospholipids (such as phosphotidylserine and phosphotidylinositol) and involved the head region of synapsin I. In contrast, the tail region, although very basic (pI ~ 12) was not able to bind phospholipids, suggesting that electrostatic interactions were not sufficient to explain the interactions observed between the head region of synapsin I with acidic phospholipids. Even though synapsin I is a peripheral protein easily removed from the membrane by salt solutions (Huttner et al., 1983; Bennett et al., 1985), photolabeling of synapsin I indicated that part of the head domain of synapsin I bound close the hydrophobic core of the lipid bilayer of SSV membranes (Benfenati et al., 1989a, b). The amphipathic regions in domain C have a high potential for surface and hydrophobic interactions (Sudhof et al., 1989) and may account for these findings.

Binding studies using purified chemically generated tail fragments of synapsin I indicate that this region interacts with a membrane protein component of SSVs. The affinity of this interaction is considerably reduced when synapsin I is phosphorylated on sites 2 and 3 (Benfenati et al., 1989b). As this interaction was dependent on the ionic concentration of the medium, this indicates that electrostatic interactions are involved in this interaction. Photolabeling cross-linking experiments have identified the a subunit of calcium/calmodulin protein kinase II as the major synaptic vesicle binding protein for synapsin I (Benfenati et al., 1992b). This enzyme, therefore, appears to have a dual role: 1) to bind synapsin I to the membrane of SSVs and 2) to phosphorylate it on sites 2 and 3 and so regulate its activities.

Synapsin II has a stronger hydrophobic interaction for SSVs than synapsin I (Thiel et al., 1990). The production of several fusion proteins which contain different fragments of the amino-terminal domain of synapsin II (domain B) have at least a single binding site for SSVs. As this interaction is trypsin-sensitive it would appear that synapsin II can interact with a SSV membrane protein component (Thiel et al., 1990).

Synapsin I and microfilaments (F-actin). Microfilaments are abundant in the presynaptic nerve terminal (Drenkhahn and Kaiser, 1983) and depolarization of synaptosomes has been linked to actin filament reorganization (Bernstein and Bamberg, 1985). *In vitro*, synapsin I cross-linked and bundled F-actin (Bahler and Greengard, 1987), indicating that synapsin I binds to the sides of actin filaments, although by its sequence it does not belong to any known major superfamily of actin-binding proteins (Sudhof et al., 1989). The affinity for the binding of synapsin I to F-actin is in the range of K_d 1.5–2.0 μM (Bahler and Greengard, 1987; Petrucci and Morrow, 1987). F-actin bundling is reduced when synapsin I is phosphorylated by cAMP-dependent protein kinase (site 1) and almost abolished by phosphorylation by calcium/calmodulin-dependent protein kinase II (sites 2 and 3; Bahler and Greengard, 1987). Synapsin I phosphorylated on sites 2 and 3 decreases the maximum binding (B_{max}) by almost 50%.

The actin binding sites on synapsin I have been mapped. The reagent, 2-nitro-5-thiobenzoic acid (NTCB), can be used to cleave polypeptides on the amino-terminal side of cysteine (Jacobson et al., 1973). The sequences of bovine, human, and rat synapsins all show that synapsins Ia and Ib have only three cysteines and that these are conserved—in the bovine protein, these are at residues 223, 360, and 370 (Sudhof et al., 1989; Sudhof, 1990). Cleavage of bovine synapsin I with NTCB, therefore, results in the formation of three groups of fragments designated N-terminus (residues 1–222), middle (residues 223–359/369), and tail (residues 360/370–706 from synapsin Ia and 360/370–670 from synapsin Ib; Bahler et al., 1989). In vitro binding assays with purified NTCB fragments showed that the amino-terminal end of synapsin I contains an actin binding site (Bahler et al., 1989). An additional high-affinity binding site was found by Bahler et al. (1989) in the middle region of the molecule. Although no actin binding or bundling activity was found in the NTCB-generated tail fragment, the tail region appears to be involved in bundling. With the use of photoactive cross-linkers, Petrucci and Morrow (1991) have found an actin-binding region in the tail region of synapsin I. It therefore seems that synapsin I has two binding sites in the head/middle region of the molecule (which are hypothesized to be spatially near to each other in the intact molecule) and a single site in the carboxy-tail region (Petrucci et al., 1991). The two sites in the head/middle domains may contribute to a single high-affinity binding site and the third binding site in the tail domain may constitute a different site required for bundling in the intact molecule (Petrucci et al., 1991).

Synapsin I and microtubules. "Synapsin I shaped" molecules have been observed cross-linking microtubules and synaptic vesicles in the presynaptic terminal (Hirokawa et al., 1989; Gotow et al., 1991). It has also been proposed that synapsin I may interact with microtubules during axonal transport. Synapsin I is synthesized in the perikaya and transported along the axon to the presynaptic nerve terminal. During axonal transport, it is associated with both fast (vesicular) and slow (cytoskeletal) transport components (Baitinger and Willard, 1987). The slow cytoskeletal fraction may represent synapsin I bound to microtubules, since a subpopulation of synapsin I copurifies with microtubules in vitro (Goldenring et al., 1986; Farrell and Keates, 1990).

Synapsin I has many of the characteristics of a MAP (microtubule-associated protein). It co-cycles with tubulin through several cycles of warm polymerization and cold depolymerization and binds saturably to taxol-stabilized microtubules ($K_d \sim 4.5$ µM; Baines and Bennett, 1986; Goldenring et al., 1986). Synapsin I also bundles microtubules (half maximal bundling at ~0.6 µM; Baines and Bennett, 1986; Bennett and Baines, 1992). As with the actin bundling activity, the question arises as to whether synapsin I monomers are oligovalent for tubulin, or whether univalent synapsin I molecules self-associate to form oligovalent complexes. There is evidence for each. Font and Aubert Foucher (1989) have demonstrated that native synapsin I in solution can interact with itself in a head-to-head association to give a dimer, which should be able to cross-link microtubules. On the other hand, synapsin I monomers are at least bivalent for microtubules. A tubulin binding site was found in a 44kD trypsin-generated fragment that came from the amino-head region of synapsin I (Aubert-Foucher and Font, 1990). Recently, two separate tubulin binding sites have been found in the head and a single site in the tail domain of synapsin I (Petrucci and Morrow, 1991). These results raise the possibility that three tubulin binding sites may be present in the intact synapsin I molecule. No bundling activity was observed with isolated intact head or tail domains (Bennett et al., 1991), indicating that both head and tail regions are required for bundling. Furthermore, synapsin I monomers can bundle microtubules (Bennett and Baines, 1992). Tubulin, which has been proteolytically digested so as to remove the C-terminus, will only bind to the head region of synapsin. This indicates that the synapsin I tail binds to the same site in tubulin as MAPs I and 2. As yet, there is no information available on the tubulin-binding activity of synapsin II, although as the head domain in both synapsins I and II are highly homologous (Sudhof et al., 1989), it would be predicted that the two tubulin binding sites in the head domain will be present.

Synapsin I and spectrin. Several techniques have shown that synapsin I can bind spectrin *in vitro* (Baines and Bennett, 1985; Okabe and Sobue, 1987). The binding is saturable and to a single high affinity site ($K_d \sim 24$ nM); and the binding is regulated by phosphorylation of synapsin I by calmodulin-dependent protein kinase II (Sikorski et al., 1991). Rotary shadowing electron microscopic techniques indicate that synapsin I binds to the ends of spectrin tetramers (Krebs et al., 1987). This may explain the observation that brain spectrin binds end-on to SSVs (Goodman et al., 1988). Filaments with a similar size to spectrin appear to link synaptic vesicles to the plasma membrane (Landis et al., 1988; Hirokawa et al., 1989) and it is, therefore, postulated that spectrin associated with the presynaptic membrane may have a role in positioning synaptic vesicles near to the presynaptic membrane.

Synapsin I and neurofilaments. The interaction of synapsin I with neurofilaments is not extensively characterized. Immunocytochemical studies showed that synapsin I was present in a partially purified preparation of neurofilaments and that it cosedimented with the light-chain of the neurofilament triplet (NF-L; Goldenring et al., 1986; Steiner et al., 1987a, 1987b). This binding to purified NF-L was saturable and the affinity was reduced by phosphorylation (Steiner et al., 1987a).

Regulation of Synapsin Activities

Synapsin I has multiple binding sites for a large number of cytoskeletal proteins and SSVs (see following). These interactions require rapid and precise regulation to enable synapsin I to play its role during neurotransmitter release. The binding activities of synapsin I are controlled by several factors including: phosphorylation, the direct interaction of synapsin I with calmodulin, and perhaps O-glycosylation.

By phosphorylation. Synapsin I undergoes phosphorylation at several sites and is a substrate for at least four different protein kinases (Huttner and Greengard, 1979; Huttner et al., 1981; Huang et al., 1982; Kennedy et al., 1983a; Nairn and Greengard, 1987; Trimble and Scheller, 1988). Both synapsin I and II are good substrates for cAMP-dependent protein kinase I and calcium/calmodulin-dependent protein kinase I (on site 1 which is on ser^9 on synapsin I and ser^{10} on synapsin II) with a $K_m \sim 2$ μM for both kinases. As predicted from the amino acid sequence, synapsin I, but not synapsin II, is a substrate for calcium/calmodulin protein kinase II (on sites 2 and 3 which are on ser^{566} and ser^{603}, respectively). Synapsin I is the best known substrate for this kinase ($K_m \sim 0.4$ μM). The amino acid sequence of site 1 is Arg–Arg–Leu–Ser–P, which conforms to the consensus sequence, Arg–Arg–X–Ser–P, which is found in many cAMP-dependent protein kinase substrates (Czernic et al., 1987). The amino acid sequence of both sites 2 and 3 (Arg–Gln–Ala–Ser–P) conforms to the consensus sequence, Arg–X–Y–Ser–P/Thr–P, which is found in many calcium/calmodulin protein kinase II substrates (Czernic et al., 1987). An additional phosphorylation site has been mapped to ser^{551} in synapsin I. This growth factor-sensitive proline-directed protein kinase phosphorylates the Ala–Ser–Pro–Ser sequence. This kinase has been shown to phosphorylate synapsin I both *in vivo* (in PC 12 cells) (Vulliet et al., 1989) and *in vitro* (Hall et al., 1990).

It has been demonstrated that synapsin I is a poor substrate *in vitro* for protein kinase C (PKC; Severin et al., 1989). Although this phosphorylation activity has not been shown to occur *in vivo* (Wang et al., 1988), a recent report indicates that activators of PKC increase the phosphorylation of synapsin I on sites 1, 2, and 3, and on synapsin II on site 1 (Browning and Dudek, 1992). This phosphorylation event may be a part of the complex regulatory mechanism required to control the activity of the synapsins.

By glycosylation. Both synapsin I and II contain N-acetylglucosamine (GlcNAc) residues (Luthi et al., 1991). However, only a small proportion (0.06 GlcNAc residues/molecule of synapsin I) of synapsin I molecules appear to contain these O-glycosylation sites. This does not mean that glycosylation is insignificant, especially since, by analogy, purified synapsin I only contains ~0.1 mol/mol phosphate (Hirokawa et al., 1989). Cytoplasmic O-glycosylation is a dynamic process (Haltiwanger et al., 1992), and the low stoichiometry of synapsin I glycosylation may be a consequence of turnover. It has been proposed that the role of glycosylation may be analogous to phosphorylation in controlling protein activities (Hart, 1992). Addition and removal of O-GlcNAc is responsive to cellular stimuli (Kearse and Hart, 1991; Chou and Omary, 1993), and it will be of interest to determine whether the activities of synapsin I are modulated by O-glycosylation.

By binding to calmodulin. The regulation of neuronal exocytosis requires the specific binding of calcium to effector molecules. Several findings suggest that calmodulin may be one of these effector molecules. Calmodulin is present in all eukaryotic cells where regulated exocytosis has been studied. Inhibitors of calmodulin are general inhibitors of regulated exocytosis (Douglas and Nemeth, 1982; Burgoyne et al., 1982; Kenigsberg et al., 1982) and anti-calmodulin antibodies prevent exocytosis both after microinjection in intact cells and during cell free exocytosis (for example, Steinhardt and Alderton, 1982 and Kenigsberg and Trifaro, 1985).

Calmodulin interacts directly with SSVs (Hooper and Kelly, 1984); several synaptic vesicle-associated proteins including calcium/calmodulin-dependent protein kinase II, synaptotagmin, and synapsin Ia and Ib are also calmodulin-binding proteins. Okabe and Sobue (1987) proposed that synapsin I and calmodulin had a 1:1 stoichiometry, but recent data (Goold and Baines, 1994) suggests that a 2:1 stoichiometry exists. Calmodulin interacts with the head domain of synapsin I (Hayes et al., 1991), which contains several positively charged amphiphilic helical regions, which are good candidates for calmodulin-binding sites (O'Neill and DeGrado, 1990). The significance for this binding activity in the regulation of neuronal exocytosis is, at present, under investigation but probably relates to control of actin binding.

B.   Calmodulin-dependent Protein Kinase II

Calmodulin-dependent protein kinase II (CaM-KII) is an abundant protein kinase in nervous tissue, and exists in a number of cellular compartments, including post-synaptic densities (Kennedy et al., 1983b), the cell cytoplasm (Bennett et al., 1983), and bound to the cytoplasmic face of SSV membranes (Huttner et al., 1983). The role of vesicle bound CaM-KII has recently been suggested to be a membrane-binding site for synapsin I, regulated by CaM-KII itself phosphorylating synapsin I (Benfenati et al., 1992b). In this respect, not all populations of CaM-KII seem to be functionally equivalent. Soluble CaM-KII isolated from brain cytosolic extracts does not co-isolate with synapsin I bound with high affinity to it (Bennett et al., 1983). Since additional sites for synapsin interaction with SSV membranes exist (see preceding), the full *in vivo* role of CaM-KII on the SSV membranes remains elusive. The nature of CaM-KII with the SSV membrane is not known, but since there is no evidence for it being an integral membrane protein, it may be assumed that there is a receptor for CaM-KII specifically associated with the SSV membrane.

C.   Rab3a (or smg p25A)

Five subfamilies of low molecular mass GTPases related to the Ras oncogene product are known—Ras, Rab, Rac, Ran, and Arf. The Rab family has more than 20 members, and they seem to have roles in directing vesicle traffic. In rat brain, one member of the Rab family, Rab3a (otherwise known as smg p25a), has been found in both soluble and membrane bound forms—the membrane bound form is associated with SSVs and the presynaptic membrane/active zones (Mizoguchi et al., 1989, 1992; Fischer von Mollard et al., 1990).

There is accumulating evidence to suggest that Rab3a has a function related to either neuronal exocytosis or endocytotic recycling. In synaptosomes, Rab3a quantitatively dissociates from SSVs during calcium-dependent exocytosis and partially re-associates with them in the recovery period after stimulation (Fischer von Mollard et al., 1991). Microinjection of Rab3a peptides into cultured neurons, which form synapses, increases the spontaneous frequency of miniature synaptic currents (Richmond and Haydon, 1993).

The primary structure of Rab3a does not predict a transmembrane region, instead it is anchored to the membrane by a hydrophobic post-translational modification; the molecule is isoprenylated at two carboxy-terminal cysteine residues by two geranylgeranyl groups (Araki et al., 1991; Johnston et al., 1991; Musha et al., 1992).

The role of Rab3a in directing exocytosis is discussed in Section VIIB.

D. SVAPP-120

Synaptic vesicle associated phospho-protein (SVAPP-120) is a peripheral membrane protein of synaptic vesicles (Bahler et al., 1991). On SDS gels it is a doublet of 119kD and 124kD. During postnatal development, its expression in rat cortex and cerebellum correlated with synaptogenesis. This protein is brain-specific and is possibly only located on a subpopulation of synaptic vesicles. SVAPP-120 is phosphorylated both *in vivo* and *in vitro* on serine residues, and to a lesser extent on threonine residues, by several protein kinases including protein kinase C, cAMP-dependent protein kinase, and an endogenous calcium/calmodulin-dependent protein kinase.

E. Amphiphysin

Amphiphysin is an acidic peripheral synaptic vesicle associated protein (Lichte et al., 1992). The predicted polypeptide chain size is 75kD, but SDS-PAGE gave 115kD, possibly indicating extensive post-translational modification. There are three structural domains; a hydrophilic N-terminal region, followed by a proline rich stretch, and a negatively charged C-terminal domain. There are many potential consensus sequences for phosphorylation. Amphiphysin has a similar distribution to synaptophysin; immunocytochemical studies reveal that it is localized to SSVs. mRNA expression is confined to the brain and adrenal gland only. The function of this protein was, at the time of this writing, unknown.

F. Other SSV Associated Proteins

In their study of otherwise unidentified SSV proteins, Bahler et al. (1991) determined the N-terminal sequences of a number of SSV proteins. This revealed the presence of two adaptins. It may be the case that these represent adaptor proteins that remain bound to SSV membrane proteins after they have been retrieved from the membrane and recycled via clathrin coated vesicles (see Section VIII).

A form of the protein kinase pp60c-src is bound to SSV (Linstedt et al., 1992). Its major SSV substrate is synaptophysin (Barnekow et al., 1990; Section IIIA). Several other proteins become associated with the SSV membrane as part of a docking/fusion complex. These are dealt with in Section VIIB.

It has been calculated that (assuming an average molecular weight of 50kD) the membrane of a single SSV cannot hold more than 10–15 different proteins in addition to the vacuolar proton pump (Jahn and Sudhof, 1993), so it seems unlikely that many more SSV-associated proteins have yet to be found.

V. EVIDENCE FOR TWO POOLS OF SMALL SYNAPTIC VESICLES

The organization of SSVs prior to exocytosis is critical for the process of neurotransmission. Several studies have suggested that SSVs are organized into two different pools (a reserve pool and a releasable pool) within the presynaptic terminal, each pool of vesicles having a different function (Llinas et al., 1985, 1991; DeCamilli and Jahn, 1990; Lin et al., 1990).

Electrophysiological evidence for the two pools suggested that dephosphorylated synapsin I could "cage" or restrain SSVs, while synapsin phosphorylated by calcium/calmodulin protein kinase II can "decage" or overcome the restraint of SSVs to allow exocytosis to occur (Lin et al., 1990). The caged pool is known as the reserve pool and the uncaged pool as the releasable pool.

There is also electron microscopic evidence available which supports the existence of two pools. Rotary shadowing of freeze-fractured presynaptic terminals of rat cerebellar mossy fibers showed that a population of SSVs were attached to 30 nm strands (presumably synapsin I) which connected to microfilaments and microtubules. Another population of SSVs was observed attached to the presynaptic membrane by 100 nm strands (maybe spectrin; Hirokawa et al., 1989). Similar results were observed with glutamate-containing SSVs (Landis et al., 1988).

The releasable pool seems to consist of clusters of SSVs docked at the active zones. The releasable pool can undergo exocytosis as soon as depolarization occurs. After depolarization, SSVs are recruited from the reserve pool to the releasable pool (Torri-Tarelli et al., 1985). The reserve pool is likely to consist of vesicles held in the peripheral axoplasm in the presynaptic terminal close to the presynaptic membrane, but not in contact with it. Additionally, a small population of SSVs are found in the center of the nerve terminal in a region that is enriched in mitochondria (Gotow et al., 1991).

A. The "Reserve Pool" of SSV

The presynaptic-terminal cytoplasm of mammalian CNS can be characterized as having two domains. The mitochondrial domain is situated in the middle of the

terminal and contains mainly mitochondria and a few synaptic vesicles. The peripheral synaptic vesicle domain is filled with SSVs, and contains relatively few mitochondria (Gotow et al., 1991). Quick-freeze deep-etch electron microscopy of unfixed nerve terminals revealed that in the mitochondrial domain, the numerous mitochondria were connected to each other by filamentous strands. There were a small number of SSVs present in this domain and large (80–100 nm in diameter) vesicles, possibly LDCVs, were also present. Both types of vesicles were linked to mitochondria and microtubules (Gotow et al., 1991). The cytoplasm of the peripheral synaptic vesicle domain had a more fibrillar appearance than the mitochondrial domain, due to the large number of fibers associated with the SSVs. These fibers were thicker and longer than those that link mitochondria to each other.

Conventional electron microscopy has shown that SSVs are cross-linked by filaments (DeCamilli et al., 1983). The quick-freeze deep-etch rotary shadowing technique has identified the main presynaptic cytoskeletal filaments as F-actin and microtubules (Landis and Reese, 1983; Hirokawa et al., 1989; Gotow et al., 1991). Filaments 30 nm in length were observed between actin and SSVs, and between microtubules and SSVs. Frequently, a spherical structure was seen in the middle of the filaments. These fine filaments had a similar morphology to synapsin I, neuron-specific phospho-protein (Hirokawa et al., 1989). The simplest conclusion is that synapsin I (and by implication synapsin II) link SSV to cytoskeletal elements in the synaptic vesicle domain, and so generate the reserve pool of vesicles.

B. SSVs at the Active Zone—An Immediately Available Pool of Vesicles

The concept of an active zone for exocytosis was derived from observations on neuromuscular junctions that exocytosis took place only at a limited portion of the presynaptic membrane (Couteaux and Pecout-Dechavassine, 1970). Conventional thin section electron microscopy delineates the active zone as a proteinaceous thickening on the cytoplasmic face of the presynaptic face of the plasma membrane (reviewed in Heuser, 1989), where SSV seem to cluster. It is also close to the point where endocytotic membrane retrieval takes place (e.g., Heuser and Reese, 1973). SSVs docked at the active zone can seem to be partially buried in the presynaptic thickening, indicating that in the process of docking, SSV proteins are likely to bind to active zone proteins. By the same token that there are SSV-specific proteins, it is likely that there should be a unique population of proteins at the active zone.

Considerations of the rapidity of exocytosis in response to Ca^{2+} influx indicated that only vesicles docked at the active zone should undergo exocytosis as the first wave of response to an action potential (reviewed in Llinas, 1987); these would be a pool of SSV immediately available for exocytosis. An extension of this idea was that Ca^{2+} channels should be located at or near the active zone itself. A key element in support of this suggestion was the observation of Ca^{2+} channels as intramembrane particles (IMPs) within the membrane of the active zone. Freeze–fracture electron microscopy revealed that intramembrane particle distribution under the presynaptic thickening is highly organized. Typically, round IMPs (7.5–15 nm diameter) on the P face of the presynaptic membrane can be detected, sometimes arranged in lines (Heuser and Reese, 1973) or bar-like aggregates (Saito, 1990). The IMPs are often associated with sites of exocytosis (Pfenninger et al., 1971; Heuser and Reese, 1973). The density of IMPs is also correlated with changes in presynaptic morphology—invaginations can be formed where they are densest, possibly representing sites of SSV association with the active zone. The suggestion arose that the IMPs were, in fact, Ca^{2+} channels (Pumplin et al., 1981), an idea that fitted well with electrophysiological evidence that the effect of the action potential was to generate a well-timed entry of Ca^{2+} into the presynaptic terminal close to the active zone (Llinas et al., 1982; Llinas, 1987). This concept was extended to include the suggestion that Ca^{2+} channels might be part of the docking apparatus itself (Smith and Augustine, 1988). Ca^{2+} channels as part of the docking apparatus would be supposed to allow the delivery of Ca^{2+} directly to the proteins that control the exocytotic process.

Ca^{2+} channels are not the only proteins specifically associated with the active zone. Other active zone proteins include syntaxin (Bennett et al., 1992) and Rab3a (Mizoguchi et al., 1992), considered to be components of the docking and fusion apparatus. The roles of these proteins at the active zone are dealt with in Section VII.

The active zone then can be considered to be a compact structural unit whose role is to couple, with accuracy and rapidity, the arrival of the action potential to the exocytosis of neurotransmitter.

VI. ORGANIZATION OF THE RESERVE POOL OF SSVS BY SYNAPSIN I AND THE CYTOSKELETON

Synapsin I has been implicated in the short-term regulation of neurotransmitter release for almost a decade. Within the last few years, evidence has accumulated to strongly suggest that continuous cycles of phosphorylation and dephosphorylation of synapsin I regulate the movement of SSVs from the reserve pool to the releasable pool by changing the affinity of synapsin I for both F-actin and SSVs. Some of this evidence is described below.

When dephosphorylated synapsin I was microinjected into the squid giant synapse (Llinas et al., 1985, 1991) or goldfish Mauthner neurons (Hackett et al., 1990), stimulus-evoked neurotransmitter release was inhibited. This inhibition was reversed if synapsin I, which had been phosphorylated by calcium/calmodulin-dependent protein kinase II (i.e., on sites 2 and 3), was subsequently injected (Llinas et al., 1985). A similar result has been found in a mammalian model. The introduction of dephosphorylated synapsin I into rat brain synaptosomes resulted in a potassium-dependent decrease in the release of glutamate, while microinjection of phosphorylated synapsin I had no effect (Nichols et al., 1992).

In addition, the phosphorylation of synapsin I on sites 2 and 3 lowered its affinity for actin and SSVs (Schiebler et al., 1986; Bahler and Greengard, 1987; Benfenati et al., 1992a; Valtorta et al., 1992b). Dephosphorylated synapsin I was found to increase the initial rate of actin polymerization, induce the growth of a large number of short actin filaments, and induce actin polymerization in physiologically unfavorable conditions, that is, in the absence of $MgCl_2$ and KCl (Benfenati et al., 1992a; Fesce et al., 1992; Valtorta et al., 1992b). In contrast, phosphorylated synapsin I reversed these effects, suggesting that dephosphorylated synapsin I has a nucleating effect on actin polymerization. This is in agreement with Lin et al. (1990) who suggest that dephosphorylated synapsin I restrains or cages synaptic vesicles while phosphorylated synapsin I decages them. Furthermore, synapsin I has been shown to partially dissociate from SSVs after electrical stimulation at frog neuromuscular junctions (Torri-Tarelli et al., 1992). Shira et al. (1989) demonstrated in depolarized synaptosomes that as the proportion of synapsin I phosphorylation increases, there is a rapid translocation of synapsin I from a membrane-bound compartment into a cytosolic compartment.

These results have led to the following hypothesis for the role of the cytoskeleton during neurotransmitter release (Baines, 1987; De Camilli et al., 1990) (see Figure 1). Dephosphorylated synapsin I tethers SSVs to actin and, therefore, restrains their

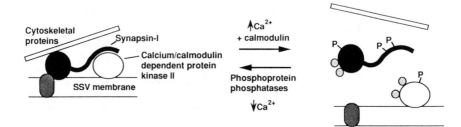

Figure 1. A schematic view of control of synapsin-I by calcium and calmodulin. Synapsin-1 (the black "tadpole" shaped molecule) interacts in its dephosphorylated state with protein and lipid sites on the cytoplasmic face of the SSV membrane; one of the protein sites is calcium/calmodulin dependent protein kinase II. Synapsin-I can also interact with a number of cytoskeletal proteins (including spectrin, actin, and microtubules—shown as a rod) and thus link SSV to the cytoskeleton. Upon the opening of calcium channels, calmodulin is activated. Activated calmodulin then binds to and activates calcium/calmodulin dependent protein kinase II which, in turn, phosphorylates the synapsin-I in its "tail" region. Possibly, another kinase, calcium/calmodulin dependent protein kinase I, phosphorylates synapsin-I in its head domain. Calmodulin also binds directly to the head region. Phosphorylated calmodulin-bound synapsin-I has a reduced affinity for both the membrane and the cytoskeleton. The interactions are therefore diminished, thereby freeing SSV to move up to the active zone for docking. See text for details.

movement within the presynaptic terminal, thereby preventing them from reaching the presynaptic plasma membrane. SSVs, which are already docked at the active zones and are available for immediate exocytosis, are not under this restraint. On depolarization, the increase in intracellular calcium leads to: (1) the exocytosis of neurotransmitter from those SSVs which are already docked and (2) to the activation of calcium/calmodulin-dependent protein kinase I and II and, therefore, to the phosphorylation of synapsin I. This phosphorylation allows the dissociation of synapsin I from either actin filaments, SSVs, or both which results in the release of a subpopulation of SSVs which are now available for docking at the active zone. As there are now more SSVs available for exocytosis, prolonged stimulation will increase the amount of neurotransmitter released at each round of exocytosis.

VII. SMALL SYNAPTIC VESICLES AT THE ACTIVE ZONE

A. SSV Docking with the Active Zone

How does docking of SSV at the active zone occur? Several hypotheses have been put forward for docking mechanisms and between them, they involve most of the integral membrane proteins of SSVs. Two hypotheses are outlined below—the first involves the proteins synaptophysin and synaptoporin, the second involves a complex of proteins, including synaptotagmins. The first idea antedates the second, and seems recently to have fallen from view.

Synaptophysin and Synaptoporin as Part of a Docking Complex

The discovery that synaptophysin was a hexameric protein with transmembrane segments that were arranged analogously to the connexin proteins of gap junctions, led to speculation that synaptophysin (and the related protein synaptoporin) might form a complex with a homologous protein in the presynaptic membrane (Knaus et al., 1990). The structure of such a complex would, in this concept, be similar in arrangement to the connexon structure of gap junctions. Since synaptophysin appears to be a voltage-sensitive channel, it could be imagined that opening of a pseudo-gap junction would follow from the depolarization of the presynaptic membrane. A corollary of this hypothesis would be that a synaptophysin binding protein would have to be present in presynaptic membranes. Such a protein has indeed been found—it has been termed physophylin, to indicate its affinity for synaptophysin (Thomas and Betz, 1990). However, whether a complex of synaptophysin/synaptoporin and physophilin can form a pseudo-gap junction, or perhaps open to form a "fusion pore" along the lines suggested by Almers (1990; and see following) awaits further investigation. However, the fact that synaptophysin and physophilin were in corresponding compartments of the nerve terminal is an indication that synaptophysin may take part in docking.

Synaptotagmins as Part of a Multi-member Docking Complex

Several lines of evidence indicate a role for synaptotagmin in SSV docking. Synaptotagmin interacts with a number of putative presynaptic plasma membrane proteins including the α-latrotoxin receptor (Petrenko et al., 1991), the ω-cono-toxin-sensitive or N-type calcium channel (Jackson, 1987), and two proteins variously named syntaxins, HPC-1, or epimorphin (Bennett et al., 1992; Inoue et al., 1992; Yoshida et al., 1992).

The neurotoxin, α-latrotoxin, is obtained from the venom of the black widow spider. This toxin is known to cause synaptic vesicle exocytosis and massive neurotransmitter release from presynaptic nerve terminals (e.g., Clark et al., 1970). Affinity chromatography studies with the presynaptic α-latrotoxin receptor showed that it binds specifically and with a high affinity to synaptotagmin (Petrenko et al., 1991).

The syntaxins are two homologous integral 35kD proteins which are probably located at or near to the active zone (Bennett et al., 1992). The syntaxins and synaptotagmin I can be co-immunoprecipitated from a preparation enriched in synaptic vesicle and presynaptic membranes, but not from a synaptic vesicle preparation which indicates that the syntaxins are not associated with synaptic vesicle membranes, but with another membrane compartment, probably synaptic sites (Bennett et al., 1992).

Anti-synaptotagmin antibodies have been found in immunoglobulin fractions from patients with the autoimmune disease, Lambert–Eaton myasthenia. They immunoprecipitated ω-conotoxin receptors (Leveque et al., 1992) which indicates that synaptotagmin is associated with ω-conotoxin-sensitive or N-type calcium channels. These calcium channels are thought to be responsible for the increase in intracellular calcium that accompanies synaptic vesicle exocytosis (for a review see Tsien and Tsien, 1990). In addition, antibodies to synaptotagmin formed a complex with synaptotagmin and the ω-conotoxin receptor (Yoshida et al., 1992). Taken together, these results suggest that synaptotagmin may form part of a docking complex which would also include syntaxin near N-type calcium channels at presynaptic active zones. Further components of this complex may include neur-exins, a group of transmembrane proteins which include the α-latrotoxin receptor and the extracellular matrix protein, laminin (Ushkaryov et al., 1992). The cyto-plasmic domain of neurexins binds to synaptotagmin (Hata et al., 1993). A potential docking complex, the "synaptosecretosome" which includes all these proteins has recently been isolated, thus establishing a case that all the interactions predicted individually can occur simultaneously (O'Connor et al., 1993).

Despite the large number of *in vitro* experiments implicating synaptotagmin as a component of the exocytotic apparatus, doubt has recently been cast on its supposed role from several lines of experimentation *in vivo* which have used either cultured cells or whole organisms in which the synaptotagmin gene has been manipulated. Synaptotagmin-defective mutants of the neuro-endocrine cell line,

PC12 (rat adrenal pheochromocytoma), continued to release catecholamines in a calcium-dependent manner (Shoji-Kasai et al., 1992). Although many aspects of calcium-dependent catecholamine release in neuro-endocrine cells have been shown to be similar to that of regulated exocytosis in neurons, the catecholamine-containing organelles were not identified in this study so the significance of this result is as yet uncertain. On the other hand, microinjection of fusion proteins containing cytoplasmic fragments of synaptotagmin into nerve growth factor-treated PC12 cells impaired the function of the catecholamine containing vesicles (Bommert et al., 1993), indicating that synaptotagmin may, indeed, have a modulatory role in the secretory process. These larger catecholamine-containing dense core vesicles were distinct from SSVs and, therefore, additional evidence is still required to elucidate the role of synaptotagmin in regulated exocytosis of SSVs in neurons.

Mutants of *Caenorhabditis elegans* and *Drosophila* have been constructed that were deficient in synaptotagmin (Nonet et al., 1993; Diantonio et al., 1993). In both organisms, limited coordinated motor movements and response to stimuli were observed, indicating that SSV continued to dock and fuse at presynaptic termini. This argues that synaptotagmin is not absolutely required for neurotransmission. It is still possible to conjecture that its role may be modulatory, since although both mutant organisms had at least some synaptic transmission, their nervous function was impaired; in any case, studies of the efficacy of synaptic transmission in these mutants have not been published at the time of writing.

A further complication to understanding the role of synaptotagmin comes with the finding that synaptotagmin stably transfected into Chinese hamster ovary (CHO) cells did not associate with vesicles that undergo membrane trafficking (Feany and Buckley, 1993). Whereas transfected synaptophysin in fibroblasts associated with small vesicles (Linstedt and Kelly, 1991), synaptotagmin associated with the actin cytoskeleton. The significance of this in the context of SSV exocytosis is not clear, and in any case, CHO cells lack the other components of the synaptosecretosome.

B. Exocytosis

How does fusion of SSV with the presynaptic membrane occur? Once docked at the plasma membrane close to Ca^{2+} channels, SSV must be kept ready for the fusion event which is signalled by depolarization of the plasma membrane and Ca^{2+} influx. A second requirement for exocytosis is GTP. While Ca^{2+} seems to be the signal for exocytosis, GTP seems to be required to make SSV "competent" for exocytosis. Any account of a mechanism for SSV exocytosis must recognize the role of both Ca^{2+} and GTP as regulators of SSV exocytosis.

The Site and Mechanism of Action of GTP

GTP potentiates most forms of regulated exocytosis, and in certain cases (mast cells, for example), activation of GTP-binding proteins alone can induce exocytosis (reviewed by Gomperts, 1990). In his review, Gomperts postulated that a GTP-binding protein would modulate exocytosis in many systems of regulated exocytosis and he designated the hypothetical G protein G_e to indicate its role. A small GTP binding protein, Rab3a, is present on SSV, and may fulfill the role of G_e.

Rab3a proteins co-isolate with SSV from many organisms (e.g., Chin and Goldman, 1992), and are detectable at the active zone of rat neuromuscular junctions (Mizoguchi et al., 1992). A connection between Rab3a and SSV exocytosis was detected with the discovery that it quantitatively dissociated from SSV at the point of exocytosis (Fischer von Mollard et al., 1991), and the finding that it only associates with SSV at late stages of the exocytotic pathway (Matteoli et al., 1991). Its target has been suggested to be a protein related to synaptotagmin, designated rabphillin (Shirataki et al., 1993).

How Rab3a regulates, or makes vesicles competent for exocytosis, is a matter for conjecture. Small GTP-binding proteins of the ras family generally have two states: an "active" state in which GTP is bound, and an "inactive" state with GDP bound. A slow intrinsic rate of GTP hydrolysis converts the active state to the inactive state, and the active state can be regained by exchange of bound GDP for GTP. The slowness of the intrinsic GTPase and nucleotide exchange indicate that for ras family proteins to regulate physiological events, the intrinsic rates must be subject to regulation, that is, be driven fast when events require. In this context, regulators of ras family function are known—GTPase activating proteins (GAPs), proteins activating nucleotide exchange (NEPs), and GDP dissociation inhibitors (GDI). Thus, by balancing the relative activities of the cognate GAP, NEP, and GDI proteins, the activity of any specific ras family member could, in principle, be controlled. Consistent with this, Rab3a, GDI, NEP, and GAP have been characterized (Burstein et al., 1991; Sasaki et al., 1991; Burstein and Macara, 1992). GDI interaction with Rab3a C-terminus regulates Rab3a interaction with membranes (Araki et al., 1990; Musha et al., 1992). GDI interaction could be postulated to keep Rab3a "inactive" or in an "inhibitory" state by retention of GDP.

One possible view of Rab3a and its regulators would be as follows. Vesicles could be prevented from fusion by GDI activity. Inhibition of GDI and subsequent nucleotide exchange mediated by NEP would make SSV competent for exocytosis by allowing binding of active GTP-Rab3a. Upon exocytosis, Rab3a could be inactivated by a specific GAP, which would render any membranes that Rab3a attached to incompetent for exocytosis (and possibly for Rab3a binding) until a further round of nucleotide exchange. The next stage in understanding how Rab3a controls synaptic exocytosis will be to explore the controls on Rab3a regulatory proteins, which at the time of writing are not well understood. The activities of Rab3a are discussed further following.

A Fusion Pore in SSV Exocytosis?

The existence of a fusion pore was predicted by electron microscopy of (first) mast cells and (subsequently) many other cell types. Chandler and Heuser (1980) showed that in degranulating mast cells, membrane-lined pores, 20–100 nm in diameter, provided water-filled channels between the lumen of the secretory granule and the extracellular environment. Patch–clamp electrophysiology experiments were consistent with this: measurement of membrane conductances in stimulated mast cells indicate an initial opening of the fusion pore of approximately 1 nm diameter, followed by rapid opening to more than 20 nm (reviewed in Monck and Fernandez, 1992).

Valtorta et al. (1990) have adopted the fusion pore model to suggest that a fusion pore might open transiently between docked synaptic vesicles and the presynaptic membrane. The SSV contents would diffuse out, and the pore would then close. Undocking of the vesicles would allow recycling of the vesicles for refilling with transmitter, and subsequent rounds of discharge.

The model of the fusion pore put forward by Almers (1990) shows a pair of membranes that are about to fuse linked together. The linkage is provided by interactions between multimeric channel proteins in each membrane. The signal for exocytosis results in channel opening. Small molecules can diffuse through the open channel in a manner analogous to the opening of a gap junction. However, in the Almers scheme, not only do the vesicle contents diffuse, but the lipids in the two membranes start to mingle together by diffusing along amphiphilic faces of the channel protein subunits. Eventually, the two membranes fuse fully and the exocytotic event is complete. In principle, the full fusion need not take place, and the channels could reclose before full membrane fusion occurs. As indicated in Section VIIA, synaptophysin and synaptoporin are candidates for hexameric channel proteins in SSV membranes; physophilin may be a candidate for the corresponding protein in presynaptic membranes. In a fusion pore scheme, one possibility might be that a fusion pore would form when docked SSV open the synaptophysin/synaptoporin–physophilin pseudo-gap junction. Such a scheme is entirely hypothetical, and arguments can be made against it. Sudhof and Jahn (1991) argued against synaptophysin as a fusion pore component on the basis that the voltage dependency of its channel would close the channel during exocytosis. At a true gap junction, homophilic interactions of connexon units via their extracellular domains establish membrane contact (Lowenstein, 1987). In the case of the hypothesized SSV junction, the interactions would be heterophilic and mediated via cytoplasmic domains—the structural analogy cannot be very close. The electron microscopic evidence for a fusion pore in mast cells is good; this is not the case for SSV exocytosis. Moreover, the time course of mast cell exocytosis is much longer than that of SSV exocytosis (Almers, 1990). The existence of a fusion pore in SSV exocytosis remains unproven.

SNAPs, SNAREs, and the Synaptosecretosome

An alternative view of proteins that direct SSV fusion has come from a combination of two sets of data. The formation of a docking complex, the synaptosecretosome, has already been dealt with (Section VIIA). A second line of experimentation has been pursued by Rothman's group, which arose from an investigation of cytoplasmic proteins required for vesicle fusion in the transport of Golgi vesicles. In a recent and remarkable convergence of ideas, the proteins that this group identified in Golgi transport now appear to take part in SSV fusion. These proteins seem to mediate a constitutive transport (i.e., requiring fusion without Ca^{2+} signaling). SSV fusion must be regulated. Rothman's proteins now seem able to generate SSV fusion, which might be regulated by proteins of the synaptosecretosome.

In studying intra-Golgi transport, Rothman's group identified a cytosolic protein that was required for membrane fusion, and whose activity could be abolished by alkylation with N-ethylmaleimide (NEM)—they termed this protein NSF (NEM-sensitive fusion protein; Block et al., 1988; Malhotra et al., 1988). A soluble NSF acceptor or attachment protein (SNAP) was identified, and the SNAP–NSF complex was found to bind to integral membrane proteins of the two membranes involved in the fusion event (Weidman et al., 1989; Whiteheart et al., 1992). These SNAP–NSF receptors were designated SNAREs; it was presumed that there would be SNAREs on both membranes undergoing fusion. Arising from this, it seemed possible that each SNARE might act as a recognition marker for its membrane compartment and so direct the specific fusion of two membranes (reviewed in Wilson et al., 1991; Rothman, 1991; Rothman and Orci, 1992). Rothman's group have now identified four putative SNAREs in brain membranes, and they turn out to be involved in synaptic exocytosis (Whiteheart et al., 1993; Sollner et al., 1993). One of the putative SNAREs was the SSV protein, synaptobrevin II. The other SNAREs were associated with presynaptic membranes—syntaxins A and B, and the otherwise little characterized protein SNAP-25 (synaptosomal associated protein, 25kD; Oyler et al., 1989). The SSV protein could be conveniently designated vesicle (v-) SNAREs, while the presynaptic membrane proteins could be designated target (t-) SNAREs. A complex of NSF and SNAP could join the cognate v- and t-SNAREs to generate a complex competent to direct fusion of SSV and presynaptic membranes. The participation of additional cytosolic proteins and other cofactors cannot be excluded.

Since NSF and SNAP were originally defined in a system where membrane fusion did not require Ca^{2+}, how is regulation to be added to the NSF–SNAP–v-SNARE–t-SNARE system? It is notable that one of the t-SNARES is syntaxin. This is also identified as part of the synaptosecretosome complex of proteins involved in vesicle docking (see Section VIIA). Within this complex are the Ca^{2+}-binding protein synaptotagmin, neurexin, and Ca^{2+} channels (see Figure 2). Synaptotagmin may possibly have an inhibitory role in controlling synaptic exocytosis; synaptotagmin C2-domain peptides inhibit neurotransmitter release (Bommert et al., 1993).

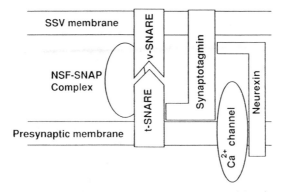

Figure 2. Docking of SSV at the presynaptic active zone. In this scheme, two classes of transmembrane proteins found at the presynaptic membrane, neurexins and t-SNAREs, can act as "receptors" for the SSV. t-SNARE class proteins include syntaxins and SNAP-25. Neurexins can interact with the SSV transmembrane protein synaptotagmin; t-SNAREs can interact with synaptotagmin, and also with their counterparts v-SNAREs as part of a complex with NSF (the N-ethyl maleimide-sensitive membrane fusion protein) and SNAP (the soluble NSF acceptor protein). The v-SNARE class of protein includes synaptobrevin. Ca^{2+} channels are present at the presynaptic membrane and open in response to an action potential. Synaptotagmin is a Ca^{2+} binding protein which has been proposed to be the presynaptic Ca^{2+} sensor—binding of Ca^{2+} to synaptotagmin promotes its interaction with phospholipids. One possibility is that synaptotagmin inhibits membrane fusion driven by the SNARE–NSF–SNAP interaction until it binds Ca^{2+}, whereupon it allows fusion to occur. After models shown in O'Connor et al. (1993) and Kelly (1993), see text for details.

It might be hypothesized that synaptotagmin's constraint on vesicle exocytosis is overcome by its binding to Ca^{2+}, and that the constraint is exercised by its interaction with a SNARE. Note also that Rab3a interacts with a protein related to synaptotagmin (Shirataki et al., 1993)—whether physophilin interacts directly with t- or v-SNAREs is unknown, but it could be imagined that the exocytosis competency that GTP-Rab3a confers on SSV could be mediated by regulation of a SNARE. A possibility that has been discussed by Zerial and Stenmark (1993) is that the role of Rab3a may be to proofread the interaction of SNAREs. Starting with the premise that rabphilin is related to synaptotagmin (a syntaxin-binding protein), they postulated that perhaps rabphilin too could bind syntaxin or a related v- or t-SNARE. However, perhaps rabphilin can only recognize its target SNARE when the correct complex of v- and t-SNARE has formed. Incorrect v-SNARE–t-SNARE complexes would not bind rabphilin, so the membranes carrying them would not become competent to fuse. Binding of the (GTP–Rab3a–rabphilin) complex to the (SNARE–NSF–SNAP) complex would allow fusion to occur, but would not trigger it (see Figure 3). In this scheme, the trigger for exocytosis would still be binding of Ca^{2+} to synaptotasmin, as described above and in Figure 2.

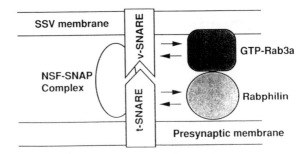

Figure 3. A model for Rab3a activity: proofreading t-SNARE and v-SNARE interaction. The GTPase, Rab3a, interacts with a synaptotagmin-like protein, rabphilin. In this model, the GTP–Rab3a–rabphilin complex recognizes specific v- and t-SNAREs, and only permits binding of the NSF-SNAP complex when the "correct" complex is formed (i.e., SSV–presynaptic membrane). Based on a model given in Zerial and Stenmark (1993), see text for details.

SSV Protein Structure and Function as Related to Other Forms of Vesicle Traffic : Implications for the Evolution of Membrane Traffic

It is interesting to note that both syntaxin and synaptobrevin belong to an unusual class of membrane protein (Baumert et al., 1989; Trimble et al., 1991; Bennett et al., 1992). They do not have N-terminal signal sequences and most of their mass is in the cytoplasm, they have a single transmembrane segment near the C terminus, and few, if any, amino acid residues in compartments topologically equivalent to the endoplasmic reticulum lumen. In this class of protein are a large number of yeast proteins related in sequence to syntaxin or synaptobrevin; these proteins are required for secretion from yeast (c.f. Warren, 1993 for review).

The observation of the closeness of mammalian syntaxin/synaptobrevin to yeast proteins raises the intriguing prospect that SSV exocytosis is a function that has evolved from the membrane traffic that came with the development of simple eukaryotic organisms. This idea is supported by numerous recent findings. For example, there are indications of functional conservation of membrane traffic control between brain and yeast; both components of the GTP-binding and regulatory apparatus (Sasaki et al., 1991; Burton et al., 1993) and the NSF/SNAP system (Wilson et al., 1989; Griff et al., 1992) contain evolutionarily conserved functions.

Botulinum and Tetanus Toxins as Inhibitors of Exocytosis

Recent data obtained from several studies involving neurotoxins has begun to aid the dissection of the exocytotic pathway in nerve cells. The *Clostridial* neurotoxins which include tetanus and the seven serotypes of botulinum toxin (A–G) inhibit presynaptic neurotransmitter release. Further evidence that synaptobrevin plays a role in neurotransmitter release comes from the finding that tetanus toxin, a zinc protease (Schiavo et al., 1992), degrades the putative v-SNARE synaptobrevin II (Link et al., 1992; McMahon et al., 1993). Both tetanus and botulinum serotype B toxins specifically cleaved rat synaptobrevin II at a single site (gln^{76}– phe^{77}), but synaptobrevin I was not cleaved at val^{76}–phe^{77}. A ubiquitous synaptobrevin homologue, cellubrevin, has been described (McMahon et al., 1993) which was also cleaved *in vitro* and *in vivo* by tetanus toxin. It has been proposed that cellubrevin is involved in constitutive membrane trafficking, although cleavage of cellubrevin by tetanus toxin does not seem to inhibit endosomal traffic (Link et al., 1993). An additional substrate for tetanus toxin was a transglutaminase, which can be constitutively activated by toxin-catalyzed proteolysis: Activation of the transglutaminase was suggested to covalently modify synapsin I (Facchiano et al., 1993). Tetanus toxin also affected synapsin I phosphorylation (Presek et al., 1992).

Botulinum toxin A seems to confirm the role of SNAP-25 in synaptic exocytosis. This toxin is a zinc protease, and its substrates include SNAP-25 (Blasi et al., 1993). Of a wide range of synaptic proteins tested, only SNAP-25 was efficiently cleaved by this protease, leading to the suggestion that selective cleavage of SNAP-25 is sufficient to inhibit exocytosis.

Botulinum neurotoxin D catalyzes the ADP-ribosylation of its target proteins. In SSV, its target is Rab3a. The activity of this toxin in inhibiting neurotransmission confirms the importance of Rab3a in SSV exocytosis.

VIII. SYNAPTIC VESICLE RECYCLING

For every SSV that fuses with the presynaptic membrane, an equivalent area of membrane has to be retrieved from it if the area of the membrane is not to increase. It also makes intuitive sense, in terms of cellular economy, to recycle SSV components for use in subsequent rounds of exocytosis. The details of SSV recycling are as yet sketchy, but an overall view can be given.

Presynaptic termini are the richest area in the brain for clathrin-coated vesicles (Maycox et al., 1992). A subpopulation of nervous tissue clathrin-coated vesicles carries some SSV proteins including the transmembrane proteins synaptophysin, synaptotagmin, and SV2 (Pfeffer and Kelly, 1985). Furthermore, antibodies against the domain of synaptotagmin that resides in the SSV lumen can be endocytosed into living neurons in culture (Matteoli et al., 1992). All these data indicate that clathrin-coated vesicles recycle membrane components back into the cell after exocytosis.

Several GTPases may have a role in endocytotic recycling of membrane components as well as in exocytosis (Carter et al., 1993). The *Drosophila* mutant, *shibire*, is defective in synaptic vesicle recycling, and seems not to form coated vesicles well. The mutation is in the GTPase, dynamin (Vanderbliek and Meyerowitz, 1991), first recognized as a microtubule motor protein (Obar et al., 1990, 1991; Shpetner and Vallee, 1991, 1992). Dynamin has also been identified with dephosphin, a protein that is dephosphorylated upon exocytosis (Robinson, 1991a, b; Robinson et al., 1993). It is a substrate for both protein kinase C and casein kinase II, which differentially regulate its activity (Robinson et al., 1993).

Sorting of the components must occur, both to prevent the endocytosis of proteins that must be retained in the presynaptic membrane, and to regenerate a fully formed SSV from uncoated endocytotic vesicles. Neither event is well understood. It is notable that there is at least one synapse-specific adaptor protein that can link integral membrane proteins to clathrin (Ahle and Ungewickell, 1986; Perry et al., 1992; Puszkin et al., 1992; Morris et al., 1993; Zhou et al., 1993); it may be that their recognition specificity covers proteins that require endocytosis, and that they would correspondingly "ignore" those that do not. In the pathways of endocytotic traffic mediated by clathrin-coated vesicles in non-neuronal cells, the sorting of endocytosed membrane components takes place in endosomal compartments (reviewed in Pearse and Robinson, 1990 and Keen, 1990). There is limited evidence for endosome-like compartments in some types of nerve ending (Heuser and Reese, 1973; see also Kelly, 1988), but the universality and the nature of synaptic endosomes is a matter for debate.

NOTE ADDED IN PROOF

Since this manuscript was completed there have been a large number of significant advances. Most of these relate to the use of genetic approaches to the function of proteins described above, to the investigation of their dynamics, interaction and regulation, and to identification of novel members of the exocytotic complex. A very few will be mentioned in this note.

- *Synaptotagmins.* A large family of synaptotagmins has been discovered. They are confirmed as Ca^{2+} sensors for exocytosis in neurons (Kelly, 1995; Littleton and Bellen, 1995), but some members of the family seem not to bind Ca^{2+} (Li et al., 1995a). Ca^{2+} appears to directly modulate the interaction of synaptotagmins with syntaxins (Chapman et al., 1995). Synaptotagmin interacts with β-SNAP to form a complex that recruits NSF, suggesting that these proteins link the process of membrane fusion to calcium entry (Schiavo et al., 1995). Synaptotagmin is now established as a receptor for clathrin adaptor proteins in the process of membrane retrieval (Li et al., 1995a): this process involves both clathrin and dynamin in membrane budding (Takei et al., 1996).
- *Synaptobrevins and syntaxins.* The use of tetanus toxin to specifically eliminate synaptobrevin from cultured neurons has revealed that it is essential for secretion, but not to the development of synaptic processes (Ahnerthilger et al., 1996). Synaptobrevin seems to exist in an intramembrane complex with the V-ATPase and synaptophysin, and this complex is distinct from the synaptic core complex of synaptobrevin/syntaxin/SNAP-25 (Galli et al., 1996). The interactions of syntaxins are more complex than previously

considered: complexes of syntaxin with N-type Ca^{2+} channels have been isolated, and the interaction of these channels with the synaptic core complex is Ca^{2+}-dependent (Sheng et al., 1996). A novel family of regulatory proteins interacting with the core complex has been uncovered (McMahon et al., 1995). NO has now been suggested to modulate the formation of the core complex too (Meffert et al., 1996). Homologues of the yeast *sec*-1 and *C. elegans unc*-18 gene products interact with syntaxins are modulators of exocytosis (Halachmi and Lev, 1996).

- *Synaptophysin.* Transgenic approaches have not revealed an essential role for synaptophysin in neuronal exocytosis (McMahon et al., 1996).
- *p29.* The sequence of p29 (now known as synaptogyrin) has been determined: it probably crosses the membrane four times, and may have a role in membrane recycling (Stenius et al., 1995).
- *Synapsins.* Transgenic approaches have revealed that the synapsins have a role in the supply and organization of synaptic vesicles, and in biogenesis of synapses, but not in long-term potentiation (Li et al., 1995b; Rosahl et al., 1995; Spillane et al., 1995). *Drosophila* synapsins have now been molecularly cloned, and the conserved C-domain has been identified, although not the phosphorylation sites (Klagges et al., 1996). The role of calmodulin in controlling actin binding by synapsins, in concert with phosphorylation, has been established (Goold et al., 1995). MAP-kinases stimulated by neurotrophins catalyze phosphorylation of synapsins (Jovanovic et al., 1996).

REFERENCES

Agoston, D. V., Ballmann, M., Conlon, J. M., & Dowe, G. H. C. (1985). Isolation of neuropeptide-containing vesicles from the guinea pig ileum. J. Neurochem. 45, 398–406.

Ahle, S., & Ungewickell, E. (1986). Purification and properties of a new clathrin assembly protein. EMBO J. 5, 3143–3149.

Ahle, S., & Ungewickell, E. (1990). Auxilin, a newly identified clathrin associated protein in coated vesicles from bovine brain. J. Cell Biol. 111, 19–29.

Ahnerthilger, G., Kutay, U., Chahoud, I., Rapoport, T., & Wiedenmann, B. (1996). Synaptobrevin is essential for secretion but not for the development of synaptic processes. Eur. J. Cell Biol. 70, 1–11.

Alder, J., Lu, B., Valtorta, F., Greengard, P., & Poo, M. M. (1992). Calcium-dependent transmitter secretion reconstituted in *Xenopus* oocytes—requirement for synaptophysin. Science 257, 657–661.

Almers, W. (1990). Exocytosis. Ann. Rev. Physiol. 52, 607–624.

Anderson, D. C., King, S. C., & Parsons, S. M. (1982). Proton gradient linkage to active uptake of [H^3]-labeled acetylcholine by torpedo electric organ synaptic vesicles. Biochemistry 21, 3037–3043.

Araki, S., Kikuchi, A., Hata, Y., Isomura, M., & Takai, Y. (1990). Regulation of reversible binding of smg p25a, a ras p21-like GTP-binding protein, to synaptic plasma-membranes and vesicles by its specific regulatory protein, GDP dissociation inhibitor. J. Biol. Chem. 265, 13007–13015.

Araki, S., Kaibuchi, K., Sasaki, T., Hata, Y., & Takai, Y. (1991). Role of the C-terminal region of smg p25a in its interaction with membranes and the GTP/GDP exchange protein. Mol. Cell Biol. 11, 1438–1447.

Archer, B. T., Ozcelik, T., Jahn, R., Francke, U., & Sudhof, T. C. (1990). Structures and chromosomal localizations of 2 human genes encoding synaptobrevin-1 and synaptobrevin-2. J. Biol. Chem. 265, 17267–17273.

Aubert-Foucher, E., & Font, B. (1990). Limited proteolysis of synapsin 1. Identification of the region of the molecule responsible for its association with microtubules. Biochemistry 29, 5351–5357.

Bahler, M., & Greengard, P. (1987). Synapsin-I bundles F-actin in a phosphorylation-dependent manner. Nature 326, 704–707.

Bahler, M., Benfenati, F., Valtorta, F., Czernic, A. J., & Greeengard, P. (1989). Characterization of synapsin 1 fragments produced by cysteine specific cleavage: A study of their interactions with F-actin. J. Cell Biol. 108, 1841–1849.

Bahler, M., Klein, R. L., Wang, J., Benfenati, F., & Greengard, P. (1991). A novel synaptic vesicle-associated phosphoprotein—SVAPP-120. J. Neurochem. 57, 423–430.

Bahr, B.A., & Parsons, S. M. (1992). Purification of the vesamicol receptor. Biochemistry 31, 5763–5769.

Baines, A. J. (1987). Synapsin 1 and the cytoskeleton. Nature 326, 646.

Baines, A. J., & Bennett, V. (1985). Synapsin I is a spectrin binding protein immunologically related to erythrocyte 4.1. Nature 315, 410–413.

Baines, A. J., & Bennett, V. (1986). Synapsin I is a microtubule bundling protein. Nature 319, 145–147.

Baitinger, C., & Willard, M. (1987). Axonal transport of synapsin I-like proteins in rabbit retinal ganglion cells. J. Neurosci. 7, 3723–3735.

Bajjalieh, S. M., Peterson, K., Shinghal, R., & Scheller, R. H. (1992). SV2, a brain synaptic vesicle protein homologous to bacterial transporters. Science 257, 1271–1273.

Barnekow, A., Jahn, R., & Schartl, M. (1990). Synaptophysin—a substrate for the protein tyrosine kinase Pp60C-SRC in intact synaptic vesicles. Oncogene 5, 1019–1024.

Baumert, M., Maycox, P. N., Navone, F., De Camelli, P., & Jahn, R. (1989). Synaptobrevin, an integral membrane protein of 18000 Dalton present in small synaptic vesicles of rat brain. EMBO J. 8, 379–384.

Baumert, M., Takei, K., Hartinger, J., Burger, P. M., Fisher von Mallard, G., Maycox, P. R., De Camilli, P., & Jahn, R. (1990). p29, A novel tyrosine-phosphorylated membrane protein present in small clear vesicles of neurones and endocrine cells. J. Cell Biol. 110, 1285–1294.

Bell, R. M., & Burns, A. J. (1991). Lipid activation of protein kinase C. J. Biol. Chem. 266, 4661–4664.

Benfenati, F., Greengard, P., Brunner, J., & Bahler, M. (1989a). Electrostatic and hydrophobic interactions of synapsin I and synapsin I fragments with phospholipid bilayers. J. Cell Biol. 108, 1851–1862.

Benfenati, F., Bahler, M., Jahn, R., & Greengard, P. (1989b). Interactions of synapsin I with small synaptic vesicles: Distinct sites in synapsin I bind to vesicle phospholipids and vesicle proteins. J. Cell Biol. 108, 1863–1872.

Benfenati, F., Neyroz, P., Bahler, M., Masotti, L., & Greengard, P. (1990). Time-resolved fluorescence study of the neuron-specific phosphoprotein synapsin-I—evidence for phosphorylation-dependent conformational changes. J. Biol. Chem. 265, 12584–12595.

Benfenati, F., Valtorta, F., Chieregatti, E., & Greengard, P. (1992a). Interaction of free and synaptic vesicle bound synapsin-I with F-actin. Neuron 8, 377–386.

Benfenati, F., Valtorta, F., Rubenstein, J. L., Gorelick, F. S., Greengard, P., & Czernik, A. J. (1992b). Synaptic vesicle-associated Ca^{2+}/Calmodulin-dependent protein kinase-II is a binding protein for synapsin-I. Nature 359, 417–420.

Bennett, A. F., Hayes, N. V. L., & Baines, A. J. (1991). Site specificity in the interactions of synapsin-1 with tubulin. Biochem. J. 276, 793–799.

Bennett, A. F., & Baines, A. J. (1992). Bundling of microtubules by synapsin 1: Characterisation of bundling and interaction of distict sites in synapsin 1 head and tail domains with different sites in tubulin. Eur. J. Biochem. 206, 783–792.

Bennett, M. K., Erondu, N. E., & Kennedy, M. B. (1983). Purification and characterisation of a calmodulin-dependent protein kinase that is highly concentrated in brain. J. Biol. Chem. 258, 12735–12744.

Bennett, M. K., Calakos, N., & Scheller, R. H. (1992). Syntaxin—A synaptic protein implicated in docking of synaptic vesicles at presynaptic active zones. Science 257, 255–259.

Bennett, V., Baines, A. J., & Davis, J. (1985). Purification of brain analogs of RBC membrane skeletal proteins: Ankyrin, protein 4.1, (synapsin) spectrin and spectrin subunits. Methods Enzymol. 134, 55–68.

Bernstein, B. W., & Bamberg, J. R. (1985). Reorganization of actin in depolarizing synaptosomes. J. Neurosci. 5, 2565–2563.

Blasi, J., Chapman, E. R., Link, E., Binz, T., Yamasaki, S., Decamilli, P., Sudhof, T. C., Niemann, H., & Jahn, R. (1993). Botulinum neurotoxin-a selectively cleaves the synaptic protein SNAP-25. Nature 365, 160–163.

Block, M. R., Glick, B. S., Wilcox, C. A., Wieland, F. T., & Rothman, J. E. (1988). Purification of an N-ethylmaleimide-sensitive protein catalyzing vesicular transport. Proc. Natl. Acad. Sci. USA 85, 7852–7856.

Bommert, K., Charlton, M. P., Debello, W. M., Chin, G. J., Betz, H., & Augustine, G. J. (1993). Inhibition of neurotransmitter release by C2-domain peptides implicates synaptotagmin in exocytosis. Nature 363, 163–165.

Brodsky, F. M., Hill, B. L., Acton, S. L., Nathke, I., Wong, D., Ponnambalam, S., & Parham, P. (1991). Clathrin light chains: Arrays of protein motifs that regulate coated vesicles dynamics. TIBS 16, 208–213.

Brose, N., Petrenko, A., Sudhof, T. C., & Jahn, R. (1992). Synaptotagmin. A calcium sensor on the synaptic vesicle surface. Science 256, 1021–1025.

Browning, M. D., & Dudek, E. M. (1992). Activators of protein-kinase-C increase the phosphorylation of the synapsins at sites phosphorylated by cAMP-dependent and Ca^{2+} calmodulin-dependent protein-kinases in the rat hippocampal slice. Synapse 10, 62–70.

Buckley, K. M., Floor, E., & Kelly, R. B. (1987). Cloning and sequence-analysis of cDNA-encoding p38, a major synaptic vesicle protein. J. Cell Biol. 105, 2447–2456.

Buckley, K., & Kelly, R. B. (1985). Identification of a transmembrane glycoprotein specific for secretory vesicles of neural and endocrine cells. J. Biol. Chem. 100, 1284–1294.

Burger, P. M., Hell, J., Mehl, E., Krasel, C., Lottspeich, F., & Jahn, R. (1991). GABA and glycine in synaptic vesicles—storage and transport characteristics. Neuron 7, 287–293.

Burgoyne, R. D., Geisow, M. J., & Barron, J. (1982). Dissection of the stages in exocytosis in adrenal chromaffin cells with the use of trifluoperazine. Proc. Roy. Soc. Lond. B 216, 111–115.

Burstein, E. S., Linko-Stentz, K., Lu, Z. J., & Macara, I. G. (1991). Regulation of the GTPase activity of the ras-like protein p25rab3A. Evidence for a rab3A-specific GAP. J. Biol. Chem. 266, 2689–2692.

Burstein, E. S., & Macara, I. G. (1992). Characterization of a guanine nucleotide-releasing factor and a GTPase-activating protein that are specific for the ras-related protein p25rab3A. Proc. Natl. Acad. Sci. USA 89, 1154–1158.

Burton, J., Roberts, D., Montaldi, M., Novick, P., & DeCamilli, P. (1993). A mammalian guanine-nucleotide-releasing protein enhances function of yeast secretory protein sec4. Nature 361, 464–467.

Carlson, S. S., Wagner, J. A., & Kelly, R. B. (1978). Purification of synaptic vesicles from elasmobranch electric organ and the use of biophysical criteria to demonstrate purity. Biochemistry 17, 1188–1199.

Carlson, S. S., & Kelly, R. B. (1983). A highly antigenic proteoglycan-like component of cholinergic synaptic vesicles. J. Biol. Chem. 258, 1082–1091.

Carter, L. L., Redelmeier, T. E., Woollenweber, L. A., & Schmid, S. L. (1993). Multiple GTP-binding proteins participate in clathrin-coated vesicle-mediated endocytosis. J. Cell Biol. 120, 37–45.

Chandler, D. E., & Heuser, J. E. (1980). Arrest of membrane fusion events in mast cells by quick freezing. J. Cell Biol. 86, 666–674.

Chapman, E. R., Hanson, P. I., An, S., & Jahn, R. (1995). Ca^{2+} regulates the interaction between synaptotagmin and syntaxin-1. J. Biol. Chem. 270, 23667–23671.

Chin, G. J., & Goldman, S. A. (1992). Purification of squid synaptic vesicles and characterization of the vesicle-associated proteins synaptobrevin and rab3a. Brain Res. 571, 89–96.

Chou, C. F., & Omary, M. B. (1993). Mitotic arrest-associated enhancement of O-linked glycosylation and phosphorylation of human keratin-8 and keratin-18. J. Biol. Chem. 268, 4465–4472.

Christensen, H., Fykse, E. M., & Fonnum, F. (1990). Uptake of glycine into synaptic vesicles isolated from rat spinal-cord. J. Neurochem. 54, 1142–1147.

Clark, A. W., Mauro, A., Langenecker, H. E. J., & Hurlbut, W. P. (1970). Effects on the fine structure of the frog neuromuscular junction. Nature 225, 703–705.

Couteaux, R., & Pecout-Dechavassine, M. (1970). Vesicules synaptiques et poches au niveau des zones actives de la jonction neuromusculaire. C. R. Soc. Biol. (Paris) 271, 2346–2349.

Czernic, A. J., Pang, D. T., & Greengard, P. (1987). Amino acid sequences surrounding the cAMP-dependent and Ca^{2+}/calmodulin-dependent phosphorylation sites in rat and bovine synapsin 1. Proc. Natl. Acad. Sci. USA 84, 7518–7522.

Davletov, B., Sontag, J.-M., Hata, Y., Petrenko, A., Fuske, E. M., Jahn, R., & Sudhof, T. C. (1993). Phosphorylation of synaptotagmin I by caesin kinase II. J. Biol. Chem. 268, 6816–6822.

De Camilli, P., Cameron, R., & Greengard, P. (1983). Synapsin 1 (Protein 1), a nerve terminal specific phosphoprotein. 1. Its general distribution in synapses of the central and peripheral nervous system demonstrated by immunofluorescence in frozen and plastic sections. J. Cell Biol. 96, 1337–1354.

De Camilli, P., & Jahn, R. (1990). Pathways to regulated exocytosis in neurons. Ann. Rev. Physiol. 52, 625–645.

De Camilli, P., Benfenati, F., Valtorta, F., & Greengard, P. (1990). The synapsins. Ann. Rev. Cell Biol. 6, 433–460.

Diantonio, A., Parfitt, K. D., & Schwarz, T. L. (1993). Synaptic transmission persists in synaptotagmin mutants of *Drosophila*. Cell 73, 1281–1290.

Douglas, W. W., & Nemeth, E. F. (1982). On the calcium receptor activating exocytosis—inhibitory effects of calmodulin-interacting drugs on rat mast-cells. J. Physiol. 323, 229–244.

Drenkhahn, D., & Kaiser, H. W. (1983). Evidence for the concentration of F-actin and mysoin in synapses and in the plasmalemmal zone of axons. Eur. J. Cell Biol. 31, 235–240.

Elferink, L. A., Trimble, W. S., & Scheller, R. H. (1989). Vesicle-associated membrane protein genes are differentially expressed in the rat central nervous-system. J. Biol. Chem. 264, 11061–11064.

Facchiano, F., Benfenati, F., Valtorta, F., & Luini, A. (1993). Covalent modification of synapsin I by a tetanus toxin-activated transglutaminase. J. Biol. Chem. 268, 4588–4591.

Farrell, K. P., & Keates, R. (1990). Synapsin-1 is found in a microtubule-associated complex of proteins isolated from bovine brain. Biochimie Et Biologie Cellulaire 68, 1256–1261.

Feany, M. B., & Buckley, K. M. (1993). The synaptic vesicle protein synaptotagmin promotes formation of filopodia in fibroblasts. Nature 364, 537–540.

Feany, M. B., Lee, S., Edwards, R. H., & Buckley, K. M. (1992). The synaptic vesicle protein SV2 is a novel type of transmembrane transporter. Cell 70, 861–867.

Fesce, R., Benfenati, F., Greengard, P., & Valtorta, F. (1992). Effects of the neuronal phosphoprotein synapsin-I on actin polymerization .2. Analytical interpretation of kinetic curves. J. Biol. Chem. 267, 11289–11299.

Fischer von Mollard, G., Mignery, G. A., Baumert, M., Perin, M. S., Hanson, T. J., Burger, P. M., Jahn, R., & Sudhof, T. C. (1990). rab3 is a small GTP-binding protein exclusively localised to synaptic vesicles. Proc. Natl. Acad. Sci. USA 87, 1988–1992.

Fischer von Mollard, G., Sudhof, T. C., & Jahn, R. (1991). A small GTP-binding protein dissociates from synaptic vesicles during exocytosis. Nature 349, 79–81.

Floor, E., Leuenthal, P. S., & Schaeffer, S. F. (1990). Partial purification and characterisation of the vacuolar H$^+$ATPase of mammalin synaptic vesicles. J. Neurochem. 55, 1663–1670.

Font, B., & Aubert-Foucher, E. (1989). Detection by chemical cross-linking of bovine brain synapsin I self-association. Biochem. J. 264, 893–899.

Galli, T., McPherson, P. S., & De Camilli, P. (1996). The V-O sector of the V-ATPase, synaptobrevin, and synaptophysin are associated on synaptic vesicles in a Triton X-100-resistant, freeze-thawing sensitive, complex. J. Biol. Chem. 271, 2193–2198.

Geppert, M., Archer, B. T., & Sudhof, T. C. (1991). Synaptotagmin-II—a novel differentially distributed form of synaptotagmin. J. Biol. Chem. 266, 13548–13552.

Goelz, S. E., Nestler, E. J., Chehrazi, B., & Greengard, P. (1981). Distribution of protein I in mammalian brain as determined by a detergent based radioimmunoassay. Proc. Natl. Acad. Sci. USA 78, 63–72.

Goldenring, J. R., Lasher, R. S., Vallano, M. L., Ueda, T., Naito, S., Sternberger, N. H., Sternberger, L. A., & De Lorenzo, R. J. (1986). Association of synapsin 1 with the neuronal cytoskeleton. Identification in cytoskeletal preparations *in vitro* and immunocytochemical localization in brain of synapsin 1. J. Biol. Chem. 261, 8495–8504.

Gomperts, B. D. (1990). G$_e$: A GTP binding protein mediating exocytosis. Ann. Rev. Physiol. 52, 591–606.

Goodman, S. R., Krebs, K. E., Whitfield, C. F., Riederer, B. M., & Zagon, I. S. (1988). Spectrin and related molecules. CRC Crit. Rev. Biochem. 23, 171–234.

Goold, R., & Baines, A. J. (1994). Evidence that two non-overlapping high affinity calmodulin binding sites are present in the head region of synapsin 1. Eur. J. Biochem. 224, 229–240.

Goold, R., Chan, K. M., & Baines, A. J. (1995). Coordinated regulation of synapsin-I interaction with F-actin by Ca^{2+}/calmodulin and phosphorylation: Inhibition of actin-binding and bundling. Biochemistry 34, 1912–1920.

Gotow, T., Miyaguchi, K., & Hashimoto, P. H. (1991). Cytoplasmic architecture of the axon terminal—filamentous strands specifically associated with synaptic vesicles. Neuroscience 40, 587–598.

Griff, I. C., Schekman, R., Rothman, J. E., & Kaiser, C. A. (1992). The yeast sec17 gene-product is functionally equivalent to mammalian alpha-SNAP protein. J. Biol. Chem. 267, 12106–12115.

Hackett, J. T., Cochran, S. L., Greenfield, L. J., Brosius, D. C., & Ueda, T. (1990). Synapsin-I injected presynaptically into goldfish mauthner axons reduces quantal synaptic transmission. J. Neurophys. 63, 701–706.

Halachmi, N., & Lev, Z. (1996). The Sec1 Family—a novel family of proteins involved in synaptic transmission and general secretion. J. Neurochem. 66, 889–897.

Hall, F. L., Mitchell, J. P., & Vulliet, P. R. (1990). Phosphorylation of synapsin I at a novel site by proline-directed protein kinase. J. Biol. Chem. 265, 6944–6948.

Haltiwanger, R. S., Kelly, W. G., Roquemore, E. P., Blomberg, M. A., Dong, L., Kreppel, L., Chou, T. Y., & Hart, G. W. (1992). Glycosylation of nuclear and cytoplasmic proteins is ubiquitous and dynamic. Biochem. Soc. Trans. 20, 264–269.

Hart, G. W. (1992). Glycosylation. Curr. Opin. Cell Biol. 4, 1017–1023.

Hata, Y., Davletov, B., Petrenko, A. G., Jahn, R., & Sudhof, T. C. (1993). Interaction of synaptotagmin with the cytoplasmic domains of neurexins. Neuron 10, 307–315.

Hayes, N. V. L., Bennett, A. F., & Baines, A. J. (1991). Selective Ca^{2+}-dependent interaction of calmodulin with the head domain of synapsin-1. Biochem. J. 275, 93–97.

Hell, J. W., Maycox, P. R., Stadler, H., & Reihard, J. (1988). Uptake of GABA by rat brain synaptic vesicles isolated by a new procedure. EMBO J. 7, 3023–3029.

Henderson, P. F. J. (1993). The 12-transmembrane helix transporters. Curr. Opin. Cell Biol. 5, 708–721.

Heuser, J. E., & Reese, T. (1973). Evidence for recycling synaptic vesicle membrane during neurotransmitter release at the frog neuromuscular junction. J. Cell Biol. 57, 315–344.

Heuser, J.E. (1989). Review of electron-microscopic evidence favoring vesicle exocytosis as the structural basis for quantal release during synaptic transmission. Quart. J. Expl. Physiol. Cogn. Med. Sci. 74N7, 1051–1069.

Hirokawa, N., Glicksman, M. A., & Willard, M. B. (1984). Organisation of mammalian neurofilament polypeptides within the neuronal cytoskeleton. J. Cell Biol. 98, 1523–1536.

Hirokawa, N., Sobue, K., Kanda, K., Harada, A., & Yorifuji, H. (1989). The cytoskeletal architecture of the presynaptic terminal and molecular architecture of synapsin 1. J. Cell Biol. 108, 111–126.

Ho, M. F., Bahler, M., Czernik, A. J., Schiebler, W., Kezdy, F. J., Kaiser, E. T., & Greengard, P. (1991). Synapsin-I Is a highly surface-active molecule. J. Biol. Chem. 266, 5600–5607.

Hooper, J. E., & Kelly, R. B. (1984). Calmodulin is tightly associated with synaptic vesicles independent of calcium. J. Biol. Chem. 259, 148–153.

Howland, D. S., Hemmendinger, L. M., Estes, P. S., & DeGennaro, L. J. (1990). Sequence and functional analysis of the rat synapsin I gene promoter. J. Cell Biol. 111, 2849a.

Howland, D. S., Hemmendinger, L. M., Carroll, P. D., Estes, P. S., Melloni, R. H., & DeGennaro, L. J. (1991). Positive-acting and negative-acting promoter sequences regulate cell type-specific expression of the rat synapsin-I gene. Molecular Brain Research 11, 345–353.

Huang, C.-K., Browning, M. D., & Greengard, P. (1982). Purification and characterisation of protein IIIb a mammalian brain phosphoprotein. J. Biol. Chem. 257, 6524–6528.

Huttner, W. B., & Greengard, P. (1979). Multiple phosphorylation sites in protein 1 and their differential regulation by cAMP and calcium. Proc. Natl. Acad. Sci. USA 76, 5402–5406.

Huttner, W. B., Degennaro, L. J., & Greengard, P. (1981). Differential phosphorylation of multiple sites in purified protein-I by cyclic AMP-dependent and calcium-dependent protein-kinases. J. Biol. Chem. 256, 1482–1488.

Huttner, W. B., Schiebler, W., Greengard, P., & De Camilli, P. (1983). Synapsin 1, a nerve terminal specific phosphoprotein. 3. Its association with synaptic vesicles studied in a highly purified synaptic vesicle preparation. J. Cell Biol. 96, 1384–1388.

Inoue, A., Obata, K., & Akagawa, K. (1992). Cloning and sequence-analysis of cDNA for a neuronal cell membrane antigen, HPC-1. J. Biol. Chem. 267, 10613–10619.

Isambert, M. F., Gasnier, B., Botton, D., & Henry, J. P. (1992). Characterization and purification of the monoamine transporter of bovine chromaffin granules. Biochemistry 31, 1980–1986.

Jackson, A. P. (1987). Clathrin light chain contains brain specific insertion sequences and a region of homology with intermediate filaments. Nature 326, 154.

Jackson, A. P. (1992). Endocytosis in the brain—the role of clathrin light-chains. Biochem. Soc. Trans. 20, 653–655.

Jacobson, G. R., Schaffer, M. H., Stark, G. R., & Vanaman, T. C. (1973). Specific chemical cleavage in high yield at the amino peptide bonds of cysteine and cysteine residues. J. Biol. Chem. 248, 6583–6591.

Jahn, R., Schiebler, W., Ouimet, C., & Greengard, P. (1985). A 38,000-dalton membrane-protein (p38) present in synaptic vesicles. Proc. Natl. Acad. Sci. USA 82, 4137–4141.

Jahn, R., & Sudhof, T. C. (1993). Synaptic vesicle traffic—rush hour in the nerve-terminal. J. Neurochem. 61, 12–21.

Johnson, D. A., Jahn, R., & Sudhof, T. C. (1989). Transmembrane topography and evolutionary conservation of synaptophysin. J. Biol. Chem. 264, 1268–1273.

Johnston, P. A., & Sudhof, T. C. (1990). The multisubunit structure of synaptophysin. J. Biol. Chem. 265, 8869–8873.

Johnston, P. A., Archer, B. T., Robinson, K., Mignery, G. A., Jahn, R., & Sudhof, T. C. (1991). rab3A attachment to the synaptic vesicle membrane mediated by a conserved polyisoprenylated carboxy-terminal sequence. Neuron 7, 101–109.

Jovanovic, J. N., Benfenati, F., Siow, Y. L., Sihra, T. S., Sanghera, J. S., Pelech, S. L., Greengard, P., & Czernik, A. J. (1996). Neurotrophins stimulate phosphorylation of synapsin-I by map kinase and regulate synapsin-I actin interactions. Proc. Natl. Acad. Sci. USA 93, 3679–3683.

Katz, B. (1969). *The Release of Neural Transmitter Substances*. Liverpool University Press, Liverpool.

Kearse, K. P., & Hart, G. W. (1991). Lymphocyte-activation induces rapid changes in nuclear and cytoplasmic glycoproteins. Proc. Natl. Acad. Sci. USA 88, 1701–1705.

Keen, J. H. (1990). Clathrin and associated assembly and disassembly proteins. Ann. Rev. Biochem. 59, 415–438.

Kelly, R. B. (1988). The cell biology of the nerve terminal. Neuron 1, 431–438.

Kelly, R. B. (1993). Much ado about docking. Curr. Biol. 3, 474–476.

Kelly, R. B. (1995). Neural transmission—synaptotagmin is just a calcium sensor. Curr. Biol. 5, 257–259.

Kenigsberg, R. L., Cote, A., & Trifaro, J. M. (1982). Trifluoperazine, a calmodulin inhibitor, blocks secretion in cultured chromaffin cells at a step distal from calcium entry. Neuroscience 7, 2277–2286.

Kenigsberg, R. L., & Trifaro, J. M. (1985). Microinjection of calmodulin antibodies into cultured chromaffin cells blocks catecholamine release in response to stimulation. Neuroscience 14, 335–347.

Kennedy, M. B., McGuinness, T., & Greengard, P. (1983a). A calcium calmodulin-dependent protein-kinase from mammalian brain that phosphorylates synapsin-I—partial-purification and characterization. J. Neurosci. 3, 818–831.

Kennedy, M. B., Bennett, M. K., & Erondu, N. K. (1983b). Biochemical and immunochemical evidence that the major postsynaptic density protein is a subunit of a calmodulin dependent protein kinase. Proc. Natl. Acad. Sci. USA 80, 7357–7361.

Klagges, B. R. E., Heimbeck, G., Godenschwege, T. A., Hofbäuer, A., Pflugfelder, G. O., Reifegerste, R., Reisch, D., Schaupp, M., Buchner, S., & Buchner, E. (1996). Invertebrate synapsins—a single-gene codes for several isoforms in Drosophila. J. Neurosci. 16, 3154–3165.

Klevin, R., Lagercrantz, H., & Zimmerman, H. (1982). *Neurotransmitter Vesicles*. Academic Press, New York.

Knaus, P., Marquezepouey, B., Scherer, H., & Betz, H. (1990). Synaptoporin, a novel putative channel protein of synaptic vesicles. Neuron 5, 453–462.

Krebs, K. E., Prouty, S. M., Zagon, I. S., & Goodman, S. R. (1987). Structural and functional relationship of red blood cell protein 4.1 to synapsin I. CRC Crit. Rev. Biochem. 23, 171–234.

Landis, D. M. D., & Reese, T. S. (1983). Cytoplasmic organization in cerebellar dendritic spines. J. Cell Biol. 97, 1169–1178.

Landis, D. M. (1988). Membrane and cytoplasmic structure at synaptic junctions in the mammalian central nervous system. J. Electron Microsc. Tech. 10, 129–151.

Landis, D. M., Hall, A. K., Weinstein, L. A., & Reese, T. S. (1988). The organization of cytoplasm at the presynaptic active zone of a central nervous system synapse. Neuron 1, 201–209.

Leube, R. E., Kaiser, P., Seiter, A., Zimbelmann, R., Franke, W. W., Rehm, H., Knaus, P., Prior, P., Betz, H., Reinke, H., Beyreuther, K., & Wiedenmann, B. (1987). Synaptophysin—molecular-organization and messenger-RNA expression as determined from cloned cDNA. EMBO J. 6, 3261–3268.

Leveque, C., Hoshino, T., David, P., Shoji-Kasai, Y., Leys, K., Omori, A., Lang, B., El-For, O., Sato, K., Martin-Mautot, N., Newsam-Davies, J., Takahashi, M., & Seager, M. J. (1992). The synaptic vesicle protein synaptotagmin associates with calcium channels and is a putative Lambert-Eaton myasthetic syndrome antigen. Proc. Natl. Acad. Sci. USA 89, 3625–3629.

Li, L. A., Suzuki, T., Mori, N., & Greengard, P. (1993). Identification of a functional silencer element involved in neuron-specific expression of the synapsin-I gene. Proc. Natl. Acad. Sci. USA 90, 1460–1464.

Li, C., Ullrich, B., Zhang, J. Z., Anderson, R. G. W., Brose, N., & Sudhof, T. C. (1995a). Ca^{2+}-dependent and Ca^{2+}-independent activities of neural and nonneural synaptotagmins. Nature 375, 594–599.

Li, L., Chin, L. S., Shupliakov, O., Brodin, L., Sihra, T. S., Hvalby, O., Jensen, V., Zheng, D., McNamara, J. O., Greengard, P., & Andersen, P. (1995b). Impairment of synaptic vesicle clustering and of synaptic transmission, and increased seizure propensity, in synapsin I- deficient mice. Proc. Natl. Acad. Sci. USA 92, 9235–9239.

Lichte, B., Veh, R. W., Meyer, H. E., & Kilimann, M. W. (1992). Amphiphysin, a novel protein associated with synaptic vesicles. EMBO J. 11, 2521–2530.

Lin, J. W., Sugimori, M., Llinas, R. R., McGuinness, T. L., & Greengard, P. (1990). Effects of synapsin-I and calcium/calmodulin-dependent protein kinase-II on spontaneous neurotransmitter release in the squid giant synapse. Proc. Natl. Acad. Sci. USA 87, 8257–8261.

Link, E., Edelmann, L., Chou, J. H., Binz, T., Yamasaki, S., Eisel, U., Baumert, M., Sudhof, T. C., Niemann, H., & Jahn, R. (1992). Tetanus toxin action—inhibition of neurotransmitter release linked to synaptobrevin proteolysis. Biochem. Biophys. Res. Commun. 189, 1017–1023.

Link, E., Mcmahon, H., Vonmollard, G. F., Yamasaki, S., Niemann, H., Sudhof, T. C., & Jahn, R. (1993). Cleavage of cellubrevin by tetanus toxin does not affect fusion of early endosomes. J. Biol. Chem. 268, 18423–18426.

Linstedt, A. D., & Kelly, R. B. (1991). Synaptophysin is sorted from endocytotic markers in neuroendocrine PC12-cells but not transfected fibroblasts. Neuron 7, 309–317.

Linstedt, A. D., Vetter, M. L., Bishop, J. M., & Kelly, R. B. (1992). Specific association of the protooncogene product pp60(c-src) with an intracellular organelle, the pc12 synaptic vesicle. J. Cell Biol. 117, 1077–1084.

Littleton, J. T., & Bellen, H. J. (1995). Synaptotagmin controls and modulates synaptic-vesicle fusion in a Ca^{2+}-dependent manner. Trends Neurosci. 18, 177–183.

Llinas, R. (1987). Functional compartments in synaptic transmission. In: *Synaptic Function* (Edelman, G. M., Gall, W. E., & Cowan, W. M., Eds.). Wiley, New York, pp. 7–20.

Llinas, R., Sugimori, M., & Simon, S. (1982). Transmission by presynaptic spike-like depolarisation in the squid giant synapse. Proc. Natl. Acad. Sci. USA 79, 2415–2419.

Llinas, R., McGuinness, T. L., Leonard, C. S., Sugimori, M., & Greengard, P. (1985). Intraterminal injection of synapsin 1 or calcium/calmodulin dependent protein kinase 2 alters neurotransmitter release at the squid giant synapse. Proc. Natl. Acad. Sci. USA 82, 3035–3039.

Llinas, R., Gruner, J. A., Sugimori, M., McGuinness, T. L., & Greengard, P. (1991). Regulation by synapsin-I and Ca^{2+}-calmodulin-dependent protein kinase II of transmitter release in squid giant synapse. J. Physiol. 436, 257–282.

Lowenstein, W. R. (1987). The cell-to-cell channel of gap junctions. Cell 48, 725–726.

Lui, Y., Peter, D., Roghani, A., Schulidner, P., Prive, G. C., Eisenberg, D., Brecha, N., & Edwards, R. H. (1992). A cDNA that supresses MPP^+ toxicity encodes a vesicular amine transporter. Cell 70, 539–551.

Luthi, T., Haltiwanger, R. S., Greengard, P., & Bahler, M. (1991). Synapsins contain O-linked N-acetylglucosamine. J. Neurochem. 56, 1493–1498.

Malhotra, V., Orci, L., Glick, B. S., Block, M. R., & Rothman, J. E. (1988). Role of an N-ethylmaleimide-sensitive transport component in promoting fusion of transport vesicles with cisternae of the golgi stack. Cell 54, 221–227.

Matteoli, M., Takei, K., Cameron, R., Hurlbut, P., Johnston, P. A., Sudhof, T. C., Jahn, R., & Decamilli, P. (1991). Association of rab3a with synaptic vesicles at late stages of the secretory pathway. J. Cell Biol. 115, 625–633.

Matteoli, M., Takei, K., Perin, M. S., Sudhof, T. C., & DeCamilli, P. (1992). Exo-endocytotic recycling of synaptic vesicles in developing processes of cultured hippocampal-neurons. J. Cell Biol. 117, 849–861.

Matthew, W. D., Tsavaler, L., & Reichardt, L. F. (1981). Identification of a synaptic vesicle-specific membrane-protein with a wide distribution in neuronal and neurosecretory tissue. J. Cell Biol. 91, 257–269.

Maycox, P., Hell, J., & Jahn, R. (1990). Amino acid neurotransmission: Spotlight on synaptic vesicles. Trends Neurosci. 13, 83–87.

Maycox, P. R., Deckwerth, T., Hell, J. W., & Jahn, R. (1988). Glutamate uptake by brain synaptic vesicles—energy-dependence of transport and functional reconstitution in proteoliposomes. J. Biol. Chem. 263, 15423–15428.

Maycox, P. R., Link, E., Reetz, A., Morris, S. A., & Jahn, R. (1992). Clathrin-coated vesicles in nervous tissue are involved primarily in synaptic vesicle tecycling. J. Cell Biol. 118, 1379–1388.

McMahon, H. T., & Nicholls, D. G. (1991). The bioenergetics of neurotransmitter release. Biochim. Biophys. Acta 1059, 243–264.

McMahon, H. T., Ushkaryov, Y. A., Edelmann, L., Link, E., Binz, T., Niemann, H., Jahn, R., & Sudhof, T. C. (1993). Cellubrevin is a ubiquitous tetanus-toxin substrate homologous to a putative synaptic vesicle fusion protein. Nature 364, 346–349.

McMahon, H. T., Missler, M., Li, C., & Sudhof, T. C. (1995). Complexins—cytosolic proteins that regulate snap receptor function. Cell 83, 111–119.

McMahon, H. T., Bolshakov, V. Y., Janz, R., Hammer, R. E., Siegelbaum, S. A., & Sudhof, T. C. (1996). Synaptophysin, a major synaptic vesicle protein, is not essential for neurotransmitter release. Proc. Natl. Acad. Sci. USA 93, 4760–4764.

Meffert, M. K., Calakos, N. C., Scheller, R. H., & Schulman, H. (1996). Nitric-oxide modulates synaptic vesicle docking/fusion reactions. Neuron 16, 1229–1236.

Mitschulat, H. (1989). Dynamic properties of the Ca^{2+}/calmodulin-dependent protein kinase in *Drosophila*: Identification of a synapsin I-like protein. Proc. Natl. Acad. Sci. USA 86, 5988–92.

Mizoguchi, A., Ueda, T., Ikeda, K., Shiku, H., Mizoguti, H., & Takai, Y. (1989). Localization and subcellular distribution of cellular ras gene products in rat brain. Brain Res. Mol. Brain Res. 5, 31–44.

Mizoguchi, A., Arakawa, M., Masutani, M., Tamekane, A., Yamaguchi, H., Minami, N., Takai, Y., & Ide, C. (1992). Localization of smg p25A/rab3A p25, a small GTP-binding protein, at the active zone of the rat neuromuscular junction. Biochem. Biophys. Res. Commun. 186, 1345–1352.

Monck, J. R., & Fernandez, J. M. (1992). The exocytotic fusion pore. J. Cell Biol. 119, 1395–1404.

Morris, S. A., Schroder, S., Plessmann, U., Weber, K., & Ungewickell, E. (1993). Clathrin assembly protein-ap180—primary structure, domain organization and identification of a clathrin binding-site. EMBO J. 12, 667–675.

Musha, T., Kawata, M., & Takai, Y. (1992). The geranylgeranyl moiety but not the methyl moiety of the smg-25a rab3a protein is essential for the interactions with membrane and its inhibitory GTP/GDP exchange protein. J. Biol. Chem. 267, 9821–9825.

Nagy, A., Varady, G., Joo, F., Rakanczay, Z., & Pik, A. (1977). Separation of acetylcholine and catecholamine containing synaptic vesicles from brain cortex. J. Neurochem. 29, 449–459.

Nairn, A. C., & Greengard, P. (1987). Purification and characterization of Ca^{2+} calmodulin-dependent protein kinase-I from bovine brain. J. Biol. Chem. 262, 7273–7281.

Navone, F., Greengard, P., & DeCamilli, P. (1984). Synapsin I in neurotranmission: Selective association with small synaptic vesicles. Science 226, 1209–1211.

Navone, F., Jahn, R., Digioia, G., Stukenbrok, H., Greengard, P., & DeCamilli, P. (1986). Protein-p38—an integral membrane-protein specific for small vesicles of neurons and neuro-endocrine cells. J. Cell Biol. 103, 2511–2527.

Navone, F., Digioia, G., Jahn, R., Browning, M., Greengard, P., & Decamilli, P. (1989). Microvesicles of the neurohypophysis are biochemically related to small synaptic vesicles of presynaptic nerve terminals. J. Cell Biol. 106, 3425–3433.

Nichols, R. A., Chilcote, T. J., Czernik, A. J., & Greengard, P. (1992). Synapsin-I regulates glutamate release from rat-brain synaptosomes. J. Neurochem. 58, 783–785.

Nonet, M. L., Grundahl, K., Meyer, B. J., & Rand, J. B. (1993). Synaptic function is impaired but not eliminated in *C-elegans* mutants lacking synaptotagmin. Cell 73, 1291–1305.

O'Connor, V. M., Shamotienko, O., Grishin, E., & Betz, H. (1993). On the structure of the synaptosecretosome—evidence for a neurexin synaptotagmin syntaxin Ca^{2+} channel complex. FEBS Lett. 326, 255–260.

O'Neill, K. T., & DeGrado, W. F. (1990). How calmodulin binds its targets: Sequence independent recognition of amphiphilic helices. TIBS 15, 59–64.

Obar, R. A., Collins, C. A., Hammarback, J. A., Shpetner, H. S., & Vallee, R. B. (1990). Molecular cloning of the microtubule-associated mechanochemical enzyme dynamin reveals homology with a new family of GTP-binding proteins. Nature 347, 256–261.

Obar, R. A., Shpetner, H. S., & Vallee, R. B. (1991). Dynamin—a microtubule-associated GTP-binding protein. J. Cell Sci. (S14), 143.

Okabe, T., & Sobue, K. (1987). Identification of a new 84/82 kDa calmodulin binding protein which interacts with actin filaments, tubulin and spectrin as synapsin 1. FEBS Lett. 213, 184–188.

Oyler, G. A., Higgins, G. A., Hart, R. A., Battenberg, E., Billingsley, M., Bloom, F. E., & Wilson, M. C. (1989). The identification of a novel synaptosomal-associated protein, SNAP-25, differentially expressed by neuronal subpopulations. J. Cell Biol. 109, 3039–3052.

Pang, D. T., Wang, J. K. T., Valtorta, F., Benfenati, F., & Greengard, P. (1988). Protein tyrosine phosphorylation in synaptic vesicles. Proc. Natl. Acad. Sci. USA 88, 762.

Pearse, B. M. F., & Robinson, M. S. (1990). Clathrin, adaptors and sorting. Ann. Rev. Cell Biol. 6, 151–171.

Perin, M. S., Fried, V. A., Mignery, G. A., Jahn, R., & Sudhof, T. C. (1990). Phospholipid binding by a synaptic vesicle protein homologous to the regulatory region of protein kinase C. Nature 345, 260–263.

Perin, M. S., Brose, N., Jahn, R., & Sudhof, T. C. (1991). Domain structure of synaptotagmin (p65). J. Biol. Chem. 266, 623–629.

Perry, D. G., Li, S., Hanson, V., & Puszkin, S. (1992). Neuromuscular-junctions contain np185—the multifunctional protein is located at the presynaptic site. J. Neurosci. Res. 33, 408–417.

Peters, A., Palay, S. L., & Webster, H. D. F. (1991). The neurones and supporting cells. In *The Fine Structure of the Nervous System*. Oxford University Press, Oxford.

Petrenko, A. G., Perin, M. S., Davletov, B. A., Ushkaryov, Y. A., Geppert, M., & Sudhof, T. C. (1991). Binding of synaptotagmin to the alpha-latrotoxin receptor implicates both in synaptic vesicle exocytosis. Nature 353, 65–68.

Petrucci, T. C., & Morrow, J. S. (1987). Synapsin-I—an actin-bundling protein under phosphorylation control. J. Cell Biol. 105, 1355–1363.

Petrucci, T. C., & Morrow, J. S. (1991). Actin and tubulin binding domains of synapsins-Ia and synapsins-Ib. Biochemistry 30, 413–422.

Petrucci, T. C., Macioce, P., & Paggi, P. (1991). Axonal transport kinetics and posttranslational modification of synapsin-I in mouse retinal ganglion cells. J. Neurosci. 11, 2938–2946.

Pfeffer, S. R., & Kelly, R. B. (1985). The subpopulation of brain coated vesicles that carries synaptic vesicle proteins contains two unique polypeptides. Cell 40, 949–957.

Pfenninger, K., Akert, K., Moor, H., & Sandri, C. (1971). Freeze-etching of presynaptic membranes in the central nervous systems. Proc. Roy. Soc. London Ser. B 261, 387–389.

Ponnambalam, S., Robinson, M. S., Jackson, A. P., Peiper, L., & Parnham, P. (1990). Conservation and diversity in families of coated vesicle adaptins. J. Cell Biol. 265, 4814–4820.

Popoli, M. (1993). Synaptotagmin is endogenously phosphorylated by Ca^{2+}-calmodulin protein kinase-II in synaptic vesicles. FEBS Lett. 317, 85–88.

Presek, P., Jessen, S., Dreyer, F., Jarvie, P. E., Findik, D., & Dunkley, P. R. (1992). Tetanus toxin inhibits depolarization-stimulated protein phosphorylation in rat cortical synaptosomes—effect on synapsin-I phosphorylation and translocation. J. Neurochem. 59, 1336–1343.

Pumplin, D. W., Reese, T. S., & Llinas, R. (1981). Are the presynaptic membrane particles calcium channels? Proc. Natl. Acad. Sci. USA 78, 7210–7213.

Puszkin, S., Perry, D., Li, S. W., & Hanson, V. (1992). Neuronal protein np185 is developmentally regulated, initially expressed during synaptogenesis, and localized in synaptic terminals. Molecular Neurobiology 6, 253–283.

Rhem, H., Wiedenmann, B., & Betz, H. (1986). Molecular characterisation of synaptophysin, a major calcium binding protein of the synaptic vesicle membrane. EMBO J. 5, 535–541.

Richmond, J., & Haydon, P. G. (1993). Rab effector domain peptides stimulate the release of neurotransmitter from cell cultured synapses. FEBS Lett. 326, 124–130.

Robinson, M. S. (1989). Cloning of cDNA encoding two related 100kD vesicle proteins (α-adaptins). J. Cell Biol. 108, 833–842.

Robinson, P. J. (1991a). The role of protein-kinase-C and its neuronal substrates dephosphin, b-50, and MARKS in neurotransmitter release. Molecular Neurobiology 5, 87–130.

Robinson, P. J. (1991b). Dephosphin, a 96000-Da substrate of protein kinase-C in synaptosomal cytosol, is phosphorylated in intact synaptosomes. FEBS Lett. 282, 388–392.

Robinson, P. J., Sontag, J. M., Liu, J. P., Fykse, E. M., Slaughter, C., McMahon, H., & Sudhof, T. C. (1993). Dynamin GTPase regulated by protein-kinase-C phosphorylation in nerve-terminals. Nature 365, 163–166.

Rosahl, T. W., Spillane, D., Missler, M., Herz, J., Selig, D. K., Wolff, J. R., Hammer, R. E., Malenka, R. C., & Sudhof, T. C. (1995). Essential functions of synapsin-I and synapsin-II in synaptic vesicle regulation. Nature 375, 488–493.

Rothman, J. E. (1991). Enzymology of intracellular membrane-fusion. Klinische Wochenschrift 69, 98–104.

Rothman, J. E., & Orci, L. (1992). Molecular dissection of the secretory pathway. Nature 355, 409–415.

Rubenstein, J. L., Greengard, P., & Czernik, A. J. (1993). Calcium-dependent serine phosphorylation of synaptophysin. Synapse 13, 161–172.

Saito, K. (1990). Freeze-fracture organization of hair cell synapses in the sensory epithelium of guinea pig organ of Corti. J. Electron Microsc. Tech. 15, 173–186.

Sasaki, T., Kaibuchi, K., Kabcenell, A. K., Novick, P. J., & Takai, Y. (1991). A mammalian inhibitory GDP/GTP exchange protein (GDP dissociation inhibitor) for smg p25a is active on the yeast sec4 protein. Mol. Cell. Biol. 11, 2909–2912.

Scharrer, E., & Scharrer, B. (1954). Hormones produced by neurosecretory cells. Recent Prog. Hormone Res. 10, 183–240.

Schiavo, G., Gmachl, M. J. S., Stenbeck, G., Sollner, T. H., & Rothman, J. E. (1995). A possible docking and fusion particle for synaptic transmission. Nature 378, 733–736.

Schiavo, G., Poulain, B., Rossetto, O., Benfenati, F., Tauc, L., & Montecucco, C. (1992). Tetanus toxin is a zinc protein and its inhibition of neurotransmitter release and protease activity depend on zinc. EMBO J. 11, 3577–3583.

Schiebler, W., Jahn, R., Doucet, J.-P., Rothlein, J., & Greengard, P. (1986). Characterisation of synapsin 1 binding to small synaptic vesicles. J. Biol. Chem. 261, 8383–8390.

Scranton, T. W., Iwata, M., & Carlson, S. S. (1993). The SV2 protein of synaptic vesicles is a keratan sulfate proteoglycan. J. Neurochem. 61, 29–44.

Severin, S. J., Moskvitina, E. L., Bykova, E. V., Lutzenko, S. V., & Shvets, V. I. (1989). Synapsin I from human brain. Phosphorylation by Ca^{2+}, phospholipid-dependent protein kinase. FEBS Lett. 258, 223–226.

Sheng, Z. H., Rettig, L., Cook, T., & Catterall, W. A. (1996). Calcium-dependent interaction of N-type calcium channels with the synaptic core complex. Nature 379, 451–454.

Shira, T. S., Wang, J. K. T., Gorelick, F. S., & Greengard, P. (1989). Translocation of synapsin I in response to depolarisation of isolated nerve terminals. Proc. Natl. Acad. Sci. USA 86, 8108–8112.

Shirataki, H., Kaibuchi, K., Sakoda, T., Kishida, S., Yamaguchi, T., Wada, K., Miyazaki, M., & Takai, Y. (1993). Rabphilin-3a, a putative target protein for smg p25a rab3a p25 small GTP-binding protein related to synaptotagmin. Mol. Cell. Biol. 13, 2061–2068.

Shoji-Kasai, Y., Yoshida, A., Sato, K., Hoshino, T., Oqura, A., Kando, S., Fujimoto, Y., Kuwahara, R., Kato, R., & Takahashi, M. (1992). Neurotransmitter release from synaptotagmin-deficient clonal variants of PC 12 cells. Science 256, 1820–1823.

Shpetner, H. S., & Vallee, R. B. (1991). Purification and characterization of dynamin. Methods Enzymol. 196, 192–201.

Shpetner, H. S., & Vallee, R. B. (1992). Dynamin is a GTPase stimulated to high-levels of activity by microtubules. Nature 355, 733–735.

Sikorski, A. F., Terlecki, G., Zagon, I. S., & Goodman, S. R. (1991). Synapsin-I-mediated interaction of brain spectrin with synaptic vesicles. J. Cell Biol. 114, 313–318.

Smith, S. J., & Augustine, G. J. (1988). Ca^{2+} ions, active zones and synaptic transmitter release. Trends Neurosci. 11, 458–464.

Sollner, T., Whitehart, S. W., Brunner, M., Erdjumentbromage, H., Geromanos, S., Tempst, P., & Rothman, J. E. (1993). SNAP receptors implicated in vesicle targeting and fusion. Nature 362, 318–324.

Spillane, D. M., Rosahl, T. W., Sudhof, T. C., & Malenka, R. C. (1995). Long-term potentiation in mice lacking synapsins. Neuropharmacology 34, 1573–1579.

Stadler, H., & Dowe, G. H. C. (1982). Identification of a heparan sulfate-containing proteoglycan as a specific core component of cholinergic synaptic vesicles from *Torpedo-marmorata*. EMBO J. 1, 1381–1384.

Stamm, S., Casper, D., Dinsmore, J., Kaufmann, C. A., Brosius, J., & Helfman, D. M. (1992). Clathrin light chain-b—gene structure and neuron-specific splicing. Nucleic Acids Research 20, 5097–5103.

Steiner, J. P., Gardner, K., Baines, A. J., & Bennett, V. (1987b). Synapsin 1: A regulated synaptic vesicle organizing protein. Brain Res. Bull. 18, 777–786.

Steiner, J. P., Ling, E., & Bennett, V. (1987a). Nearest neighbour analysis for synapsin 1. J. Biol. Chem. 262, 905–932.

Steinhardt, R. A., & Alderton, J. M. (1982). Calmodulin confers calcium sensitivity on secretory exocytosis. Nature 295, 154–155.

Stenius, K., Janz, R., Sudhof, T. C., & Jahn, R. (1995). Structure of synaptogyrin (P-29) defines novel synaptic vesicle protein. J. Cell. Biol. 131, 1801–1809.

Sternbach, Y., Greenbergofrath, N., Flechner, I., & Schuldiner, S. (1990). Identification and purification of a functional amine transporter from bovine chromaffin granules. J. Biol. Chem. 265, 3961–3966.

Stone, D. K., Crider, B. P., & Xie, X. S. (1990). Structural properties of vacuolar proton pumps. Kidney Int. 38, 649–653.

Sudhof, T. C. (1990). The structure of the human synapsin I gene and protein. J. Biol. Chem. 265, 7849–7852.

Sudhof, T. C., Lottspeich, F., Greengard, P., Mehl, E., & Jahn, R. (1987). The cDNA and derived amino-acid sequences for rat and human synaptophysin. Nucleic Acids Research 15, 9607.

Sudhof, T. C., Baumert, M., Perin, M., & Jahn, R. (1989). A synaptic vesicle protein is conserved from mammals to *Drosophilia*. Neuron 2, 1475–1481.

Sudhof, T. C., Czernik, A. J., Kao, H. T., et al. (1989). Synapsins: Mosaics of shared and individual domains in a family of synaptic vesicle phosphoproteins. Science 245, 1474–1480.

Sudhof, T. C., & Jahn, R. (1991). Proteins of synaptic vesicles involved in exocytosis and membrane recycling. Neuron 6, 665–677.

Tabb, J. S., Kish, D. E., Van Dyke, R., & Ueda, T. (1992). Glutamate transport into synaptic vesicles: Roles of membrane potential, pH gradient and intravesicular pH. J. Biol. Chem. 267, 15412–15418.

Takei, K., Mundigl, O., Daniell, L., & Decamilli, P. (1996). The synaptic vesicle cycle—a single vesicle budding step involving clathrin and dynamin. J. Cell. Biol. 133, 1237–1250.

Tashiro, T., & Stadler, H. (1978). Chemical composition of cholinergic synaptic vesicles from *Torpedo marmorata* based on improved purification. Eur. J. Biochem. 90, 479–487.

Thiel, G., Sudhof, T. C., & Greengard, P. (1990). Synapsin-II—Mapping of a domain in the NH_2-terminal region which binds to small synaptic vesicles. J. Biol. Chem. 265, 16527–16533.

Thiel, G., Greengard, P., & Sudhof, T. C. (1991). Characterization of tissue-specific transcription by the human synapsin-I gene promoter. Proc. Natl. Acad. Sci. USA 88, 3431–3435.

Thomas, L., Hartung, K., Langosch, D., Rehm, H., Bamberg, E., Franke, W. W., & Betz, H. (1988). Identification of synaptophysin as a hexameric channel protein of the synaptic vesicle membrane. Science 242, 1050–1053.

Thomas, L., & Betz, H. (1990). Synaptophysin binds to physophilin, a putative synaptic plasma-membrane protein. J. Cell Biol. 111, 2041–2052.

Torri-Tarelli, F., Grohovaz, F., Fesce, R., & Ceccarelli, B. (1985). Temporal coincidence between synaptic vesicle fusion and quantal secretion of acetylcholine. J. Cell Biol. 101, 1386–1399.

Torri-Tarelli, F., Bossi, M., Fesce, R., Greengard, P., & Valtorta, F. (1992). Synapsin I partially dissociates from synaptic vesicles during exocytosis induced by electrical stimulation. Neuron 9, 1143–1153.

Trimble, W. S., & Scheller, R. H. (1988). Molecular biology of synaptic vesicle associated proteins. Trends Neurosci. 11, 241–242.

Trimble, W. S., Cowan, D. M., & Scheller, R. H. (1988). VAMP-1—a synaptic vesicle-associated integral membrane-protein. Proc. Natl. Acad. Sci. USA 85, 4538–4542.

Trimble, W. S., Linial, M., & Scheller, R. H. (1991). Cellular and molecular biology of the presynaptic nerve terminal. Ann. Rev. Neurosci. 14, 93–122.

Tsien, R. W., & Tsien, R. Y. (1990). Calcium channels, stores and oscillations. Ann. Rev. Cell Biol. 6, 715–760.

Tuazon, P. T., & Trough, J. A. (1991). Casein kinase-I and kinase -II- multipotential serine protein kinase, structure, function and regulation. Advances in Second Messenger and Phosphoprotein Research 23, 123–164.

Ueda, T., & Greengard, P. (1977). Adenosine 3'-5'-monophosphate regulated phosphoprotein system of neuronal membranes. J. Biol. Chem. 252, 5155–5163.

Ushkaryov, Y. A., Petrenko, A. G., Geppert, M., & Sudhof, T. C. (1992). Neurexins—synaptic cell-surface proteins related to the alpha-latrotoxin receptor and laminin. Science 257, 50–56.

Valtorta, F., Fesce, R., Grohovaz, F., Haimann, C., Hurlbut, W. P., Iezzi, N., Tarelli, F. T., Villa, A., & Ceccarelli, B. (1990). Neurotransmitter release and synaptic vesicle recycling. Neurosci. 35, 477–489.

Valtorta, F., Benfenati, F., & Greengard, P. (1992a). Structure and function of the synapsins. J. Biol. Chem. 267, 7195–7198.

Valtorta, F., Greengard, P., Fesce, R., Chieregatti, E., & Benfenati, F. (1992b). Effects of the neuronal phosphoprotein synapsin-I on actin polymerization .1. Evidence for a phosphorylation-dependent nucleating effect. J. Biol. Chem. 267, 11281–11288.

Vanderbliek, A. M., & Meyerowitz, E. M. (1991). Dynamin-like protein encoded by the drosophila-shibire gene associated with vesicular traffic. Nature 351, 411–414.

Vulliet, P. R., Hall, F. L., Mitchell, J. P., & Hardie, D. G. (1989). Identification of a novel proline-directed serine/threonine protein kinase in rat pheochromocytoma. J. Biol. Chem. 264, 16292–16298.

Wang, J. K. T., Walaas, S. I., & Greengard, P. (1988). Protein-phosphorylation in nerve-terminals—comparison of calcium calmodulin-dependent and calcium diacylglycerol-dependent systems. J. Neurosci. 8, 281–288.

Warren, G. (1993). Bridging the gap. Nature 362, 297–298.

Weidman, P. J., Melancon, P., Block, M. R., & Rothman, J. E. (1989). Binding of an N-ethylmaleimide sensitive fusion protein to Golgi membranes requires both a soluble protein(s) and an integral membrane-receptor. J. Cell Biol. 108, 1589–1596.

Wendland, B., Miller, K. G., Schilling, J., & Scheller, R. H. (1991). Differential expression of the p65 gene family. Neuron 6, 993–1007.

Whiteheart, S. W., Brunner, M., Wilson, D. W., Wiedmann, M., & Rothman, J. E. (1992). Soluble N-ethylmaleimide-sensitive fusion attachment proteins (SNAPS) bind to a multi-SNAP receptor complex in golgi membranes. J. Biol. Chem. 267, 12239–12243.

Whiteheart, S. W., Griff, I. C., Brunner, M., Clary, D. O., Mayer, T., Buhrow, S. A., & Rothman, J. E. (1993). SNAP family of NSF attachment proteins includes a brain-specific isoform. Nature 362, 353–355.
Whittaker, V. P. (1984). The structure and function of cholinergic synaptic vesicles. Biochem. Soc. Trans. 12, 561–576.
Whittaker, V. P., Michaelson, I. A., & Kirkland, R. J. A. (1964). The separation of synaptic vesicles from nerve ending particles ("synaptosomes"). J. Biochem. 90, 293–303.
Wiedenmann, B., & Franke, W. W. (1985). Identification and localization of synaptophysin, an integral membrane glycoprotein of Mr 38,000 characteristic of presynaptic vesicles. Cell 41, 1017–1028.
Wilson, D. W., Wilcox, C. A., Flynn, G. C., Chen, E., Kuang, W. J., Henzel, W. J., Block, M. R., Ullrich, A., & Rothman, J. E. (1989). A fusion protein required for vesicle-mediated transport in both mammalian-cells and yeast. Nature 339, 355–359.
Wilson, D. W., Whiteheart, S. W., & Rothman, J. E. (1991). Intracellular membrane-fusion. TIBS 16, 334–337.
Xie, X. S., Crider, B. P., & Stone, D. K. (1989). Isolation and reconstitution of the chloride transporter of clathrin-coated vesicles. J. Biol. Chem. 264, 18870–18873.
Yoshida, A., Oho, C., Omori, A., Kuwahara, R., Ito, T., & Takahashi, M. (1992). HPC-1 is associated with synaptotagmin and omega-conotoxin receptor. J. Biol. Chem. 267, 24925–24928.
Zerial, M., & Stenmark, H. (1993). Rab GTPases in vesicular transport. Curr. Opin. Cell Biol. 5, 613–620.
Zhou, S. B., Tannery, N. H., Yang, J., Puszkin, S., & Lafer, E. M. (1993). The synapse-specific phosphoprotein-f1-20 is identical to the clathrin assembly protein ap-3. J. Biol. Chem. 268, 12655–12662.

MEMBRANE ATTACK BY COMPLEMENT:
ASSEMBLY AND BIOLOGY OF TERMINAL COMPLEMENT COMPLEXES

Moon L. Shin, Horea G. Rus, and Florin I. Niculescu

Biomembranes
Volume 4, pages 123–149.
Copyright © 1996 by JAI Press Inc.
All rights of reproduction in any form reserved.
ISBN: 1-55938-661-4.

123

I. INTRODUCTION TO THE COMPLEMENT SYSTEM

The complement system comprises more than 14 soluble proteins in body fluid, a number of receptors, and several soluble and membrane proteins performing regulatory functions. The complement system, when activated, serves to defend the host against infection and to function as immune effectors and regulators. The complement system was first described a century ago by Hans Buchner as a heat-labile "bactericidal principle" in the blood. Jules Bordet, then demonstrated that the bactericidal activity of serum, destroyed by heating, could be restored by adding fresh serum, and named it "alexine." Paul Ehrlich, called this serum factor "Komplement." The complement fixation assay was developed to measure the antigen–antibody reaction by Bordet and Gengou in 1901. The hemolytic assay, which determines the complement activation end-point by lysing erythrocytes, has been used effectively for nearly half a century to unravel the activation cascade and the protein–protein and protein–membrane interactions (see these reviews: Mayer, 1961; Müller-Eberhard, 1988; Frank and Fries, 1989; Walport and Lachmann, 1993). The complement system is activated by the classical pathway, which includes the proteins C1, C4, C2, and C3 and also by the alternative pathway with the participation of protein C3 and protein factors B, D, and P. Both pathways lead to activation of C5, C6, C7, C8, and C9 proteins, resulting in sequential assembly of C5b-7, C5b-8, and C5b-9 complexes on target cell lysis. The activation cascade of the complement system is shown in Figure 1.

A.   Activation by the Classical Pathway

The classical pathway activation occurs when a minimum of two globular regions of C1q interact with the C_H2 domains of an IgG duplex or with the C_H3 domains of a single IgM in immune complexes. In the absence of antibody, other substrates such as viral envelope membranes, gram-negative bacterial cell wall, C-reactive protein, cardiolipin, DNA, cytoskeletal intermediate filaments, and central nervous system myelin can also activate the classical pathway. The C1q bound to the activator allows the ternary complexes $C1r_2–C1s_2$ to become active when they are bound to the collagenous portion of C1q in the presence of Ca^{2+}. Both C1r and C1s are proenzymes possessing serine esterase catalytic domains and the sequential auto-activation of C1r and C1s can lead to activation of C4 and C2 through cleavage

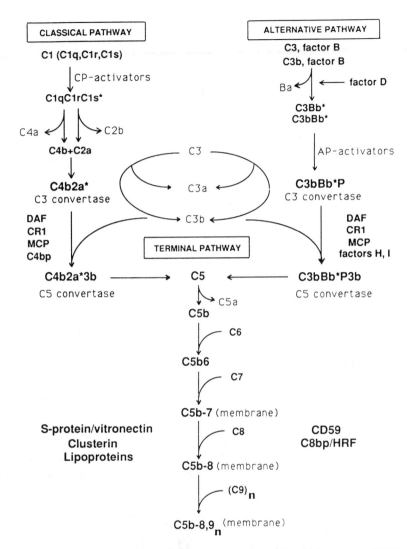

Figure 1. Activation of the complement system. Activation of the classical (CP) and alternative (AP) pathways through specific activators leads to generation of enzyme complexes with serine-protease activity. C3-convertases cleave C3 and C5-convertases cleave C5. Generation of C5b fragment allows activation and assembly of the terminal pathway leading to formation of C5b–9 complexes and cell death. DAF, CR1, MCP, and C4bp for CP, and factors H and I for AP are regulators of C3/C5 convertases. S-protein/vitronectin, clusterin, and lipoproteins are inhibitors of the terminal pathway in the fluid phase, while CD59 and C8bp/HRF are inhibitors anchored to the plasma membrane.

by C1s. Cleavage of the α chain of C4 releases a 9 kD peptide, C4a, while the remaining 190 kD C4b binds covalently to target molecules when hydrolysis of a reactive thioester present in C4α occurs, and reacts with nucleophilic species such as hydroxyl- or amino-groups. C1s cleaves C2 to produce C2b and C2a. C2a, which has the catalytic site, complexes with C4b to form C4b2a, a C3 cleaving enzyme "C3-convertase." C3 is cleaved by C4b2a thereby generating C3b and C3a. C3b also binds covalently to nucleophiles present in the molecules, such as C4b, and antibodies present near the C3b generating site. C3b and its split products interact with specific complement receptors, CR1 for C3b, CR2 for C3dg or C3d, and CR3 for iC3b. When a C3b binds to C3-convertase, C4b2a3b complex "C5-convertase" is generated. By binding to C3b in a C4b2a3b complex, C5 is then cleaved into C5a and C5b by C2a enzyme (Frank and Fries, 1989).

B. Activation by the Alternative Pathway

Activation of the alternative pathway starts when factor D, a serine protease in serum, cleaves factor B, but only when factor B is bound to C3b derived from the classical pathway or when factor B is associated with C3(H$_2$O), a spontaneously hydrolyzed C3. Cleavage of factor B generates C3bBb or C3(H$_2$O)Bb, "alternative pathway C3 convertase," in which Bb, as C2a, has a catalytic site. This enzyme complex is inactivated rapidly by factors H and I in serum. Factor H competes with Bb in binding to C3b, and dissociates Bb from C3b, then allowing factor I enzyme to cleave C3b and produce iC3b. Properdin (P) increases the stability of this enzyme complex by forming C3bBbP. C3bBbP is converted into a C5 convertase with an additional C3b, which provides a binding site for C5. These convertases, when formed in a fluid-phase, are not as effective as when assembled on a solid-phase like cell membranes. Activators of the alternative pathway include zymosan, high molecular weight dextrans, plastic surfaces, peripheral nerve myelin, endotoxin, and other surface structures of microbial and tumor cells. The alternative pathway activators function by protecting the activator-bound C3-convertases from factors H and I and by increasing the binding affinity of Bb to C3b (Pangburn, 1988).

II. ACTIVATION OF C5–C9 AND ASSEMBLY OF TERMINAL COMPLEMENT COMPLEXES

Assembly of the C5b-9 complex in a membrane lipid bilayer starts with cleavage of an Arg–Leu bond of the C5 α-chain to generate C5a and C5b. Although association between native C5 and C6 occurs, C5b6 complexes are formed when C5b fragments associate with C6. C5b6 complex can bind membranes reversibly through ionic as well as hydrophobic interactions. Subsequent interaction of C7, C8, and C9 with C5b6 and the membrane results in heteropolymeric transmembrane pores. This pore assembly takes place through distinct phases of intermediate formation, namely C5b-7, C5b-8, and C5b-9. These complexes are collectively

Table 1. Properties of Terminal Complement Proteins and Their Regulators

Components	Subunit	Molecular Weight	Plasma Concentration, µg/ml
C5		190,000	80
	C5α	115,000	
	C5β	75,000	
C6	single-chain	120,000	70
C7	single-chain	110,000	70
C8		151,000	50
	C8-α	64,000	
	C8-β	64,000	
	C8-γ	22,000	
C9	single-chain	71,000	50

Regulator	Subunit	Molecular Weight	Location	Site of Action
C8-bp (HRF)	single-chain	65,000	membrane	C8 and C9; decrease tubular poly-C9 formation in human
S-Protein/Vitronectin	single-chain	80,000	plasma	C5b–7; inhibit C5b–7 interaction with membrane
Clusterin/Sp 40,40	heterodimer	70,000	plasma, seminal fluid	Binds to C5b–7, C5b–8, C5b–9; modulates TCC formation
CD59	single-chain	18,000–21,000	membrane	Inhibit reactive lysis; blocks C7 and C8 binding to C5b6 preventing TCC formation; blocks C9 polymerization

referred as terminal complement complexes (TCC), while C5b-9, the final complex, and the most effective complex in inducing cell death, is referred to as the membrane attack complex (MAC; Mayer et al., 1981; Bhakdi and Tranum-Jensen, 1987; Müller-Eberhand, 1988; Shin and Carney, 1988). The properties of the proteins involved in TCC assembly and the mechanisms of the TCC forming steps will be reviewed. The properties of complement proteins and inhibitors participating in the assembly of TCC are summarized in Table 1.

A.   Proteins of The Terminal Complement Cascade

C5

C5 is a glycoprotein of 190 kD, composed of two disulfide-linked peptides, α (115 kD) and β (75 kD), and is structurally homologous to C3, C4, and α2-macro-

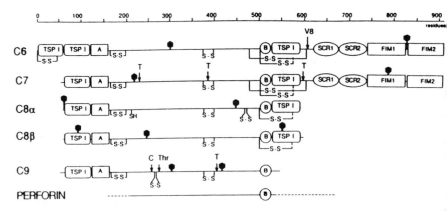

Figure 2. The structural organization of the terminal complement proteins. Five different types of structural modules are common to proteins of human terminal components and to perforin: A = LDL receptor; B = EGF receptor; TSPI = trombospondin I; SCR = short consensus repeat; FIM = factor I module; Site cleavage by trypsin (T), chemotypsin (C), thrombin (Thr), and V8 protease (V8) are marked by arrows. Black hexagonal symbols represent the asparagine-linked glycosylation sites. (after Haeflinger et al., 1987).

globulin. The predicted cDNA sequence of human and mouse C5 revealed that the pro-C5 molecule begins with the β-chain at the N-terminus (Wetsel et al., 1987, 1988; Haviland et al., 1991). Although most C5 synthesis occurs in hepatocytes, C5 is also made by macrophages and alveolar epithelial cells. The C5 gene of 78 kb, located on chromosome 2 of mouse and chromosome 9 of human, consists of 42 exons and 41 introns and is transcribed and processed to a mature 6.0 kb mRNA. The 5′-flanking region, similar to that of IL-4 and IL-6, contains a dexamethasone-response element and an NF-κB site. The cis-acting sequences for C5 gene transcription have not been completely delineated, though C5 is transcriptionally up-regulated during inflammation. C5 has 2.7% Asn-linked carbohydrate in the α-chain. The major polypeptides, C5a and C5b, are produced by cleavage of the C5α chain by C5 convertases. In vitro, C5b can be further fragmented into C5c and C5d. C5a, a potent inflammatory mediator, has one glycosylation site and 50% of C5a is in an α-helical conformation (DiScipio et al., 1983).

C6 and C7

C6 is a single-chain 120 kD glycoprotein with 913 amino acids and 4% Asn-linked carbohydrate. C7, a single-chain 110 kD glycoprotein, consists of 821 amino acids (DiScipio et al., 1988). C6 has a unique structure of a mosaic protein, consisting of several homology modules existing in many other proteins (Figure 2; Haeflinger et al., 1989; Hobart et al., 1993). C6 has a C-terminal Cys-rich domain that is partly homologous to regions of thrombospondin I, the LDL-receptor, and the EGF-receptor. The Cys-poor central region of C6 is partly homologous to perforin. C6 and C7 have several domains with homologies to the "short consensus repeats," a characteristic motif shared among large numbers of proteins including complement regulatory proteins and receptors. The Cys-rich C-terminal region of C6 binds to the α-chain of C5b to form C5b6.

C8

C8 consists of three polypeptides: the α-chain (64 kD, 533 amino acids) and the β-chain (64 kD, 537 amino acids) share extreme homology and are noncovalently linked, and the γ-chain (22 kD, 182 amino acids) is linked to the α-chain by a disulfide bond. C8-α, -β, and -γ are coded by three different genes. C8-α and -β share significant homologies with other TCC proteins, including a Cys-poor central region and Cys-rich C-terminal domains (Rao et al., 1987; Sodetz, 1988). C8-γ belongs to a family of proteins that includes α1-microglobulin, protein HC, retinol binding protein, and the family of lipocalins which binds lipophilic ligands (Haeflinger et al., 1987).

C9

C9, a single-chain globular protein of 71 kD and 537 amino acids, consists of predominantly hydrophobic N-terminal and hydrophilic C-terminal segments. The C9 sequence, which is similar to the C6 structure from residue 56 to 540, also reveals conserved domains that are shared with other TCC proteins (Marazziti et al., 1988; Haeflinger, 1989). The C9 sequence homologies shared with C6 and C8α/C8β are 27% and 34%, respectively. C9 can be cleaved at His–Gly, residues 244–245, by α-thrombin to yield noncovalently linked peptides of 34 kD (C9a) and 37 kD (C9b). C9b has cytolytic activity via interaction with the membrane lipid bilayer (Ishida et al., 1982). The 80 kb C9 gene, located on chromosome 5 of human, consists of at least 11 exons. C9 expression, which is highest in hepatocytes, is modulated under many pathologic conditions. The C9 protein has distinct domains, with domains 1 and 2 showing homologies with other serum and membrane proteins as previously mentioned (Stanley, 1988).

B. C5b–7 Complex Assembly

Upon binding to the α-chain of C5b, C6 undergoes conformational changes and acquires a capacity to interact with hydrophobic domains of the lipid bilayer (Hu et al., 1981). Interaction of C7 with the α-chain of C5b in C5b6 results in a C5b–7 complex with an amphiphilic transformation of the C7 molecule as demonstrated by the release of phospholipids from liposomes, the binding to photoreactive probes in the membrane, increased conductance across 35 Å thin planar black lipid membranes (BLM), and also by binding to non-ionic detergents (Shin et al., 1977;

Hu et al., 1981; Ramm et al., 1983; Podack, 1988). The C5b–7 cytolytically inactive complex does not form pores, nor does it induce Ca^{2+} influx (Morgan and Campbell, 1985; Shin and Carney, 1988). Insertion of C5b–7 into the membrane, however, stimulates cells, as evidenced by activation of membrane-associated phospholipases, enhanced elimination of C5b–7 from the cell surface, and the ability of C5b–7 to induce hydrolysis of myelin basic protein (Carney et al., 1985; Vanguri and Shin, 1988; Niculescu et al., 1993). By electron microscopy, C5b6 is shown as a bi-lobal structure, the C5b–7 monomer as a long hook-like slender structure of 0.45 nm in length, and the C5b–7 dimer as a supercoiled stalk. C5b–7 monomers and dimers anchor to the membrane through the stalk and allow C8 binding and C9 polymerization. C5b–7 formed in solution aggregates or complexes with inhibitory proteins in serum. This C5b–7 complex, designated SC5b–7, does not interact with the membrane (DiScipio et al., 1988).

C.   C5b–8 Complex Assembly

The C8-β-chain binds to C5b of membrane-associated C5b–7. Amphiphilic transition of C8 occurs in both α- and β-chains of C8 as in C7, which allows insertion of hydrophobic peptides from the α- and β-chains. C8-α in the C5b–8 complex serves as the binding site for C9 to form C5b–9. C8-γ is not essential for the cytolytic activity. Both C8-α- and C8-β-chains have multiple functional domains. C8-α has a β-chain binding domain, a domain to associate with γ-chain by disulfide-linkage, a membrane insertion domain, and a domain to bind and activate C9. C8-β has a C5b-binding domain, a C8-α-binding domain, and a domain which interacts with membrane lipids (Sodetz, 1988). Pore formation by C5b–8 in the membrane has been demonstrated by an increased conductance across BLM and by marker release/ion flux through resealed erythrocyte ghosts. The C5b–8 pores, ranging from 0.4 nm to 3 nm in diameter, are unstable with a finite life span (Ramm et al., 1982). To lyse an erythrocyte, a large number of C5b–8 is required (Gee et al., 1980). The kinetics of hemolysis by C5b–8 is slower than that by C5b–9. Nucleated cell death achieved by C5b–8 also requires large numbers of these complexes. Lytic C5b–8 pores have been demonstrated in M21 human melanoma cells, the U937 human histiocytic cell line, and *Giardia lamblia*. Sublytic numbers of C5b–8 are capable of activating target cells by increasing $[Ca^{2+}]_i$ and by generating other signal messengers, which will be discussed later.

D.   C5b–9 Complex Assembly

A single C9 binds to the C8-α-chain in membrane-bound C5b–8 or C5b–8 formed in solution. C9 interacts with C5b–8 rapidly and initiates transformation of a globular C9 (8 nm in length) into an elongated C9 (16 nm in length). This structural change is associated with an increase in the β-sheet conformation of C9 from 32% to 38% and membrane insertion. The rapid binding of C9 to C5b–8 produces the C5b–$8,9_1$ complex, which is then followed by slower incorporation

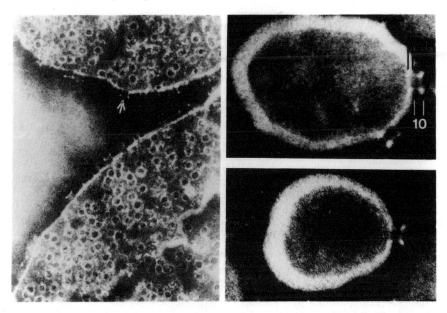

Figure 3. Ultrastructures of complement lesions. (**A**) Sheep erythrocytes lysed with antibody and human complement show C5b–9 lesions in the membrane stained with phosphotungstic acid. Side projections (arrow) as well as in tow view. (**B**) Profile views of complement channels that had been eluted from erythrocyte membranes with Triton X-100, purified by chromatography, and reincorporated into liposomes made from sheep erythrocyte membrane lipids (courtesy of Drs. Bhakdi and Tranum-Jensen).

of multiple C9 to form C5b–8,9$_n$ through C9–C9 polymerization, in which n can be as high as 16 molecules of C9 (Podack and Tschopp, 1982; Tschopp et al., 1985; Whitlow et al., 1985; Laine and Esser, 1989). C9 polymerization, not required for erythrocyte lysis or nucleated cell killing, is however, necessary for killing of gram-negative bacteria (Joiner, 1988). Although the process is inefficient, C9 polymerization can be initiated without C5b–8 by a similar mechanism of C9 polymerization (Podack and Tschopp, 1982). C9 polymerization with more than 6 molecules of C9 forms an SDS-resistant C5b–8,9$_n$ complex of tubular structure, also called poly-C9 (Bhakdi and Tranum-Jensen, 1987; Podack, 1988). As shown in Figure 3, the C5b–9 complex has an annular ring structure with external diameter of 20 nm, internal diameter of 5 nm, and a height of 15 nm. C9 polymerization with less than 6 molecules of C9 forms an SDS-dissociable C5b–8,9$_{1-6}$ complex which does not show the characteristic ultrastructure of poly-C9. The functional size of the C5b–9 channel ranges from 1 nm to 11 nm, and the pore size increases with an increasing number of C9 molecules (Dalmasso and Benson, 1981; Ramm et al.,

1985). The diameter of tubular poly-C9 without C5b–8 has been reported to be 10 nm (Young et al., 1986).

III. REGULATION OF TCC ASSEMBLY

The regulation of TCC formation operates by affecting one of the three phases of TCC assembly. Accordingly, erythrocyte lysis or nucleated cell death mediated by C5b–9 is influenced by factors regulating (1) the interaction of TCC proteins with the membrane lipid bilayer, (2) the presence of regulatory membrane proteins that can restrict TCC formation, and (3) different mechanisms which operate to eliminate TCC from the cell surface.

A. Modulation of Protein–Lipid Interaction

The membrane lipid composition which affects the efficiency of TCC assembly has been demonstrated in liposomes, erythrocytes, microorganisms, and nucleated cells (Shin et al., 1978; Dahl et al., 1979; Schlager and Ohanian, 1980). The lipid composition modulates C5b–9 assembly by affecting peptide–lipid and peptide–peptide interactions through changes in membrane fluidity and/or acyl-chain packing. The marker release from liposomes induced by C5b–9 decreases with increasing acyl-chain length, saturation of fatty acids, and increasing cholesterol concentration.

B. Inhibitors of TCC Assembly

Several membrane proteins have been shown to regulate C5b–9 assembly by affecting the number and/or the pore size. Receptor proteins such as CR1, CR2, CR3, and CR4, and regulatory proteins such as DAF and MCP down-regulate TCC formation by influencing the C3/C5 convertase activity. On the other hand, membrane proteins, especially CD59 and the C8-binding protein, (C8bp)/homologous restriction factor (HRF), that are anchored to the membrane by a glycanphosphatidylinositol (GPI)-anchor, restrict C5b–9 assembly in homologous species, thus preventing hemolysis or nucleated cell death. Bordet has reported that lysis of erythrocytes by serum complement is least efficient in homologous species (Bordet, 1900). Subsequent studies indicated that species restriction of hemolytic activity of complement is largely attributed to the steps of TCC assembly. Membrane molecules such as DAF reduce the magnitude of C3/C5 convertases, while other membrane proteins, such as CD59, directly affect TCC assembly by inhibiting C8 and C9 interaction with membrane-bound C5b–7 and C5b–8, as well as by blocking C9 polymerization. In the absence of GPI-anchored proteins on cell membranes, C5b–9 complexes are formed 3.8-fold more than on cells expressing GPI-anchored proteins, and the number of C9 molecules per C5b–8 is also 2.3-fold more in GPI-negative cells (Niculescu et al., 1993).

CD59

CD59 is an 18–20 kD glycoprotein anchored to the membrane through GPI. CD59 has been called P-18, MIRL, HRF-20, H-19, MEM-43, or "protectin" by various investigators. In paroxysmal nocturnal hemoglobinuria (PNH), membrane expression of CD59 is diminished or absent (Yamashina et al., 1990). The genetic defect of PNH cells involves abnormal transcription of the PIG-A gene, which belongs to a group of genes called PIG (*P*hosphatidyl*I*nositol*G*lycans) which are involved in the biosynthesis of GPI-anchored proteins. PIG-A encodes an early protein required for anchoring GPI to the protein backbone, near its C-terminus. Transcriptional and/or splicing defects result in small PIG-A transcripts or transcripts of normal size without function due to a mutation T→A in the coding region (Miyata et al., 1993). In addition to its expression as a membrane protein, CD59 exists also in a soluble form in blood, urine, and other body fluids (Meri et al., 1991). CD59 inhibits C9 binding to C5b–8 by affecting the association of a domain of C8-α with C9b (Ninomiya and Sims, 1992). Inhibition of C9 binding to C5b–8 by CD59 can be blocked by excess C9, or when C9 binding is allowed to occur prior to CD59 addition. These results suggest that CD59 competes with C9 to bind a nascent epitope on C8 which is exposed during C8 activation. Moreover, neither SC5b–9, inactive complexes isolated from complement-activated serum, nor poly-C9 bind to CD59. CD59, as well as other GPI-anchored proteins, are reported to be involved in signal transduction by increasing tyrosine kinase activity. Specifically, CD59 associated with CD58 exerts stimulatory effects on T cells after CD2 cross-linking (Deckert et al., 1992; Hahn et al., 1992).

C8-Binding Protein (C8bp)/Homologous Restriction Factor (HRF)

C8bp (Schönermark et al., 1986), also termed HRF (Zalman et al., 1986), is a 65 kD, GPI-anchored glycoprotein, similarly deficient in some forms of PNH diseases (Hänsch et al., 1988). C8bp binds to C8-α and inhibits the assembly of C5b–8 and C5b–9, as well as C9 polymerization. Translocation of intracellular C8bp to the plasma membrane is stimulated by IL-1β, IFN-γ, LPS, or phorbol esters in the presence of cycloheximide (Schieren et al., 1992). Increased expression of TCC-inhibitory proteins during inflammation may function to protect the host cells from the bystander C5b–9 attack (Hänsch, 1992). There are also other proteins which inhibit TCC assembly. These proteins, unlike CD59 and C8bp, function in heterologous system.

Clusterin

Clusterin, a 70 kD glycoprotein first identified in rete testis fluids by its ability to aggregate a variety of cells, is present in blood in association with lipoproteins. The clusterin gene, identical with TRPM-2, is expressed in cells involved directly in epithelial differentiation and morphogenesis (French et al., 1993). In addition, clusterin also prevents cell death induced by C5b–9 (Jenne and Tschopp, 1989; Murphy et al., 1989). Clusterin inhibits C5b–7, C5b–8, and C5b–9 assembly by interacting with a structural motif common to C7, C8-α and C9b (Tschopp et al., 1993). Clusterin is also associated with hemolytically inactive SC5b–9 complexes formed in solution together with S-protein/vitronectin.

S-protein/Vitronectin

S-protein, an 80 kD multi-functional glycoprotein, was first identified as a part of C5b–9 complexes activated in serum. Purified S-protein has inhibitory activity for hemolysis by C5b–9 via preventing C5b–7 association with the membrane (Podack and Müller-Eberhard, 1978). By cloning, it was found that S-protein is identical to "vitronectin" and "serum spreading factor" identified by inducing cell adherence and spreading on glass beads. The functional properties of the S-protein, other than its effect on TCC, are referred to in a review by Preissner (1991). S-protein/vitronectin binds to metastable sites of the nascent C5b–7 and C5b–8, and produces water soluble SC5b–7 or SC5b–8 unable to interact with the membrane. S-protein also inhibits C9 polymerization and channel formation by perforin, thus limiting not only complement pores, but also pores produced by cytotoxic lymphocytes. S-protein is abundantly present in tissue matrix including vascular walls, suggesting its possible role in protecting cells in the vessel from complement attack (Niculescu et al., 1987, 1989).

Lipoproteins

Serum lipoproteins, especially HDL, inhibit C5b–9 assembly and C9 polymerization. Apo A-I does not bind to native C9. However, Apo A-I can complex with clusterin, which interacts with TCC at earlier stages than Apo A-I alone, and it can synergize with clusterin to enhance inhibitory activity by interfering with membrane insertion processes (Hamilton et al., 1993). The binding of Apo A-I and Apo A-II to poly-C9 is saturable.

IV. MECHANISMS OF CELL DEATH INDUCED BY TCC AND RECOVERY FROM THE TCC ATTACK

The mechanisms of killing of metabolically active nucleated cell by complement differ substantially from those of erythrocyte lysis (Mayer et al., 1981; Shin and Carney, 1988). Nucleated cell death by C5b–9 is a multi-hit process requiring multiple C5b–9, whereas a single C5b–9 channel is sufficient to lyse an erythrocyte (Koski et al., 1983). In Ehlrich cells, C5b–$8,9_4$ is twice as effective in killing than C5b–$8,9_1$ (Kim et al., 1987), although the single-hit requirement of erythrocyte lysis has been shown only with the C5b–9 complex carrying 3 to 6 molecules of C9. Since limited C5b–9 complexes are efficiently eliminated, multiple channel formation is required to induce cell death in order to overcome the repair process.

Thus, the efficiency of complement in killing nucleated cells depends on the ability of the cell to repair the C5b–9 induced membrane damage, as well as on the ability of C5b–9 to initiate the killing process.

A. Repair Mechanism for Cell Survival Following Complement Attack

As discussed, cell survival from limited complement attack involves the ability of cells to respond to TCC by rapidly eliminating potentially lytic C5b–9 from the cell surface. The TCC elimination, as studied under limited complement attack, occurs by endocytotic processes or membrane shedding. The half-life of C5b–8,9$_n$ channels remaining on the cell surface ranges from 1 min to 3 min for nucleated cells to 72 h on erythrocytes (Ramm et al., 1983). C5b–8 elimination from the cell surface is slower than C5b–9 removal and C5b–7 is eliminated at the slowest rate. In Ehrlich cells, immunotracing of C5b–8,9$_n$ with colloidal gold revealed the membrane-bound TCC entering multivesicular bodies through endocytotic coated vesicles (Carney et al., 1985). TCC removal in polymorphonuclear leukocytes was 35% by endocytosis and 65% by membrane shedding (Morgan et al., 1987). Enzymatic degradation of internalized TCC was observed within 30 min and membrane vesiculation, detected as early as 30 sec, was also completed within 30 min. About 2% of total plasma membrane, which carried most of the TCC, was eliminated following C5b–9 assembly. The rate of TCC elimination increases with increasing $[Ca^{2+}]_i$. Ca^{2+} influx through the channel activates protein kinase C (PKC) and this signaling is responsible for membrane vesiculation and internalization of TCC. In platelets, vesiculation is associated with Ca^{2+}-initiated kinase activation, protein phosphorylation, and calmodulin interaction with other membrane components (Wiedmer and Sims, 1991; Weidmer et al., 1987). Comparison of PKC activity in relation to $[Ca^{2+}]_i$ in cells stimulated with C5b–8,9$_n$ or ionomycin showed that C5b–8,9$_n$ induced higher PKC activity than ionomycin, at an ionomycin concentration which increased 2-fold higher $[Ca^{2+}]_i$ than the level caused by C5b–8,9$_n$ (Carney et al., 1990). This finding indicates that additional signals that synergize with Ca^{2+} may be involved in TCC-induced PKC activation. Generation of a transient, but sustained, increase in mass levels of an endogenous PKC activator, diacylglycerol (DAG), is demonstrated in the human JY B cell line by C5b–9, as well as C5b–8 and C5b–7 (Niculescu et al., 1993). C5b–7 does not form pores nor cause Ca^{2+} influx, however, H-7, a PKC inhibitor, blocks C5b–7 elimination. DAG generation by TCC is mediated by activation of Pertussis toxin (PTX)-sensitive G-proteins, as demonstrated by the inhibition of DAG increase and TCC elimination by PTX and not by cholera toxin (Figure 4) (Niculescu et al., 1993, 1994). Phosphatidylcholine (PC) is indicated as a possible source of DAG, as evidenced by a sustained DAG increase and the temporal correlation between DAG production and PC hydrolysis following C5b–9 assembly.

Elimination of C5b–9 by membrane vesiculation was also seen in erythrocytes without hemolysis (Iida et al., 1991). Membrane vesiculation induced by Ca^{2+}

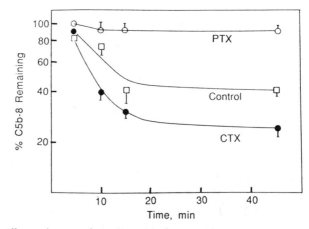

Figure 4. Effects of PTX and CTX on TCC elimination from the surface of living cells. JY25 cells treated with 500 ng/ml PTX or CTX for 3 h, or with medium, were incubated with 100 μg/ml IgG and 20% C9D for 15 min to form C5b–8. The remaining C5b–8 on the plasma membranes was determined by adding 10 μg of C9 at the time indicated to convert the C5b–8 to lytic C5b–9. Cells were further incubated for 60 min at 37°C, then cell death was determined by lactate dehydrogenase (LDH) release. Data are expressed as percent C5b–8 remaining on the plasma membrane. The LDH released from C5b–8 carrying cells treated with C9 at zero time was used as 100% and the LDH released from cells with C5b–8 at zero time was used as background control.

ionophore was 10% to 50% more in PNH erythrocytes than in normal erythrocytes, whereas TCC-induced vesiculation in GPI-deficient JY5 cells was about 50% of that in JY25 wild-type (Whitlow et al., 1993). This finding suggests an involvement of different GPI-anchored protein signaling in membrane vesiculation.

B. Mechanisms of Cell Death Induced by TCC

It has been shown that poly-C9 is not required for erythrocyte lysis nor nucleated cell death induced by C5b–9. C5b–$8,9_1$ can cause nucleated cell death, although C5b–$8,9_4$ is twice as effective and the rise in $[Ca^{2+}]_i$ during limited C5b–9 attack functions as a stimulus to eliminate TCC, as previously discussed. However, when TCC formation exceeds the elimination, the rate of cell death increases with increasing $[Ca^{2+}]_i$ (Figure 5). In the presence of a large number of C5b–9, the rate and extent of cell death are independent of cell volume regulation (Shin and Carney, 1988). Loss of adenine nucleotides (ATP, ADP, and AMP) and loss of mitochondrial membrane potential (MMP) have been demonstrated during the pre-lytic phase of complement attack (Papadimitriou et al., 1991). EGTA in the medium prevents MMP loss and acute cell death. These findings collectively indicate that Ca^{2+}-dependent acute cell death induced by C5b–9, which also affect mitochondrial

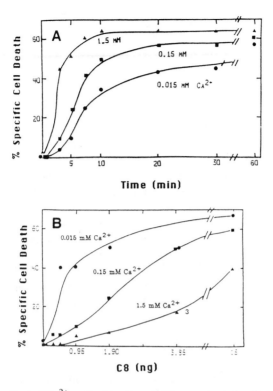

Figure 5. The effect of Ca^{2+} on the kinetics of killing nucleated cells. (**A**) Cells carrying C5b–7 complexes were mixed with varying amounts of C8 and excess C9 at different concentrations of Ca^{2+}, 0.015-, 0.15-, and 1.5-mM. The protective effect of Ca^{2+} was observed when a small number of C5b–9 complexes were assembled on the plasma cell membrane. (**B**) Cells carrying an equal number of C5b–8 complexes were incubated with an excess of C9 at 0°C for 30 min, then resuspended in buffer with Ca^{2+} concentrations of 0.015-, 0.15-, and 1.5-mM respectively. Cell death occurred about 3- to 7-fold more rapidly at higher Ca^{2+} concentrations.

function, may not be due to colloid-osmotic deregulation, as proposed previously. The mechanisms of gram-negative bacterial killing by C5b–9 are much less well understood than those of mammalian cell death, primarily due to the complex structure of the bacterial wall. Since TCC requiring at least three C9 per C5b–8, are expected to assemble on the outer membrane, it is unclear how the bacteria are killed without direct TCC attack on inner membrane. Experimental evidence indicated that bacterial cell death may occur from inner membrane dissolution induced by C5b–9 through activation of certain metabolic processes like oxidative phosphorylation (Joiner, 1988).

V. CELL ACTIVATION INDUCED BY TCC IN NUCLEATED CELLS

A. Activation of Membrane Phospholipases and Proteases

Arachidonic acid (AA) Mobilization

Multiple C5b–9 complexes induce nucleated cell death when formation exceeds elimination. Thus, surviving cells during and after elimination of limited C5b–9 may be in a modified state of cell activity which can be affected by peptide insertion and elevated $[Ca^{2+}]_i$. Following sublytic attack, Ca^{2+}-dependent mobilization of AA and formation of the AA-derived inflammatory mediators, leukotriene B4, pro-staglandin E2, and thromboxane, have been demonstrated in macrophages, poly-morphonuclear leukocytes, platelets, oligodendrocytes, and Ehrlich cells (Imagawa et al., 1983; Hänsch et al., 1987; Shirazi et al., 1987; Wiedmer et al., 1987). AA mobilization may be achieved by C5b–8 in the absence of C9 (Seeger et al., 1986; Shirazi et al., 1987) and phospholipid hydrolysis to generate AA is mediated by phospholipase A2, which is regulated by a kinase-dependent mechanism which requires Ca^{2+}. AA release, inhibited by PTX, may be also mediated by TCC-induced G-protein activation, as in DAG generation.

Generation of DAG and Ceramide

During complement activation, as studied with the human JY cell line and human complement, increases in DAG and ceramide levels have been demonstrated following TCC assembly, even at the step of C5b–7 (Niculescu et al., 1994), which does not form a channel nor cause Ca^{2+} influx. Increased levels of DAG and ceramide, both known endogenous regulators of PKC activities, are sustained and decline slowly over a 60 min period. Closely correlated kinetics of PC and sphingomyelin (SM) hydrolysis, together with DAG and ceramide generation following sublytic TCC attack, indicated that DAG and ceramide are derived from the phospholipid pools of PC and SM, respectively, by the activation of specific phospholipases (Figure 6).

Activation of PKC

Activation of PKC by Ca^{2+} and DAG is synergistic; DAG increases the binding affinity of Ca^{2+} to the enzyme. At a single cell level, formation of sublytic C5b–9, studied by digital imaging fluorescent microscopy, reveals oscillatory intracellular Ca^{2+} spikes, most likely each spike represents a C5b–9 channel (Figure 7).

PKC activity enhanced by C5b–9 is higher than that enhanced by C5b–8, and it is not detectable at the C5b–7 stage (Carney et al., 1990). Since DAG can be produced by C5b–7, the effect of C5b–7 on PKC may be too small to be detected by the assay, which measures the activity translocated from the cytosol to the

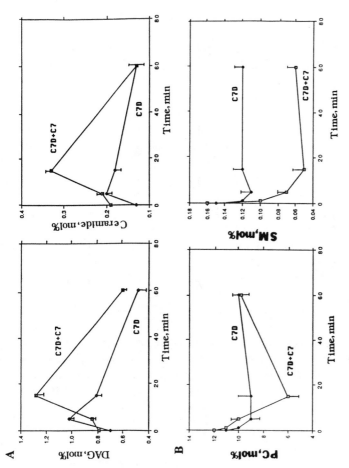

Figure 6. (A) TCC induced generation of DAG and ceramide in parallel with hydrolysis of phosphatidylcholine (PC) and sphingomyelin (SM). JY25 cells sensitized with IgG anti-MHC class II were exposed for 60 min at 37°C to C7 deficient serum (C7D) with and without C7 reconstitution. DAG and ceramide increased at 15 min about 2.5-fold when TCC assembled in the plasma membrane (C7D + C7), compared to cells without TCC (C7D). (B) TCC induced generation of DAG and ceramide in parallel with hydrolysis of phosphatidylcholine (PC) and sphingomyelin (SM). In a similar experiment, PC and SM hydrolysis were found at 15 min after TCC assembly (C7D + C7), showing a decrease of these phospholipids about 2-fold compared to C7D-treated cells.

139

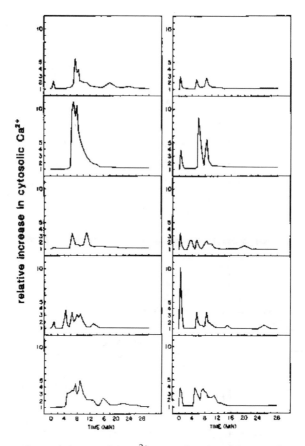

Figure 7. Oscillation of cytosolic Ca^{2+} treated with sublytic C5b–9. Ehrlich cells adherent to coverslips were loaded with Fura-2, then C5b–8 complexes were formed. After addition of limited C9, individual cells were analyzed by digital image fluorescence microscopy (Carney et al., 1990). Individual cells demonstrate a heterogenous pattern of Ca^{2+} oscillation, which appear 2 to 3 min after C9 addition, with an amplitude of 2- to 10-fold and an average 2 to 6 spikes per cell over a period of 10 min. Representative tracings of 10 cells are shown.

membrane. Activation of PKC by TCC is also supported by an increased phosphorylation of a PKC substrate, myristoylated alanine-rich C kinase substrate (Niculescu et al., 1993).

B. Interaction with G-protein

Heterotrimeric G-proteins are activated either by the ligand–receptor interaction or in a receptor-independent manner. Activation of G-proteins by TCC was indi-

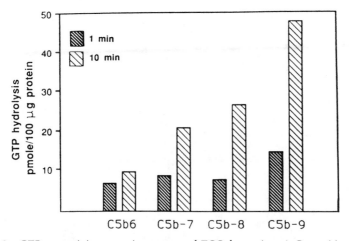

Figure 8. GTPase activity at various steps of TCC formation. IgG-sensitized JY25 membranes were incubated for 5 min at 37°C with 20% complement deficient serum. C7D, C8D, and C9D were used to form C5b–6, C5b–7, and C5b–8, respectively. C5b–9 complexes were formed by C7D + C7. GTP hydrolysis was determined at 1- and 10-min, showing 5-fold increase after TCC assembly. C5b–7 induced a 2-fold and C5b–8 induced a 3-fold increase in GTPase. C5b–9 is the strong activator since at 1 min it induced a 2-fold increase in GTP hydrolysis.

cated by the ability of PTX to inhibit TCC-mediated DAG generation. G-protein activation by TCC is directly demonstrated by the time-dependent increase in GTPγS binding and GTP-ase activity in plasma membranes carrying C5b–7, C5b–8, or C5b–9 (Figure 8; Niculescu et al., 1994).

In the absence of receptor, TCC may activate G-proteins by receptor-independent mechanisms, similar to mastoparan, directly interacting with Gα. Direct activation of G-protein was demonstrated by the association of the G-protein with membrane-inserted TCC. When C5b–7, C5b–8, or C5b–9 immunoprecipitated from detergent-solubilized cell lysates with anti-C7, anti-C8, or anti-C5b–9 neoantigen, were ADP-ribosylated with PTX, G-protein α subunits were identified on SDS-PAGE migrating as a 41 kD protein. Western blots revealed the ADP-ribosylated Gα subunit as Giα/Goα species (Figure 9).

G-proteins associated with TCC in the same complex are functionally active, since the anti-TCC immunoprecipitates also contain Gβγ subunits, and shows GTPγS binding activity (Niculescu et al., 1994).

C. Inhibition of mRNA Accumulation in Oligodendrocytes and Myotubes

TCC assembly on oligodendrocytes (OLG) selectively induces reduction of mRNA levels encoding myelin basic protein (MBP) and proteolipid protein (PLP;

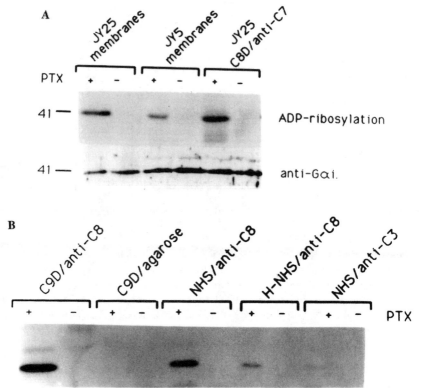

Figure 9. ADP-ribosylation by PTX of Giα-interacting with TCC. (**A**) IgG-sensitized JY25 cells treated with C8 deficient human serum (C8D) were lysed then immunoprecipitated with anti-C7 antibody and protein A agarose. These complexes were exposed to ADP-ribosylation by PTX in parallel with purified membranes of JY25 and JY5, respectively. The same blot was also immunoreacted with rabbit IgG anti-Giα$_{1,2}$ then developed by ECL. Both techniques revealed the 41 kD Gα subunit interacting with C5b–7 complexes. (**B**) TCC from JY25 cell membranes immunoprecipitated with anti-C8 and protein A agarose were exposed to ADP-ribosylation by PTX revealing the 41 kD Giα present in these complexes, but absent in immunoprecipitates with isotype IgG (C9D/agarose), in cells treated with heat-inactivated serum (H-NHS), or in immunoprecipitates with anti-C3 antibody (NHS/anti-C3).

Shirazi et al., 1993). The mRNAs encoding 2′-3′-cyclic nucleotide 3′-phosphodiesterase, a minor myelin protein, β-actin, or aldolase A are minimally affected. Reduced MBP and PLP mRNA accumulation is noted as early as 90 min and becomes significant around 3 h, and is mediated by mRNA destabilization. PLP mRNA decay is affected by extracellular Ca^{2+} concentration and MBP mRNA decay is blocked by PKA inhibitors. Sublytic TCC on myotubes, differentiated from murine C2 myoblasts *in vitro*, similarly reduced the mRNA accumulation affecting muscle-specific protein transcripts, such as α-actin, troponin Is, and aldolase A (Lang and Shin, 1991). mRNAs encoding constitutively expressed proteins, such as heat shock protein 83, are not affected by TCC, whereas β-actin mRNA is increased by C5b–9.

D. Effects on Oncogenes and Cell Cycle Kinases

The oncogenes c-*jun* and c-*fos* are induced in OLG following incubation with anti-galactocerebroside antibodies and serum complement. The specific increase attributed to TCC by using C7D reconstituted with C7 is twofold for c-*jun* and the increase, noted at 30 min, is sustained for a period of 3h (Shirazi et al., 1993). Similar behavior was noted for c-*fos* and *jun* D. As expected, these oncogenes are minimally expressed in differentiated OLG, whereas OLG progenitor cells (O-2A) cultured in growth medium express high levels of c-*jun* and c-*fos*, in the absence of MBP and PLP mRNA expression. Induction of c-*fos* and c-*jun* expression by TCC is associated with increased cdc2 and cdk4 kinase activity. The cdc2 kinase activity, low in differentiated OLG and high in O-2A progenitor cells, allows us to distinguish cycling from quiescent cells. The cdk4 kinase is markedly increased in OLG following C5b-9 attack in late G1 and G1/S transition (Rus et al., 1996). These findings indicated that differentiated cells following sublytic TCC attack undergo changes in cellular activities from the differentiated to the progenitor state. How the TCC-induced activation of c-*jun* and c-*fos* genes regulates the cell cycle of differentiated cells remains to be determined since antisense c-*jun* oligonucleotides block cell cycle activation.

VI. PARTICIPATION OF TCC IN HEALTH AND DISEASES IN HUMAN

The deposition of C5b–9 in tissues associated with the production of cytokines and reactive oxygen metabolites are among the major events responsible for tissue injury in inflammatory human diseases. The release of reactive oxygen metabolites, O_2^-, OH^-, and H_2O_2, following sublytic TCC attack requires C5b–9 assembly and the presence of Ca^{2+} in the extracellular environment (Adler et al., 1986; Hänsch et al., 1987). In addition, TCC-induced generation of leukotriene B4, PGE2, and thromboxane, can function as potent inflammatory mediators by increasing vascular permeability, attracting inflammatory cells, and activating platelets. TCC also induces the production of IL-1 and TNF, potent cytotoxic inflammatory cytokines (Lovett et al., 1987; Schönermark et al., 1991). The presence of C5b–9 neoantigens in human tissues affected by diverse diseases indicates *in situ* complement activation and subsequent TCC assembly in these tissues. Involvement of C5b–9 in various human diseases are reviewed in Table 2.

As was demonstrated in the arterial wall in atherosclerosis, C5b–9 deposits are associated with the cell debris or localized on plasma membranes of intact cells

Table 2. Presence of C5b–9 Neoantigens in Human Tissue

	Disease	Reference
Kidney	SLE[a]	Biesecker et al. (1981)
	SLE, diabetes, Amyloid	Falk et al. (1983)
	Membranous and	Rus et al. (1986)
	Membranoproliferative glomerulonephritis	Hinglais et al. (1986)
	Hemodialysis	Deppisch et al. (1990)
Muscle and Nervous Tissue	Muscle Necrosis	Engel and Biesecker (1982)
	Multiple Sclerosis	Sanders et al. (1986a)
	Guillian–Barré	Koski et al. (1987)
Skin	Cutaneous SLE	Biesecker et al. (1982)
	Pemphigus	Dahl et al. (1984)
	Dermatomyositis	Kiesell et al. (1986)
	Henoch–Schönlein purpura	Kawana et al. (1990)
Joints	Rheumatoid arthritis	Mollnes and Paus (1986)
		Sanders et al. (1986b)
Cardiovascular	Atherosclerosis	Niculescu et al. (1985)
	Myocardial infarction	Schäfer et al. (1986)
	Cardiomyopathias	Rus et al. (1987)
	Cardiopulmonary bypass	Salama et al. (1988)
Gastrointestinal Tract	Ulcerative colitis	Halstensen et al. (1990)
	Liver	Polihronis et al. (1993)
Breast	Malignant	Niculescu et al. (1992)

Note [a]SLE = systemic lupus erythematous.

adjacent to the areas of necrosis and sclerosis. Many of these cells carrying TCC complexes are monocytes/macrophages, suggesting a role for both inflammatory cells and complement activation in tissue injury (Niculescu et al., 1987, 1989). At present, it is reasonable to assume that C5b–9 complexes are directly involved in the pathogenesis of chronic inflammation and tissue healing by modulating a variety of metabolic activities of target cells.

REFERENCES

Adler, S., Baker, P., Johnson, R. J., Ochi, R., Pritzl, P., & Couser, W. G. (1986). Complement membrane attack complex stimulates production of reactive oxygen metabolites by cultured rat mesangial cells. J. Clin. Invest. 77, 762–767.

Bhakdi, S., & Tranum-Jensen, J. (1987). Damage to mammalian cells by proteins that form transmembrane pores. Rev. Physiol. Biochem. Pharmacol. 107, 148–223.

Biesecker, G., Lavin, L., Ziskind, M., & Koffler, D. (1982). Cutaneous localization of the membrane attack complex in discord and systemic lupus erythematosus. N. Engl. J. Med. 306, 264–270.

Biesecker, G., Katz, S., & Koffler, D. (1981). Renal localization of the membrane attack complex in systemic lupus erythematosus nephritis. J. Exp. Med. 159, 1779–1794.

Bordet, J. (1900). Les sérums hémolytiques, leurs antitoxines et les théories des sérums cytolytiques. Ann. Inst. Pasteur 14, 257–275.

Carney, D. F., Lang, T. J., & Shin, M. L. (1990). Multiple signal messengers generated by terminal complement complexes and their role in terminal complex elimination. J. Immunol. 145, 623–629.

Carney, D. F., Koski, C. L., & Shin, M. L. (1985). Elimination of terminal complement intermediates from the plasma membrane of nucleated cells: The rate of disappearance differs for cells carring C5b-7 or C5b-8 or a mixture of C5b-8 with a limited number of C5b-9. J. Immunol. 134, 1804–1809.

Dahl, J. S., Dahl, C. E., & Levine, R. P. (1979). Role of lipid fatty acyl composition and membrane fluidity in the resistance of Acholeplasma laidlawii to complement-mediated killing. J. Immunol. 123, 104–108.

Dahl, M. V., Falk, R. J., Carpenter, R., & Michael, A. F. (1984). Deposition of the membrane attack complex in bullous pemphigoid. J. Invest. Dermatol. 82, 132–136.

Dalmasso, A. P., & Benson, B. A. (1981). Lesions of different functional size produced by human and guinea pig complement in sheep red blood cell membranes. J. Immunol. 127, 2214–2218.

Deckert, M., Kubar, J., & Bernard, A. (1992). CD58 and CD59 molecules exhibit potentializing effects in T cell adhesion and activation. J. Immunol. 148, 672–677.

Deppisch, R., Schmitt, V., Bommer, J., Hänsch, G. M., Ritz, E., & Rauterberg, E. W. (1990). Fluid phase generation of terminal complement complex as a novel index of biocompatibility. Kidney Int. 37, 696–706.

DiScipio, R. G., Chakvarati, D. N., Müller-Eberhard, H. J., & Fey, G. H. (1988). The structure of human complement C7 and C5b-7 complex. J. Biol. Chem. 263, 549–560.

DiScipio, R. G., Smith, C. A., Müller-Eberhard, H. J., & Hugli, T. E. (1983). The activation of human complement component C5 by a fluid phase C5 convertase. J. Biol. Chem. 258, 10629–10636.

Engel, A. G., & Bieseker, G. (1982). Complement activation in muscle fiber necrosis: Demonstration of the membrane attack complex of complement in necrotic fibers. Ann. Neurol. 12, 289–296.

Falk, R. J., Dalmasso, A. P., Kim, Y., Tsai, C., Scheinmann, J. I., Gewunz, H., & Michael, A. F. (1983). Neoantigen of the polymerized ninth component of complement. Characterization of a monoclonal antibody and immunohistochemical localization in renal disease. J. Clin. Invest. 72, 560–573.

Frank, M. M., & Fries, L. F. (1989). Complement. In: *Fundamental Immunology*, 2nd Ed. (Paul, W. E., Ed.). Raven Press, New York, pp. 679–701.

French, L., Chonn, A., Ducrest, D., Baumann, B., Belin, D., Wohlwend, A., Kiss, J. Z., Sappino, A. P., Tschopp, J., & Schifferli, J. A. (1993). Murine clusterin: Molecular cloning and mRNA localization during embriogenesis. J. Cell. Biol. 122, 1119–1130.

Gee, A. P., Boyle, M. D. P., & Borsos, T. (1980). Distinction between C8 mediated and C8/C9 mediated hemolysis on the basis of independent ^{86}Rb and hemoglobin release. J. Immunol. 124, 1905–1910.

Haeflinger, J. A., Jenne, D., Stanley, K. K., & Tschopp, J. (1987). Structural homology of human complement component C8γ and plasma protein HC: Identity of the cysteine bond pattern. Biochim. Biophys. Acta 149, 750–754.

Haeflinger, J. A., Tschopp, J., Vial, N., & Jenne, D. E. (1989). Complete primary structure and functional characterization of the sixth component of the human complement system. J. Biol. Chem. 264, 18041–18051.

Hahn, W. C., Menu, E., Bothwell, A. L. M., Sims, P. J., & Bierer, B. E. (1992). Overlapping but non-identical binding sites on CD2 for CD58 and a second ligand CD59. Science 256, 1805–1807.

Halstensen, T. S., Mollnes, T. E., Garred, P., Fausa, O., & Brandtzaeg, P. (1990). Epithelial deposition of immunoglobulin G1 and activated complement (C3b and terminal complement complex) in ulcerative colitis. Gastroenterology 98, 1261–1271.

Hamilton, K. K., Zhao, J., & Sims, P. J. (1993). Interaction between apolipoproteins AI and AII and the membrane attack complex of complement. J. Biol. Chem. 268, 3632–3638.

Hänsch, G. M. (1992). The complement attack phase: Control of lysis and non-lethal effects of C5b-9. Immunopharmacology 24, 107–117.

146   MOON L. SHIN, HOREA G. RUS, and FLORIN I. NICULESCU

Hänsch, G. M., Weller, P. F., & Nicholson-Weller, A. (1988). Release of C8 binding protein (C8bp) from the cell membrane by phosphatidylinositol-specific phospholipase C. Blood 72, 1089–1092.

Hänsch, G. M., Seitz, M., & Betz, M. (1987). Effect of the late complement components C5b-9 on human monocytes; release of prostanoids, oxygen radicals and of a factor inducing cell proliferation. Int. Arch. Allergy. Appl. Immunol. 82, 317–320.

Haviland, D. L., Haviland, J. C., Fleischer, D. T., & Wetsel, R. A. (1991). Structure of the murine fifth complement component (C5) gene. J. Biol. Chem. 266, 11818–11825.

Hinglais, N., Kazatchkine, M. D., Bhakdi, S., Appay, M. D., Mandet, C., Grossetete, J., & Bariety, J. (1986). Immunohistochemical study of the C5b-9 complex of complement in human kidneys. Kidney Int. 30, 399–410.

Hobart, M. D., Fernie, B., & DiScipio, R. G. (1993). Structure of the human C6 gene. Biochemistry 32, 6198–6205.

Hu, V. W., Esser, A. F., Podack, E. R., & Wisnieski, B. J. (1981). The membrane attack mechanism of complement photolabelling reveals insertion of terminal proteins into target membranes. J. Immunol. 127, 380–386.

Iida, K., Whitlow, M. B., & Nussenzweig, V. (1991). Membrane vesiculation protects erythrocytes from destruction by complement. J. Immunol. 147, 2638–2642.

Imagawa, D. K., Osifchin, N. E., Paznekas, W. A., Shin, M. L., & Mayer, M. M. (1983). Consequence of cell membrane attack by complement: Release of arachidonate and formation of inflammatory derivatives. Proc. Natl. Acad. Sci. USA 80, 6647–6651.

Ishida, B., Wisnieski, B. J., Lavine, C. H., & Esser, A. F. (1982). Photolabelling of a hydrophobic domain of the ninth component of human complement. J. Biol. Chem. 257, 10551–10553.

Jenne, D. E., & Tschopp, J. (1989). Molecular structure and functional characterization of a human complement cytolysis inhibitor found in blood and seminal plasma: Identity to sulfated glycoprotein 2, a constituent of rat testis. Proc. Natl. Acad. Sci. USA 86, 7123–7127.

Joiner, K. A. (1988). Complement evasion by bacteria and parasites. Ann. Rev. Microbiol. 42, 201–230.

Kawana, S., Shen, G. H., Kobayashi, Y., & Nijkiyana, S. (1990). Membrane attack of complement in Henoch-Schönlein purpura skin and nephritis. Arch. Dermatol. Res. 292, 183–187.

Kiessel, J. T., Mendeli, J. R., & Rammohan, K. W. (1986). Microvascular deposition of complement membrane attack complex in dermatomyositis. N. Engl. J. Med. 314, 329–334.

Kim, S. H., Carney, D. F., Hammer, C. H., & Shin, M. L. (1987). Nucleated cell killing by complement: Effects of C5b-9 channel size and extracellular Ca^{2+} on the lytic process. J. Immunol. 138, 1530–1536.

Koski, C. L., Sanders, M. E., Swoveland, P. T., Lawley, T. J., Shin, M. L., Frank, M. M., & Joiner, K. A. (1987). Activation of terminal components of complement in patients with Guillain-Barré syndrome and other demyelinating neuropathies. J. Clin. Invest. 80, 1492–1497.

Koski, C. L., Ramm, E. L., Hammer, C. H., Mayer, M. M., & Shin, M. L. (1983). Cytolysis of nucleated cells by complement: Cell death displays multi-hit characteristics. Proc. Natl. Acad. Sci. USA 80, 3816–3820.

Laine, R. O., & Esser, A. F. (1989). Detection of refolding conformers of complement protein C9 during insertion into membranes. Nature 341, 63–65.

Lang, T., & Shin, M. (1991). Changes in levels of muscle specific mRNAs in a myotube cell line after treatment with sublytic doses of C5b-9. Complement. Inflamm. 8, 179. (Abstr.).

Lovett, D. H., Hänsch, G. M., Goppelt, M., Resch, K., & Gemsa, D. (1987). Activation of glomerular mesangeal cells by terminal membrane attack of complement. J. Immunol. 138, 2473–2480.

Marazziti, D., Eggertsen, G., Fey, G., & Stanley, K. K. (1988). Relationships between the gene and protein structure in human complement component C9. Biochemistry 27, 6529–6534.

Mayer, M. M., Michaels, D. W., Ramm, L. E., Whitlow, M. B., Willoughby, J. B., & Shin, M. L., (1981). Membrane damage by complement. CRC Crit. Rev. Immunol. 7, 133–165.

Mayer, M. M. (1961). Development of one-hit theory of immune lysis. In: *Immunochemical Approaches to Problems in Biochemistry* (Heidelberg, M., & Plescia, T., Eds.). Rutgers University Press, New Brunswick, NJ, pp. 279–288.

Meri, S., Waldmann, H., & Lachmann, P. J. (1991). Distribution of protectin (CD59) a complement membrane attack inhibitor, in normal human tissue. Lab. Invest. 65, 532–537.

Miyata, T., Takeda, J., Iida, Y., Yamada, N., Inoue, N., Takahashi, M., Maeda, K., Kitani, T., & Kinoshita, T. (1993). The cloning of PIG-A, a component in the early step of GPI-anchor biosynthesis. Science 259, 1318–1320.

Mollness, T. E., & Paus, A. (1986). Complement activation in synovial fluid and tissue from patients with juvenile rheumatoid arthritis. Arthritis Rheum. 29, 1359–1365.

Morgan, B. P., Dankert, J. R., & Esser, A. F. (1987). Recovery of human neutrophils from complement attack: Removal of membrane attack complex by endocytosis and exocytosis. J. Immunol. 138, 246–253.

Morgan, B. P., & Campbell, A. K. (1985). A recovery of human polymorphonuclear leukocytes from sublytic complement attack is mediated by changes in intracellular free calcium. Biochem. J. 231, 205–208.

Murphy, B. F., Saunders, J. R., O'Bryan, M. K., Kirsbaum, L., Walker, I. D., & d'Apice, A. J. F. (1989). Sp 40,40 is an inhibitor of C5b-6 initiated hemolysis. Int. Immunol. 1, 551–556.

Müller-Eberhard, H. J. (1988). Molecular organization and function of the complement system. Ann. Rev. Biochem. 57, 321–347.

Niculescu, F., Rus, H. G., & Shin, M. L. (1994). Receptor-independent activation of guanine nucleotide-binding proteins by terminal complement complexes. J. Biol. Chem. 269, 4417–4423.

Niculescu, F., Rus, H. G., Shin, S., Lang, T., & Shin, M. L. (1993). Generation of diacylglycerol and ceramide during homologous complement activation. J. Immunol. 150, 214–224.

Niculescu, F., Rus, H. G., Retegan, M., & Vlaicu, R. (1992). Persistent complement activation on tumor cells in breast cancer. Amer. J. Pathol. 140, 1039–1043.

Niculescu, F., Rus, H. G., Porutiu, D., Ghiurca, V., & Vlaicu, R. (1989). Immuno-electron microscopic localization of S-protein/vitronectin in human atherosclerotic wall. Atherosclerosis 78, 197–203.

Niculescu, F., Rus, H. G., & Vlaicu, R. (1987). Immunohistochemical localization of C5b-9, S-protein, C3d and apolipoprotein B in human arterial tissues with atherosclerosis. Atherosclerosis 65, 1–11.

Niculescu, F., Rus, H. G., Cristea, A., & Vlaicu, R. (1985). Localization of the terminal C5b-9 complement complex in aortic atherosclerotic wall. Immunol. Lett. 10, 109–114.

Ninomiya, H., & Sims, P. J. (1992). The human complement regulatory protein CD59 binds to the α-chain of C8 and the "b" domain of C9. J. Biol. Chem. 267, 13675–13680.

Pangburn, M. K. (1988). Initiation and activation of the alternative pathway of complement. In: *Cytotoxic Lymphocytes and Complement: Effectors of the Immune System* (Podack, E. R., Ed.). CRC Press, Boca Raton, FL, pp. 41–56.

Papadimitriou, J. C., Ramm, L. E., Drachenberg, C. B., Trump, B. F., & Shin, M. L. (1991). Quantitative analysis of adenine nucleotides during the prelytic phase of cell death mediated by C5b-9. J. Immunol. 147, 212–217.

Podack, E. R. (1988). Assembly and structure of membrane attack complex (MAC) of complement. In: *Cytolytic Lymphocyte and Complement: Effectors of the Immune System* (Podack, E. R., Ed.). CRC Press, Boca Raton, FL, pp. 174–184.

Podack, E. R., & Tschopp, J. (1982). Polymerization of the ninth component of complement (C9): Formation of poly C9 with a tubular structure resembling the membrane attack complex of complement. Proc. Natl. Acad. Sci. USA 79, 574–578.

Podack, E. R., & Müller-Eberhard, H. J. (1978). Binding of desoxycholate, phosphatidylcholine vesicles, lipoprotein and of the S-protein to complexes of terminal complement proteins. J. Immunol. 121, 1025–1030.

Polihronis, M., Machet, D., Seunders, J., O'Bryan, M., McRae, J., & Murphy, B. (1993). Immunohistological detection of C5b-9 complement complexes in normal and pathological human livers. Pathology 25, 20–23.

Preissner, K. T. (1991). Structure and biological role of vitronectin. Ann. Rev. Cell Biol. 7, 275–310.

Ramm, L. E., Whitlow, M. B., & Mayer, M. M. (1985). The relationship between channel size and the number of C9 molecules in the C5b-9 complex. J. Immunol. 134, 2594–2599.

Ramm, L. E., Michaels, D. W., Whitlow, M. B., & Mayer, M. M. (1983). On the heterogenity and molecular composition of the transmembrane channels produced by complement. In: *Biological Response Mediators and Modulators* (August, T., Ed.). Academic Press, San Diego, pp. 117–132.

Ramm, L. E., Withlow, M. B., & Mayer, M. M. (1982). Size and transmembrane channels produced by complement proteins C5b-8. J. Immunol. 122, 1143–1146.

Rao, A. G., Howard, O. M. Z., Ng, S. C., Whitehead, A. S., Colten, H. R., & Sodetz, J. M. (1987). Complementary DNA and derived aminoacid sequence of the α subunit of human complement C8: Evidence for the existence of a separate α subunit messenger RNA. Biochemistry 26, 3556–3564.

Rus, H. G., Niculescu, F., & Shin, M. L. (1996). Sublytic complement attack induces cell cycle in oligodendrocytes. S phase induction is dependent on c-jun activation. J. Immunol. 156, 4892–4900.

Rus, H. G., Niculescu, F., & Vlaicu, R. (1987). Presence of C5b-9 complement complex and S-protein in human myocardial areas with necrosis and sclerosis. Immunol. Lett. 16, 15–20.

Rus, H. G., Niculescu, F., Nanulescu, M., Cristea, A., & Florescu, P. (1986). Immunohistochemical detection of the terminal C5b-9 complement complex in children with glomerular disease. Clin. Exp. Immunol. 65, 66–72.

Salama, A., Hugo, F., Heinrich, D., Höge, R., Müller, R., Kiefel, V., Müller-Eckhart, C., & Bhakdi, S. (1988). Deposition of terminal C5b-9 complement complexes on erythrocytes and leukocytes during cardiopulmonary bypass. N. Eng. J. Med. 318, 408–414.

Sanders, M. E., Koski, C. L., Robbins, D., Shin, M. L., Frank, M. M., & Joiner, K. A. (1986a). Activated terminal complement in cerebrospinal fluid in Guillain-Barré syndrome and multiple sclerosis. J. Immunol. 136, 4456–4459.

Sanders, M. E., Kopicky, J. A., Wigley, F. M., Shin, M. L., Frank, M. M., & Joiner, K. A. (1986b). Membrane attack complex of complement in rheumatoid synovial tissue demonstrated by immunofluorescent microscopy. J. Rheumatol. 13, 1028–1034.

Schäfer, H. J., Mathey, D., Hugo, F., & Bhakdi, S. (1986). Deposition of the terminal C5b-9 complement complex in infarcted areas of human myocardium. J. Immunol. 137, 1945–1949.

Schieren, G., Janssen, O., & Hänsch, G. M. (1992). Enhanced expression of the complement regulatory factor C8 binding protein (C8bp) on U937 cells after stimulation with interleukin-1β, endotoxin, γ-interferon or phorbol ester. J. Immunol. 148, 3183–3188.

Schlager, S. I., & Ohanian, S. H. (1980). Tumor cell lipid composition and sensitivity to humoral immune killing. II. Influence of plasma membrane and intracellular lipid and fatty acid content, J. Immunol. 125, 508–517.

Schönermark, M., Deppisch, R., Riedasch, G., Rother, K., & Hänsch, G. M. (1991). Induction of mediators release from human glomerular mesangial cells by the terminal complement components C5b-9. Int. Arch. Allergy. Appl. Immunol. 96, 331–337.

Schönermark, S., Rauterberg, E. W., Shin, M. L., Löke, S., Roelcke, D., & Hänsch, G. M. (1986). Homologous species restriction in lysis of human erythrocytes: A membrane derived protein with C8-binding functions as an inhibitor. J. Immunol. 136, 1772–1776.

Seeger, W., Suttorp, N., Hellwig, A., & Bhakdi, S. (1986). Noncytolytic terminal complement complexes may serve as calcium gates to elicit leukoterine B4 generation in human polymorphonuclear leukocytes. J. Immunol. 137, 1286–1293.

Shin, M. L., & Carney, D. F. (1988). Cytotoxic action and other metabolic consequences of terminal complement proteins. Prog. Allergy 40, 44–81.

Shin, M. L., Paznekas, W. A., & Mayer, M. M. (1978). On the mechanism of membrane damage by complement: The effect of length and saturation of the acyl chains in the liposomal bilayers and the effect of cholesterol concentration in sheep erythrocytes and liposomal membranes. J. Immunol. 120, 1996–2002.

Shin, M. L., Paznekas, W. A., Abramovitz, A. S., & Mayer, M. M. (1977). On the mechanism of membrane damage by complement: Exposure of hydrophobic sites on activated complement proteins. J. Immunol. 119, 1358–1364.

Shirazi, Y., Rus, H. G., Macklin, W. B., & Shin, M. L. (1993). Enhanced degradation of messenger RNA encoding myelin proteins by terminal complement complexes in oligodendrocytes. J. Immunol. 150, 4581–4590.

Shirazi, Y., Imagawa, D. K., & Shin, M. L. (1987). Release of leukotriene B4 from sublethally injured oligodendrocytes by terminal complement complexes. J. Neurochem. 48, 271–278.

Sodetz, J. M. (1988). Structure and function of C8 in the membrane attack sequence of complement. In: *Cytotoxic Effector Mechanisms* (Podack, E. R., Ed.). Springer Verlag, Berlin, pp. 19–31.

Stanley, K. K. (1988). The molecular mechanisms of complement C9. Insertion and polymerization in biological membranes. Curr. Topics. Microbiol. Immunol. 140, 49–65.

Tschopp, J., Chonn, A., Herting, S., & French, L. E. (1993). Clusterin, the human apolipoprotein and complement inhibitor, binds to complement C7, C8β and the b domain of C9. J. Immunol. 151, 2159–2165.

Tschopp, J., Podack, E. R., & Müller-Eberhard, H. J. (1985). The membrane attack of complement: C5b-8 as accelerator of C9 polymerization. J. Immunol. 134, 10551–10553.

Vanguri, P., & Shin, M. L. (1988). Hydrolysis of myelin basic protein in human myelin by terminal complement complexes. J. Biol. Chem. 263, 7228–7234.

Walport, H. J., & Lachmann, P. J. (1993). Complement. In: *Clinical Aspects of Immunology* (Lachmann, P. J., Peters, K., Rosen, F. S., & Walport, H. J., Eds.). Blackwell Science Publishers, Oxford, pp. 347–375.

Wetsel, R. A., Lemons, R. S., LeBeau, M. M., Barnum, S., Noack, D., & Tack, B. F. (1988). Molecular analysis of human complement component C5: Localization of the structural gene to chromosome 9. Biochemistry 27, 1474–1482.

Wetsel, R. A., Ogata, R. T., & Tack, B. F. (1987). Primary structure of the fifth component of murine complement. Biochemistry 26, 737–743.

Whitlow, M., Iida, K., Marshall, P., Silber, R., & Nussenzweig, V. (1993). Cells lacking glycan phosphatidylinositol-linked protein have impaired ability to vesiculate. Blood 81, 510–516.

Whitlow, M. B., Ramm, L. E., & Mayer, M. M. (1985). Penetration of C8 and C9 in the C5b-9 complex across the erythrocyte membrane into the cytoplasmic space. J. Biol. Chem. 260, 998–1005.

Wiedmer, T., & Sims, P. J. (1991). Participation of protein kinases in complement C5b-9 induced shedding of platelet plasma membrane vesicles. Blood 78, 2880–2886.

Wiedmer, T., Ando, B., & Sims, P. J. (1987). Complement C5b-9 stimulated platelet secretion is associated with a Ca^{2+}-initiated activation of cellular protein kinases. J. Biol. Chem. 262, 13674–13682.

Yamashina, M., Ueda, E., Kinoshita, T., Takami, T., Ojima, A., Ono, H., Tanaka, H., Kondo, N., Orri, T., Okada, N., Okada, H., Inoue, K., & Kitani, T. (1990). Inherited complete deficiency of 20-kilodalton homologous restriction factor (CD59) as a cause of paroxysmal nocturnal hemoglobinuria. N. Engl. J. Med. 323, 1184–1189.

Young, D. E., Cohn, Z. A., & Podack, E. R. (1986). The ninth component of complement and the pore forming protein perforin (perforin 1) from cytotoxic T cell: Structural, immunological and functional studies. Science 233, 184–190.

Zalman, L., Wood, M., & Müller-Eberhard, H. J. (1986). Isolation of a human erythrocyte membrane capable of inhibiting expression of homologous complement transmembrane channels. Proc. Natl. Acad. Sci. USA 83, 6975–6979.

CELL MEMBRANE DAMAGE IN LYMPHOCYTE-MEDIATED CYTOLYSIS:
ROLE OF LYMPHOCYTE PORE-FORMING PROTEIN PERFORIN

Chau-Ching Liu, Pedro M. Persechini,
M. Fatima Horta, and John Ding-E Young

Biomembranes
Volume 4, pages 151–173.

151

I. INTRODUCTION

Cytotoxic T lymphocytes (CTLs) and natural killer (NK) cells play an important role in immune protection by lysing malignant and virus-infected cells, as well as other invading foreign agents (Trinchieri and Perussia, 1984; Herberman et al., 1986; Young, 1989). Lysis of a target cell by an effector lymphocyte requires an intimate contact between the two cells, which presumably triggers the delivery of a so-called "lethal hit." Conjugation between the CTL and the target cell involves a specific interaction between the T cell receptor (TCR) and a complex consisting of a unique antigen and the major histocompatibility complex (MHC) class I-molecule present on the target cell surface (Sprent and Webb, 1987; Waver and Unanue, 1990). Recognition by NK cells, on the other hand, is not MHC-restricted and involves yet unidentified receptor(s) and target cell determinants (Herbeman, 1986; Giorda and Trucco, 1991; Ryan et al., 1991; Yokoyama, 1993). Other types of cytotoxic lymphocytes such as lymphokine-activated killer (LAK) cells (Rosenberg, 1988; Zychlinsky et al., 1990), MHC class II-restricted CTLs (Ju, 1991), and γ/δ T lymphocytes with cytolytic activity (Fisch et al., 1990; Saito et al., 1990) are also thought to execute cell-killing via a contact-dependent mechanism.

According to a widely held view, conjugation with the target cell triggers degranulation by the killer cell which then results in the release of a number of putative cytotoxins into the intercellular space (Henkart and Henkart, 1982). A prominent member among these toxins is the pore-forming protein (PFP; also known as perforin, cytolysin, or C9-related protein, C9RP; Henkart, 1985; Masson and Tschopp, 1985; Podack et al., 1985; Liu et al., 1986; Young et al., 1986a; Zalman et al., 1986a). Studies have shown that perforin is able to "perforate" membranes by forming transmembrane channels and thereby mediate cell lysis.

Before describing perforin in more detail, it is pertinent to have a glance at how the biological membranes can be damaged by pore-forming toxins.

II. BIOLOGICAL MEMBRANES AS TARGETS FOR PORE-FORMING PROTEINS

The basic structural unit of biological membranes consists of amphipathic phospholipids which aggregate spontaneously into a bilayer with the polar lipid heads facing the aqueous surface and the fatty acyl chains being sequestered within the hydrophobic interior. The bilayer represents a natural barrier to the nonspecific flow of ions and other polar or high molecular weight solutes, and the selective transport of solutes across biological membranes is a process essential for living cells to maintain homeostasis of the intracellular *milieu*.

Damaging the integrity of the plasma membranes is a strategy used by both vertebrate and invertebrate organisms to attack invading foreign agents. They accomplish this task by producing proteins that are capable of forming nonselective channels/pores in the target membrane lipid bilayer. These cytolytic pore-forming toxins have been isolated from sources as diverse as bacteria, sea anemones, amoeba, higher fungi, the vertebrate immune systems, and venoms from insects and snakes (Cavard et al., 1986; Young et al., 1982; Bernheimer, 1990; Massotte et al., 1990; Ojcius and Young, 1991). In addition to perforin, the vertebrate immune systems produce other pore-forming proteins, such as leukocyte defensins (Lehrer et al., 1991) and the terminal components of the complement cascade (Mayer et al., 1981; Bhakdi and Tranum-Jensen, 1984; Müller-Eberhard, 1986). All of them are soluble proteins that can insert into biological membranes and form transmembrane channels/pores. In view of the similarities to the other pore-forming proteins, perforin not only is an important lymphocyte toxin, but also may serve as an interesting model for studying both the functional and structural features of the pore-forming cytolysis mechanism that seems to be highly conserved during evolution.

A large number of pore-forming toxins are known to create channels through a "barrel stave" mechanism (Ehrenstein and Lecar, 1977). Three discrete steps have been defined for the pore formation in this model: (1) binding of the water-soluble monomers to the membrane, (2) insertion of the monomers into the membrane, and (3) aggregation of the monomers, like barrel staves, surrounding a central pore such that the hydrophobic side of the proteins is exposed to the membrane acyl chains and the hydrophilic side lines up the pore. As a result, stable transmembrane pores are formed that allow ions and small molecules to flow passively across the bilayer. Since many of the homopolymeric channels formed by toxins are not large enough to allow the passage of cytoplasmic proteins, this model predicts that the channels would produce an ionic imbalance in the cell which then results in colloid osmotic lysis. As will be reviewed in the following sections, perforin is likely to lyse target cells by this mechanism.

III. GRANULE EXOCYTOSIS/PORE FORMATION MODEL OF LYMPHOCYTE-MEDIATED CYTOLYSIS: PERFORIN-DEPENDENT CELL KILLING MECHANISM

Several earlier studies have demonstrated that, upon exposure to CTL, the target membrane becomes leaky to ions and small molecules (Henney, 1974; Martz et al., 1974), suggestive of membrane lesions. The direct evidence for membrane lesions, however, was not available until 1980 when tubular structures with an internal diameter of 15 nm on the surface of erythrocytes attacked by lymphocytes were demonstrated (Dourmashkin et al., 1980). This initial observation of the formation of membrane pores was later confirmed by several research groups, including ours. By using cloned CTL and NK cell lines, it was shown that both cell types could generate ring-like lesions of two distinct diameters, respectively 16 nm and 5 nm, in the membrane of target cells (Podack and Dennert, 1983; Dennert and Podack, 1983). These lesions resembled the tubular lesions produced by complement (C) proteins (Mayer et al., 1981; Bhakdi and Tranum-Jensen, 1984; Muller-Eberhard, 1986).

One of the most striking morphological features of CTLs and NK cells is the presence of numerous large cytoplasmic granules, each of which contains an electron-dense core surrounded by vesicular material. The involvement of these granules in lymphocyte-mediated cell killing was first suggested by the morpho-

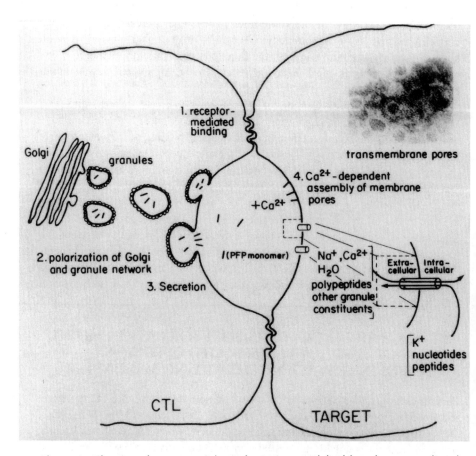

Figure 1. The granule exocytosis/pore formation model of lymphocyte-mediated cytotoxicity. After binding to its target, the lymphocyte degranulates and releases pore forming protein (PFP/perforin) monomers, as well as other granule contents, into the intercellular space. In the presence of Ca^{2+}, PFP/perforin binds to, inserts into the membrane, and polymerizes into transmembrane pores, which in turn, cause colloid-osmotic lysis of the target cell. The inset shows the membrane pores in target membrane by electron microscopy. (Reproduced, with permission, from Young and Cohn, 1986.)

logical studies showing reorientation and accumulation of granules towards the site of target cell contact, upon the conjugation between the effector cell and the target cell (Henkart and Henkart, 1982; Kupfer and Dennert, 1984; Yannelli et al., 1986). Degranulation of the effector lymphocytes was further indicated by the observation that serine esterases, normally located in the granules, were released upon contact with the target cell (Young et al., 1986b). These cytoplasmic granules were subsequently identified as the organelles responsible for causing osmotic lysis of the target cell, since purified granules were capable of forming pore lesions and lysing a variety of tumor cells *in vitro* (Martz, 1984; Millard et al., 1984; Podack and Konisberg, 1984; Young et al., 1986a; Tschopp and Nabholz, 1987). Furthermore, perforin was purified from the cytolytic granules and was shown to mediate the transmembrane channel/pore formation and the lysis of various targets including erythrocytes, nucleated cells, and lipid vesicles (reviewed by Young, 1989; Tschopp and Nabholz, 1990; Podack et al., 1991). These results have led us and others to propose the granule exocytosis/pore formation model of lymphocyte-mediated cytolysis (Figure 1a; Young and Cohn, 1986). According to this model, the contact between effector and target triggers the killer lymphocyte to degranulate and thus release various cytotoxins, namely perforin, onto the target cell surface. Perforin forms transmembrane pores in the target membrane which perturbate the selective permeability of the membrane, resulting in colloid-osmotic lysis of the target cell. The recent report that rat basophilic leukemia cells transfected with the perforin cDNA were able to lyse IgE-conjugated target cells (Shiver and Henkart, 1991) has lent additional support for the granule exocytosis/pore formation model and indicated the importance of perforin in the cell-killing process.

A. Structure of Perforin

To date, perforin has been purified from cytotoxic lymphocytes of mouse, rat, and human (Henkart, 1985; Masson and Tschopp, 1985; Podack et al., 1985; Liu et al., 1986; Young et al., 1986a; Zalman et al., 1986a), and appears to be highly conserved in all three species. Mouse perforin consists of 534 amino acids, with about 70% and 85% identity, respectively, to human and rat perforin (Kwon et al., 1989). Its molecular mass of 66–68 kD (Masson and Tschopp, 1985) or 70 kD (Podack et al., 1985; Young and Cohn, 1986), determined by SDS-PAGE under disulfide bond-reducing conditions, suggests that perforin may be glycosylated at several potential sites. Indeed, we and others have noted that N-glycanase, although not endoglycosidase H, could remove the carbohydrate moieties from the polypeptide backbone of perforin of about 60 kD (Burkhardt et al., 1989; Liu et al., unpublished data).

Perforin has been shown to be immunologically related to the various components of the complement membrane attack complex (MAC), that is, C5b, C6, C7, C8, and C9 (Young et al., 1986c, 1986d; Tschopp et al., 1986). The homology between perforin and complement proteins was further confirmed by the results obtained from the cDNA cloning/sequencing and amino acid sequence comparison (Lichtenheld et al., 1988; Shinkai et al., 1988a; Stanley, 1988; Ishikawa et al., 1989; Kwon et al. 1989; Lowrey et al., 1989). It has been reported that perforin is approximately 20–25% identical to C7, C8, and C9 within a range encompassing 270 amino acid residues located in the central region of the perforin molecule (Kwon et al., 1989). The 40 N-terminal and the 100 C-terminal amino acid residues do not share significant homology with the complement components and appear to be unique to perforin. Notably, perforin contains 20 cysteine residues which are conserved in all three different species (human, mouse, and rat). Seven out of the 8 cysteine residues present between positions 200 and 380 are also preserved in the relevant region of C9 (Kwon et al., 1989). These findings have led us to speculate that similar cysteine-rich domains with structural/functional importance may be present in these two types of pore-forming proteins.

We and others (Peitsch et al., 1990; Ojcius et al., 1991c; Persechini et al., 1992) have attempted to determine the membrane-spanning and the pore-forming domains of perforin. Hydropathic analysis of the deduced amino acid sequence has revealed perforin as a predominantly hydrophilic protein with several hydrophobic and amphipathic regions (Figure 2a). A short sequence, designated $\alpha 1$, comprising amino acid residues 189 to 218, has attracted special attention because of its predicted amphipathic α-helix configuration, a property shared by the homologous regions of C6, C7, C8, and C9. The similarity has led to a postulation that they may represent the pore-forming domains of perforin and complement components. A recent study employing an intramembrane-labeling technique has shown that this may be indeed the case for C9 (Peitsch et al., 1990), although it remains speculative for perforin. Other sequences that could potentially form amphipathic α-helix or β-sheets have also been identified in the central portion of the perforin molecule (Kwon et al., 1989). Two sequences, designated as $\beta 1$ and $\beta 2$, comprising amino acid residues 237–257 and 100–126, respectively, are expected to adopt the amphipathic β-sheet configuration, and thus may also represent candidates for the membrane-binding and/or pore-forming domains. It is noteworthy that $\beta 1$ is flanked by two cysteine residues as well as a cluster of positive and negative charges.

Although the above-mentioned putative α-helical and β-sheet structures have been proposed to be responsible for spanning the target membranes, recent attempts to locate a pore-forming domain(s) in these regions have failed. None of the synthetic peptides mimicking the amino acid sequences of the candidate structures were capable of lysing susceptible cells or lipid vesicles (Ojcius et al., 1991c). Nevertheless, a search outside the complement-homologous region has revealed that peptides comprising the first 34 amino acid residues of the N-termini of both human and murine perforin molecules were capable of lysing nucleated cells as well as lipid vesicles (Figure 2b; Ojcius et al., 1991c). These peptides were also shown to form ion channels in planar lipid bilayers (Ojcius et al., 1991c), suggesting that the N-terminal region may be an important player in membrane binding,

A

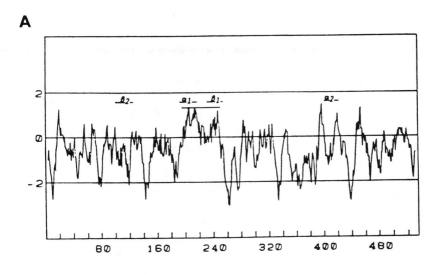

B

α_1	HAYHRLISSY GTHFITAVDL GGRISVLTAL
β_1	LNVEAQVSIG AQASVSSEYK ACEEKKKQHK
β_2	INNDWRVGLD VNPRPEANMR ASVAGSHSK
α_2	RAEHLWGDYT TATDAYLKVF F
MuNP	PCYTATRSEC KQKHKFVPGV WMAGEGMDVT TLRR
HuNP	PCHTAARSEC KRSHKFVPGA WLAGEGVDVT SLRR
CP	YAPQGLLGDP PGNRSGAVW

Figure 2. (**A**) Kyte and Doolittle hydropathy analysis of mouse PFP/perforin using a window of 9 a.a. Hydropathy values are plotted as a function of residue position in the sequence. Positive values are hydrophobic. The region of high hydrophobicity and the candidate α-helical and β-sheet domains are marked. (Reproduced, with permission, from Kwon et al., 1989.) (**B**) Amino acid sequences (one letter code) of the peptides used in the functional mapping studies of PFP/perforin. Hydrophobic residues are underlined, and charged residues are indicated by a sign above the residue. (Reproduced, with permission, from Ojcius et al., 1991c.)

insertion, and polymerization of perforin. However, it may well be that other regions of the perforin molecule, such as α_1, also participate in forming the transmembrane channel/pore structures (Persechini et al., 1992).

It is generally thought that binding of calcium ions (Ca^{2+}) may induce profound conformational change in the perforin molecule, which would render the initially hidden hydrophobic domains exposed and able to interact with the membranes. The putative Ca^{2+}-binding site(s), however, have not been identified. Since the classic Ca^{2+}-binding motif can not be discerned in the deduced amino acid sequence of perforin (Kwon et al., 1989), the Ca^{2+}-dependent perforin function may be associated with certain yet obscure structural motif(s).

B. The Perforin Pore

Perforin monomer is a soluble protein which, in the presence of Ca^{2+} ions, forms aggregates of various sizes, including supramolecular structures with molecular masses exceeding one million daltons that remain undissociated even in the presence of SDS and disulfide bond-reducing reagents (Young et al., 1986a). Under the electron microscope (EM), perforin aggregates resemble tubular structures with internal diameters ranging from 5 nm to 20 nm, with an average diameter of 16 nm (Figure 3a).

When target cells or lipid vesicles are placed in a solution containing perforin and Ca^{2+} ions, perforin molecules will bind to and polymerize in the membranes to form transmembrane tubules, as evidenced by SDS-PAGE and ultrastructural analysis (Young et al., 1986a). Binding and polymerization of the perforin molecules are separate events (Young et al., 1987; Kuta et al., 1991). It has been shown that, at 0 °C, binding of perforin molecules to red blood cell membranes occurred, but no hemolysis could be detected. As the temperature was raised, perforin molecules inserted into the membrane and polymerized to form pores; hemolysis was then observed (Young et al., 1987). The hemolytic activity of perforin as well as its release from isolated granules have been shown to be pH-dependent phenomena which occur optimally within the pH range of 7–8 (Persechini et al., 1989).

Perforin binding and pore-formation are Ca^{2+}-dependent processes that can occur even in vesicles made of purified lipids and do not require any other accessory proteinaceous molecule (Young, 1989; Tschopp and Nabholz, 1990; Podack et al.,

---

Figure 3. Electron microscopic study of PFP/perforin-mediated lysis of cells. (**A**) Tubular lesions formed by PFP/perforin on erythrocyte membranes. Arrows point to top view of tubular structures. Arrowheads point to incompletely polymerized tubules. (Reproduced, with permission, from Young et al., 1986a). (**B**) Schematic comparison of the homopolymeric pore formed by PFP/perforin and the heteropolymeric pore of the complement membrane attack complex. (Reproduced, with permission, from Young and Cohn, 1988.)

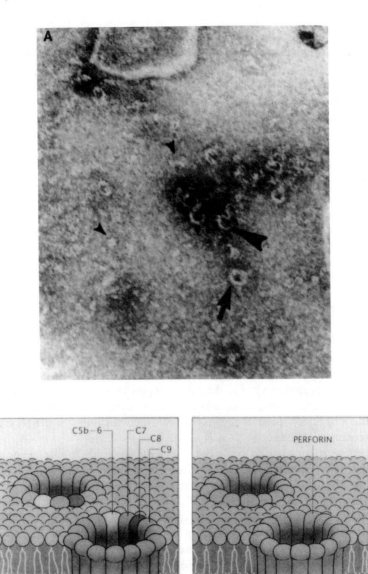

1991). This property contrasts with the action of the complement protein, C9, which usually requires a preassembled C_{5b678} complex as the receptor in membranes in order to form MAC. Another important difference between the complement MAC and the perforin pores is that the latter are composed of only perforin molecules, i.e., they are essentially homopolymers, whereas the complement MAC is a heteropolymeric complex consisting of one molecule each of C5b, C6, C7, and C8 and several C9 molecules (Figure 3b; Young and Cohn, 1988). Moreover, the MAC is smaller than the perforin pore structures, with an internal diameter of approximately 10 nm (Mayer et al., 1981; Bhakdi and Tranum-Jensen, 1984; Muller-Eberhard, 1986).

The large size of perforin pores potentially allows the passage of both low- and high-molecular weight solutes. Several lines of evidence support this view. Osmotic protection experiments showed that small carbohydrates such as glucose, maltose, and arabinose, as well as polyethylene glycol (PEG) with a molecular mass up to 3.4 kD, could not block perforin-mediated hemolysis. PEG with molecular weight of 8 kD, however, effectively blocked lysis of red blood cells by all tested concentrations of perforin, suggesting a cut-off size of about 8 kD for the perforin pores (Persechini et al., unpublished data). Recently, the leakage of several fluorescent dyes through discrete single perforin pores has been monitored using confocal laser microscopy, yielding a functional diameter of approximately 50 Å (Sauer et al., 1991). These data indicate that small intracellular molecules, such as ATP and K^+ ions, may leak from perforin-damaged cells and thus may disturb the intracellular homeostasis. This could contribute to cell death, even in cases where the osmotic stress alone is insufficient. Influx of potentially lethal extracellular molecules, such as Ca^{2+} ions, has also been detected during target–effector interaction or following incubation of targets with purified perforin (Poenie et al., 1987; Persechini et al., 1990).

Perforin pores can also be detected and studied by electrophysiological techniques. These methods can register currents conducted by small ions through high resistance biological and artificial membranes and are sensitive enough to detect the assembly of a single channel (Hille, 1984). Using the whole-cell patch–clamp technique, perforin was shown to be capable of increasing the conductance and depolarizing the cytoplasmic membranes of a number of cell types, where it formed high-conductance pores with heterogeneous sizes ranging from 0.4- to 6-nanoSiemen (nS) (Persechini et al., 1990). This was a rapid phenomenon that occurred within the first 5–30 seconds after adding perforin to the target membranes in a Ca^{2+}-containing medium. At lower concentrations, small unitary ion channels/pores formed. These channels/pores seemed to be cation- and anion-nonselective and voltage-insensitive, as observed in planar lipid bilayer studies (Figure 4) (Young et al., 1986a).

The results described above are consistent with a working model for the action of perforin previously proposed by our group and by other groups (Henkart, 1985; Young and Cohn, 1988; Podack et al., 1991). The model proposes that: (1) perforin

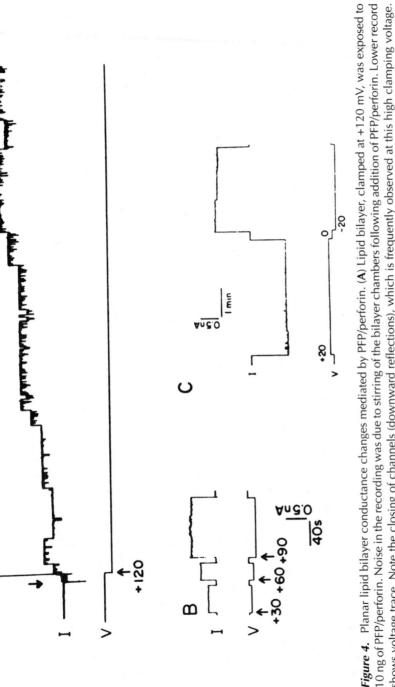

Figure 4. Planar lipid bilayer conductance changes mediated by PFP/perforin. (**A**) Lipid bilayer, clamped at +120 mV, was exposed to 10 ng of PFP/perforin. Noise in the recording was due to stirring of the bilayer chambers following addition of PFP/perforin. Lower record shows voltage trace. Note the closing of channels (downward reflections), which is frequently observed at this high clamping voltage. (**B**) A bilayer as in (**A**) exposed to three voltage pulses of +30-, +60-, and +90-mV (V). (**C**) Bilayer, exposed to 10 ng PFP/perforin, and subjected to two voltage pulses of opposite polarities. The discrete jump in current (I) represents a rare closing of PFP/perforin channel at this low voltage. (Reproduced, with permission, from Young et al., 1986a.)

undergoes conformational changes in the presence of Ca^{2+} ions and exposes the hydrophobic domain(s) to bind and insert into the membrane lipid bilayer; (2) the first membrane-inserted perforin monomers serve as nucleation sites allowing additional inserted perforin molecules to bind and polymerize, through the exposed hydrophobic domain(s), to form ion channels/aqueous pores; and (3) the selective permeability of the cell membrane breaks down eventually leading to osmotic lysis of the target cell. This model also predicts that functionally active channels may form prior to the assembly of the EM-visible pore lesions. It is, therefore, possible that target cells may become leaky or damaged even in the absence of ultrastructurally visible membrane pores. These latter lesions are expected to form only when high concentrations of perforin monomers are deposited onto the target membranes.

In the presence of purified perforin, susceptible cells undergo rapid morphological changes. At the level of light microscopy, bleb formation and swelling could be observed in less than two minutes (Persechini et al., 1990). Under the electron microscope, cytoplasmic boiling and membrane damage could be noticed during the first hour of incubation (Figure 5) (Zheng et al., 1991b). Leakage of intracellular contents (such as hemoglobin, lactate dehydrogenase, and various radiolabeled proteins), an indication of cell lysis, could be detected within 30 min. It is thus reasonable to assume that perforin is, at least partially, responsible for the target cell killing mediated by cytotoxic lymphocytes.

C.   Expression and Regulation of the Perforin Gene

Perforin protein and activity have so far been detected exclusively in cells with cytolytic functions. Distribution of perforin in various cell types has also been investigated by northern blot analysis and *in situ* hybridization using the cDNA probe specific for perforin. A 2.9 Kb transcript was detected in lymphocytes bearing CD8 (CTL marker) or CD16 (Fc receptor, NK cell marker) surface molecules, while it was not found in noncytotoxic and nonlymphocytic cells (Lichtenheld et al., 1988; Shinkai et al., 1988a; Liu et al., 1989; Lowrey et al., 1989; Smyth et al., 1990a). More recently, the perforin transcript has also been shown to be present in γ/δ T cells and some $CD4^+$-lymphocytes with cytotoxic activity (Nakata et al., 1990; Koizumi et al., 1991; Lancki et al., 1991). Several studies have shown that, although perforin transcripts were generally undetectable in resting primary lymphocytes, they could be induced by reagents known to activate cytotoxic lymphocytes, such as interleukins (IL-2, -4, -6, and -7), phorbol esters, lectins, and Ca^{2+}-ionophores (Liu et al., 1989, 1992; Joag et al., 1990; Smyth et al., 1990b, 1991). Taken together, these results suggest that perforin is expressed in a cell type-specific manner and its expression correlates with the functional status of cytotoxic lymphocytes (Kawasaki et al., 1990; Griffiths and Mueller, 1991). It is thus reasonable to consider perforin as a marker for activated cytotoxic lymphocytes. Using antibodies and nucleotide probes specific for perforin, we have been able to extend the

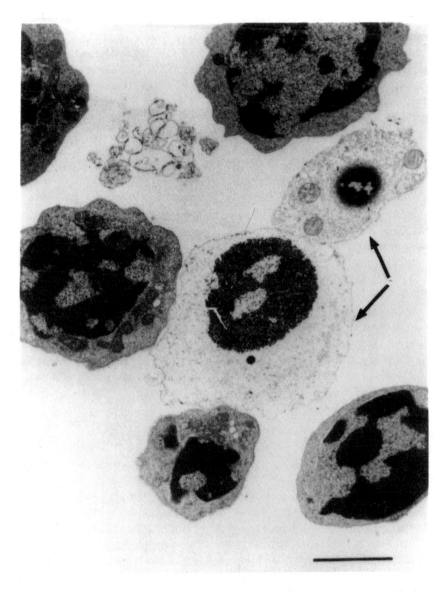

Figure 5. Mouse thymocytes were treated with PFP/perforin for 1 h at 37°C before fixation. Note the obliteration of the cytoplasmic and nuclear ultrastructure in some cells (arrows). (Reproduced, with permission, from Zheng et al., 1991b.)

investigation on perforin expression in tissues and cells derived from various pathological foci as described below.

Both mouse and human perforin genes have been cloned (Lichtenheld and Podack, 1989; Trapani et al., 1990; Youn et al., 1991; Lichtenheld and Podack, 1992). In both species, the perforin gene consists of 3 exons with the protein coding region extending across the second and third exons. Although there is some controversy regarding the 5' flanking sequence and the structure of the mouse perforin gene (Trapani et al., 1990; Youn et al., 1991; Lichtenheld and Podack, 1992), it is generally accepted that there is only one single perforin gene which is located on chromosome 10 (Trapani et al., 1990). The human perforin gene was originally mapped to chromosome 17 (Shinkai et al., 1989), but it has recently been reassigned to chromosome 10 which is syntenic to mouse chromosome 10 (Fink et al., 1992).

The 5' flanking sequences of human and mouse perforin genes share significant homology, suggesting that similar putative regulatory regions/mechanisms may be used by these two species for controlling the expression of the perforin gene. The putative promoter region for mouse perforin gene has recently been localized to the 120 nucleotide residues beyond the transcription initiation site (Lichtenheld and Podack, 1992; Liu et al., unpublished data). Since perforin is expressed exclusively in cytotoxic lymphocytes, transcriptional control appears to be important and may depend upon cell type-specific transcription factor(s) that bind to unique regulatory element(s) of the perforin gene. Several positive and negative regulatory activities have already been identified in the 5' flanking region of the mouse perforin gene (Lichtenheld and Podack, 1992; Koizumi et al., 1993). Identification of the candidate killer cell-specific regulatory motifs and their cognate binding proteins are currently being pursued.

D.   Physiological Role of Perforin

Although perforin is well recognized for its potent cytolytic activity, its precise role in lymphocyte-mediated cell killing has not been completely elucidated. Perforin has been found *in vivo* in $CD8^+$ T cells and NK cells which infiltrate the sites of viral infections (Müller et al., 1989; Young et al., 1989b; Young et al., 1990a), autoimmune reactions (Young et al., 1989a, 1990b), and graft rejections (Griffiths et al., 1991), in γ/δ-T cells circulating in peripheral blood (Nakata et al., 1990; Koizumi et al., 1991), and in peritoneal exudate lymphocytes (Takeuchi et al., 1992). Moreover, a considerable amount of perforin has been detected in a population of NK-like cells, called granulated metrial gland (GMG) cells, residing in the mouse pregnant uterus (Parr et al., 1987; Zheng et al., 1991a), which implies also an active role for this molecule in the physiology of pregnancy. The presence of perforin in the above mentioned cells provides circumstantial support for the role of perforin as a cytolytic mediator which is released by cytotoxic lymphocytes activated *in vivo*. To date, perforin-expressing cells have been found in all the investigated pathological situations where lymphocyte-mediated cytolysis is thought to be involved (Griffiths and Mueller, 1991).

The finding of perforin-expressing cells *in vivo* also argues against an earlier notion that perforin might be a product of the artificial culture conditions for CTLs where a high concentration of IL-2 is present (Berke and Rosen, 1987, 1988; Dennert et al., 1987). It should be pointed out, however, that despite its seemingly obvious importance, perforin is, at most, only partially responsible for the cell death mediated by cytotoxic lymphocytes. Studies have shown that, in addition to membrane damage, cytotoxic cells can also induce apoptosis of target cells with DNA fragmentation and other striking nuclear changes (Russell, 1983; Ju, 1991; Zychlinsky et al., 1991). Purified perforin does not seem to mediate apoptosis since it can neither induce the typical morphological changes associated with apoptosis, such as nuclear segmentation and chromatin condensation, nor cause DNA fragmentation (Duke et al., 1989; Golstein et al., 1991). With respect to these perforin-independent cell-killing pathways, several cytotoxins capable of inducing DNA fragmentation in target cells have recently been purified from cytotoxic lymphocytes (Liu et al., 1987; Tian et al., 1991; Kawakami et al., 1992; Shi et al., 1992a, 1992b). It is likely that multiple cytotoxic mediators/mechanisms are needed to render CTLs and NK cells more effective in killing a wide variety of targets.

IV. RESISTANCE OF CYTOTOXIC LYMPHOCYTES TO PERFORIN-MEDIATED CYTOLYSIS

The fact that cytotoxic lymphocytes can lyse multiple targets in successive cycles (Martz, 1976) suggests that they are refractory to self-mediated lysis. In fact, as opposed to tumor cells and noncytolytic lymphocytes, murine CTL or NK cell lines are not lysed by other killer cells which would recognize them as allogeneic targets (Kranz and Eisen, 1987; Gorman et al., 1988). Likewise, these cells are extremely resistant to either isolated cytolytic granules (Blakely et al., 1987; Verret et al., 1987; Gorman et al., 1988) or purified perforin (Jiang et al., 1988; Shinkai et al., 1988b; Persechini et al., 1990). These observations indicate that cytotoxic lymphocytes possess innate mechanisms to protect themselves from perforin-mediated lysis. Other evidence suggesting a self-protection mechanism includes the absence of morphological changes in CTL treated with high doses of perforin (Persechini et al., 1990), the lack of binding of perforin to CTL (Jiang et al., 1990), and the impaired formation of functional pores on CTL membrane, as evidenced by patch–clamp and Ca^{2+} flux measurements (Persechini et al., 1990).

That the mechanism of self-protection of cytotoxic lymphocytes is confined to the plasma membrane has been suggested by patch–clamp experiments (Persechini et al., 1990). Perforin, at doses that induce large conductance changes in membranes of all susceptible cells, does not increase the conductance in resistant cell lines, indicating the absence of pore formation. Furthermore, cytoplasts ("cells" depleted of intracellular organelles and nucleus) derived from cytotoxic lymphocytes also resist perforin-mediated membrane damage, suggesting that the intracellular components are dispensable for the resistance to perforin (Ojcius et al., 1991a). These studies, taken together, suggest that the inhibition of pore formation in the plasma membrane may be the only event that is determinative for the self-protection phenomenon of cytotoxic lymphocytes.

Since perforin and complement display some similar structural and functional features, it has been questioned whether cells resist both perforin- and complement-mediated lysis through common mechanisms. Along this line, the homologous restriction factor (HRF), a C8-binding membrane protein involved in the protection of cells against complement of the homologous species, has been implicated in perforin resistance (Zalman et al., 1986b, 1988; Martin et al., 1988). Recently, however, some evidence has been provided indicating that HRF, as well as CD59 (Sugita et al., 1988; Davies et al., 1989), another restriction factor of complement, do not account for the resistance against perforin (Jiang et al., 1989a; Meri et al., 1990). Cells that lack HRFs, for example, erythrocytes from patients with paroxysmal nocturnal hemoglobinuria, do not show increased sensitivity to perforin-mediated cytolysis, even though they are more sensitive to homologous complement (Hollander et al., 1989; Jiang et al., 1989b; Krahenbuhl et al., 1989). Furthermore, unlike resistance to complement, resistance to perforin is not restricted to homologous species; CTLs are equally resistant to both homologous and heterologous perforin (Jiang et al., 1988).

Recent studies from our laboratory suggest that a protein present on the surface of cytotoxic cells, which we have named "protectin," is responsible for the resistance of cytotoxic lymphocytes to perforin-mediated lysis (Young and Liu, 1988; Ojcius et al., 1991b). This putative "protectin" is postulated to abolish the insertion and polymerization of perforin molecules following the initial binding of the monomers to the surface. If this protection mechanism is correct, cytotoxic lymphocytes should bind (adsorb) much less perforin than the susceptible target cells in whose membranes multiple perforin molecules are inserted and aggregated. We have indeed observed that in a competition assay the susceptible target cells, but not cytotoxic cells, could adsorb perforin and, therefore, reduce the amount of perforin available for lysing erythrocytes also present in the reaction mixture (Jiang et al., 1990). Pretreatment of CTLs with proteolytic enzymes renders these cells capable of binding perforin and inhibiting lysis of erythrocytes in the same competition assay. This latter finding suggests that the putative molecule ("protectin") responsible for preventing aggregation of perforin in CTL membrane is probably a proteinaceous component. We are now in the process of identifying this "protectin" molecule.

Even though the protection mechanism discussed above may be true, we still cannot rule out the existence of other mechanisms which may involve either membrane repair by shedding or endocytosis of the pores (Morgan et al., 1987; Jones and Morgan, 1991) or some intracellular machinery yet to be identified. Moreover, considering that multiple mediators and mechanisms may be employed by cytotoxic lymphocytes, there may even be other mechanisms protecting lymphocytes against cytotoxins other than perforin.

ACKNOWLEDGMENTS

We wish to thank Drs. Ming-Cheh Liu and Leslie Leonard for critical reading of the manuscript. The work described here was supported by U.S. Public Health Service (NIH) grant CA-47307 and grants from the Cancer Research Institute, and American Heart Association-New York City Affiliate. C.-C. Liu is supported by a Career Scientist Award from Irma Hirschl Trust and an Established Investigatorship from American Heart Association-National Center. M.F. Horta was supported by Fogarty International Center, NIH and National Research Council, CNPq, Brazil.

REFERENCES

Berke, G., & Rosen, D. (1987). Are lytic granules and perforin 1 involved in lysis induced by *in vivo*-primed peritoneal exudate cytolytic T lymphocytes. Transplant. Proc. 19, 412–416.

Berke, G., & Rosen, D. (1988). Highly lytic *in vivo* primed cytolytic T lymphocytes devoid of lytic granules and BLT-esterase activity acquire these constituents in the presence of T cell growth factors upon blast transformation *in vitro*. J. Immunol. 141, 1429–1436.

Bernheimer, A. W. (1990). Cytolytic peptides of sea anemones. In: *Marine Toxins: Origin, Structure, and Molecular Pharmacology*, ACS Symposium Series 418 (Hall, S., & Strichartz, G., Eds.). American Chemical Society, Washington, DC, pp. 305–310.

Bhakdi, S., & Tranum-Jensen, J. (1984). Mechanism of complement cytolysis and the concept of channel-forming proteins. Phil. Trans. R. Soc. Lond. B 306, 311–324.

Blakely, A., Gorman, K., Ostergaard, H., Svoboda, K., Liu, C.-C., Young, J. D.-E., & Clark, W. R. (1987). Resistance of cloned cytotoxic T lymphocytes to cell-mediated cytotoxicity. J. Exp. Med. 166, 1070–1083.

Burkhardt, J. K., Hester, S., & Argon, Y. (1989). Two proteins targeted to the same lytic granule compartment undergo very different posttranslational processing. Proc. Natl. Acad. Sci. USA 86, 7128–7132.

Cavard, D., Crozel, V., Gorvel, J.-P., Pattus, F., Baty, D., & Lazdunski, C. (1980). A molecular, genetic and immunological approach to the functioning of colicin A, a pore-forming protein. J. Mol. Biol. 187, 449–459.

Davies, A., Simmons, D. L., Hale, G., Harrison, R. A., Tighe, H., Lachmann, P. J., & Waldmann, H. (1989). CD59, an Ly-6-like protein expressed in human lymphoid cells, regulates the action of the complement membrane attack complex on homologous cells. J. Exp. Med. 170, 637–654.

Dennert, G., & Podack, E. R. (1983). Cytolysis by H-2-specific T killer cells. Assembly of tubular complexes on target membranes. J. Exp. Med. 157, 1483–1495.

Dennert, G., Anderson, C. G., & Prochazka, G. (1987). High activity of N^αbenzyloxycarbonyl-L-lysine thiobenzyl ester serine esterase and cytolytic perforin in cloned cell lines is not demonstrable in *in vivo*-induced cytotoxic effector cells. Proc. Natl. Acad. Sci. USA 84, 5004–5008.

Dourmashkin, R. R., Deteix, P., Simone, C. B., & Henkart, P. A. (1980). Electron microscopic demonstration of lesions in target cell membranes associated with antibody-dependent cellular cytotoxicity. Clin. Exp. Immunol. 42, 554–560.

Duke, R. C., Persechini, P. M., Chang, S., Liu, C.-C., Cohen, J. J., & Young, J. D.-E. (1989). Purified perforin induces target cell lysis but not DNA fragmentation. J. Exp. Med. 170, 1451–1456.

Ehrenstein, G., & Lecar, H. (1977). Electrically gated ionic channels in lipid bilayers. Q. Rev. Biophys. 10, 1–34.

Fink, T. M., Zimmer, M., Weitz, S., Tschopp, J., Jenne, D. E., & Lichter, P. (1992). Human perforin (HRF1) maps to 10q22, a region that is syntenic with mouse chromosome 10. Genomics1 13, 1300–1302.

Fisch, P., Malkovsky, M., Braakman, E., Sturm, E., Bolhuis, R. L. H., Prieve, A., Sosman, J. A., Lam, V. A., & Sondel, P. M. (1990). γ/δ T cell clones and natural killer cell clones mediate distinct patterns of non-major histocompatibility complex-restricted cytolysis. J. Exp. Med. 171, 1567–1579.

Giorda, R., & Trucco, M. (1991). Mouse NKR-P1. A family of genes selectively co-expressed in adherent lymphokine-activated killer cells. J. Immunol. 147, 1701–1708.

Golstein, P., Ojcius, D. M., & Young, J. D.-E. (1991). Cell death mechanisms and the immune system. Immunol. Rev. 121, 29–65.

Gorman, K., Liu, C.-C., Blakely, A., Young, J. D.-E., Torbett, B. E., & Clark, W. R. (1988). Cloned cytotoxic T lymphocytes as target cells. II. Polarity of lysis revisited. J. Immunol. 141, 2211–2215.

Griffiths, G. M., & Mueller, C. (1991). Expression of perforin and granzymes in vivo: Potential diagnostic markers for activated cytotoxic cells. Immunol. Today 12, 415–419.

Griffiths, G. M., Namikawa, R., Muller, C., Liu, C-C., Young, J. D-E., Billingham, M., & Weissman, I. (1991). Granzyme A and perforin as markers for rejection in cardiac transplantation. Eur. J. Immunol. 21, 687–692.

Henkart, M. P., & Henkart, P. A. (1982). Lymphocyte mediated cytolysis as a secretory phenomenon. Adv. Exp. Med. Biol. 146, 227–247.

Henkart, P. A. (1985). Mechanism of lymphocyte-mediated cytotoxicity. Annu. Rev. Immunol. 3, 31–58.

Henney, C. S. (1974). Estimation of the size of a T-cell-induced lytic lesion. Nature 249, 456–458.

Herberman, R. B., Reynolds, C. W., & Ortaldo, J. R. (1986). Mechanism of cytotoxicity by natural killer (NK) cells. Annu. Rev. Immunol. 4, 651–680.

Hille, B. (1984). Ionic Channels of Excitable Membranes, 1st Ed. Sinauer Assoc., Inc., Sunderland.

Hollander, N., Shin, M. L., Rosse, W. F., & Springer, T. A. (1989). Distinct restriction of complement- and cell-mediated lysis. J. Immunol. 142, 3913–3916.

Ishikawa, H., Shinkai, Y.-I., Yagita, H., Yue, C. C., Henkart, P. A., Sawada, S., Young, H. A., Reynolds, C. W., & Okumura, K. (1989). Molecular cloning of rat cytolysin. J. Immunol. 143, 3069–3073.

Jiang, S., Persechini, P. M., Zychlinsky, A., Liu, C.-C., Perussia, B., & Young, J. D.-E. (1988). Resistance of cytolytic lymphocytes to perforin-mediated killing: Lack of correlation with complement-associated homologous species restriction. J. Exp. Med. 168, 2207–2219.

Jiang, S., Persechini, P. M., Perussia, B., & Young, J. D.-E. (1989a). Resistance of cytolytic lymphocytes to perforin-mediated killing: Murine cytotoxic T lymphocytes and human natural killer cells do not contain functional soluble homologous restriction factor or other specific soluble protective factors. J. Immunol. 143, 1453–1460.

Jiang, S., Persechini, P. M., Rosse, W. F., Perussia, B., & Young, J. D.-E. (1989b). Differential susceptibility of type III erythrocytes of paroxysmal nocturnal hemoglobinuria to lysis mediated by complement and perforin. Biochem. Biophys. Res. Commun. 162, 316–325.

Jiang, S., Ojcius, D. M., Persechini, P. M., & Young, J. D.-E. (1990). Resistance of lymphocytes to perforin-mediated killing; inhibition of perforin binding activity by surface membrane proteins. J. Immunol. 144, 998–1003.

Joag, S. V., Liu, C.-C., Kwon, B. S., Clark, W. R., & Young, J. D.-E. (1990). Expression of mRNAs for pore-forming proteins and two serine esterases in murine primary and cloned effector lymphocytes. J. Cell. Biochem. 43, 81–88.

Jones, J., & Morgan, P. (1991). Killing of cells by perforin. Resistance to killing is not due to diminished binding of perforin to the cell membrane. Biochem. J. 280, 199–204.

Ju, S.-T. (1991). Distinct pathways of CD4 and CD8 cells induce rapid target DNA fragmentation. J. Immunol. 146, 812–818.

Kawakami, A., Tian, Q., Duan, X., Streuli, M., Schoolman, S. F., & Anderson, P. (1992). Identification and functional characterization of a TIA-1-related nucleolysin. Proc. Natl. Acad. Sci. USA 89, 8681–8685.

Kawasaki, A., Shinkai, Y., Kuwana, Y., Furuya, A., Yutaka, I., Hanai, N., Itoh, S., Yagita, H., & Okumura, K. (1990). Perforin, a pore-forming protein detectable by monoclonal antibodies, is a functional marker for killer cells. Int. Immunol. 2, 677–684.

Koizumi, H., Liu, C.-C., Zheng, L. M., Joag, S. V., Bayne, N. K., Holoshitz, J., & Young, J. D.-E. (1991). Expression of perforin and serine esterases by γ/δ T-cells. J. Exp. Med. 173, 499–502.

Koizumi, H., Horta, M. F., Youn, B-C., Fu, K-C., Kwon, B. S., Young, J. D-E., & Liu, C-C. (1993). Identification of a killer cell-specific regulatory element of the mouse perforin gene: An Ets-binding site-homologous motif that interacts with Ets-related proteins. Mol. Cell. Biol. 13, 6690–6701.

Krahenbuhl, O. P., Peter, H. H., & Tschopp, J. (1989). Absence of homologous restriction factor does not affect CTL-mediated cytolysis. Eur. J. Immunol. 19, 217–219.

Kranz, D. M., & Eisen, H. N. (1987). Resistance of cytotoxic T lymphocytes to lysis by a clone of cytotoxic T lymphocytes. Proc. Natl. Acad. Sci. USA 84, 3375–3379.

Kupfer, A., & Dennert, G. (1984). Reorientation of the microtubule-organizing center and the Golgi apparatus in cloned cytotoxic lymphocytes triggered by binding to lysable target cells. J. Immunol. 133, 2762–2766.

Kuta, A., Bashford, C. L., Pasternak, C. A., Reynolds, C. W., & Henkart, P. A. (1991). Characterization of non-lytic cytolysin-membrane intermediates. Mol. Immunol. 28, 1263–1270.

Kwon, B. S., Wakulchik, M., Liu, C.-C., Persechini, P. M., Trapani, J. A., Haq, A. K., Kim, Y., & Young, J. D.-E. (1989). The structure of the lymphocyte pore-forming protein perforin. Biochem. Biophys. Res. Commun. 158, 1–10.

Lancki, D. W., Hsieh, C.-S., & Fitch, F. W. (1991). Mechanisms of lysis by cytotoxic T lymphocyte clones. Lytic activity and gene expression in cloned antigen-specific CD4+ and CD8+ lymphocytes. J. Immunol. 146, 3242–3249.

Lehrer, R. I., Ganz, T., & Selsted, M. E. (1991). Defensins: Endogenous antibiotic peptides of animal cells. Cell 64, 229–230.

Lichtenheld, M. G., Olsen, K., Lu, P., Lowrey, D. M., Hameed, A., Hengartner, H., & Podack, E. R. (1988). Structure and function of human perforin. Nature 335, 448–451.

Lichtenheld, M. G., & Podack, E. R. (1989). Structure of the human perforin gene. A simple gene organization with interesting potential regulatory sequences. J. Immunol. 143, 4267–4274.

Lichtenheld, M. G., & Podack, E. R. (1992). Structure and function of the murine perforin promoter and upstream region: Reciprocal gene activation or silencing in perforin positive and negative cells. J. Immunol. 149, 2619–2626.

Liu, C.-C., Perussia, B., Cohn, Z. A., & Young, J. D.-E. (1986). Identification and characterization of a pore-forming protein of human peripheral blood natural killer cells. J. Exp. Med. 164, 2061–2076.

Liu, C.-C., Steffen, M., King, F., & Young, J. D.-E. (1987). Identification, isolation, and characterization of a novel cytotoxin in murine cytolytic lymphocytes. Cell 51, 393–403.

Liu, C.-C., Rafii, S., Granelli-Piperno, A., Trapani, J. A., & Young, J. D.-E. (1989). Perforin and serine esterase gene expression in stimulated human T cells: Kinetics, mitogen requirements, and effect of cyclosporin A. J. Exp. Med. 170, 2105–2118.

Lowrey, D. M., Aebischer, T., Olsen, K., Lichtenheld, M., Rupp, F., Hengartner, H., & Podack, E. R. (1989). Cloning, analysis, and expression of murine perforin 1 cDNA, a component of cytolytic T-cell granules with homology to complement component C9. Proc. Natl. Acad. Sci. USA 86, 247–251.

Lu, P., Garcia-Sanz, J. A., Lichtenheld, M. G., & Podack, E. R. (1992). Perforin expression in human peripheral blood mononuclear cells. Definition of an IL-2-independent pathway of perforin induction in CD8+ T cells. J. Immunol. 148, 3354–3360.

Martin, D. E., Zalman, L. S., & Müller-Eberhard, H. J. (1988). Induction of expression of cell-surface homologous restriction factor upon anti-CD3 stimulation of human peripheral lymphocytes. Proc. Natl. Acad. Sci. USA 85, 213–217.

Martz, E., Burakoff, S. J., & Benacerraf, B. (1974). Interruption of the sequential release of small and large molecules from tumor cells by low temperature during cytolysis mediated by immune T-cells or complement. Proc. Natl. Acad. Sci. USA 71, 177–181.

Martz, E. (1976). Multiple target cell killing by the cytolytic T lymphocytes and the mechanism of cytotoxicity. Transplantation 21, 5–11.

Martz, E. (1984). Lytic granules, adhesion molecules, and other recent insights. Immunol. Today 5, 254–255.

Masson, D., & Tschopp, J. (1985). Isolation of a lytic, pore-forming protein (Perforin) from cytolytic T-lymphocytes. J. Biol. Chem. 260, 9069–9072.

Massotte, F. P. D., Wilmsen, H. U., Lakey, J., Tsernoglou, D., Tucker, A., & Parker, M. W. (1990). Colicins: Prokaryotic killer-pores. Experientia 46, 180–192.

Mayer, M. M., Michaels, D. W., Ramm, L. E., Whitlow, M. B., Willoughby, J. B., & Shin, M. L. (1981). Membrane damage by complement. CRC Crit. Rev. Immunol. 1, 133–165.

Meri, S., Morgan, B. P., Wing, G., Jones, J., Davies, A., Podack, E. R., & Lachmann, P. J. (1990). Human protectin (CD59), an 18–20-kD homologous complement restriction factor, does not restrict perforin-mediated lysis. J. Exp. Med. 172, 367–370.

Millard, P. J., Henkart, M. P., Reynolds, C. W., & Henkart, P. A. (1984). Purification and properties of cytoplasmic granules from cytotoxic rat LGL tumors. J. Immunol. 132, 3197–3204.

Morgan, P., Dankert, J. R., & Esser, A. F. (1987). Recovery of human neutrophils from complement attack: Removal of the membrane attack complex by endocytosis and exocytosis. J. Immunol. 38, 246–253.

Müller, C., Kägi, D., Aebischer, T., Odermatt, B., Held, W., Podack, E. R., Zinkernagel, R., & Hengartner, H. (1989). Detection of perforin and granzyme A mRNA in infiltrating cells during infection of mice with lymphocytic choriomeningitis virus. Eur. J. Immunol. 19, 1253–1259.

Müller-Eberhard, H. J. (1986). The membrane attack complex of complement. Annu. Rev. Immunol. 4, 503–528.

Nakata, M., Smyth, M. J., Norihisa, Y., Kawasaki, A., Shinkai, Y., Okumura, K., & Yagita, H. (1990). Constitutive expression of pore-forming protein in peripheral blood γ/δ T cells: Implication for their cytotoxic role in vivo. J. Exp. Med. 172, 1877–1880.

Ojcius, D. M., Jiang, S., Persechini, P. M., Detmers, P. A., & Young, J. D.-E. (1991a). Cytoplasts from cytotoxic T lymphocytes are resistant to perforin-mediated lysis. Mol. Immunol. 28, 1011–1018.

Ojcius, D. M., Muller, S., Hasselkus-Light, C. S., Young, J. D.-E., & Jiang, S. (1991b). Plasma membrane-associated proteins with the ability to partially inhibit perforin-mediated lysis. Immunol. Let. 28, 101–108.

Ojcius, D. M., Persechini, P. M., Zheng, L. M., Notaroberto, P. C., Adeodato, S. C., & Young, J. D.-E. (1991c). Cytolytic and ion channel-forming properties of the N terminus of lymphocyte perforin. Proc. Natl. Acad. Sci. USA 88, 4621–4625.

Ojcius, D. M., & Young, J. D.-E. (1991). Cytolytic pore-forming proteins and peptides: Is there a common structural motif? TIBS 16, 225–229.

Parr, E. L., Parr, M. B., & Young, J. D.-E. (1987). Localization of a pore-forming protein (perforin) in granulated metrial gland cells. Biol. Reprod. 37, 1327–1335.

Peitsch, M. C., Amiguet, P., Guy, R., Brunner, J., Maizel, J. V., Jr., & Tschopp, J. (1990). Localization and molecular modelling of the membrane-ionserted domain of the ninth component of human complement and perforin. Mol. Immunol. 27, 589–602.

Persechini, P. M., Liu, C.-C., Jiang, S., & Young, J. D.-E. (1989). The lymphocyte pore-forming protein perforin is associated with granules by a pH-dependent mechanism. Immunol. Lett. 22, 23–28.

Persechini, P. M., Young, J. D.-E., & Almers, W. (1990). Membrane channel formation by the lymphocyte pore-forming protein: Comparison between susceptible and resistant target cells. J. Cell Biol. 110, 2109–2116.

Persechini, P. M., Ojcius, D. M., Adeodato, S. C., Notaroberto, P. C., Daniel, C. B., & Young, J. D.-E. (1992). Channel-forming activity of the perforin N-terminus and a putative alpha-helical region homologous with complement C9. Biochemistry 31, 5017–5021.

Podack, E. R., & Dennert, G. (1983). Assembly of two types of tubules with putative cytolytic function by cloned natural killer cells. Nature 302, 442–445.

Podack, E. R., & Konigsberg, P. J. (1984). Cytolytic T cell granules. Isolation, structural, biochemical, and functional characterization. J. Exp. Med. 160, 695–710.

Podack, E. R., Young, J. D.-E., & Cohn, Z. A. (1985). Isolation and biochemical and functional characterization of perforin 1 from cytolytic T-cell granules. Proc. Natl. Acad. Sci. USA 82, 8629–8633.

Podack, E. R., Hengartner, H., & Lichtenheld, M. G. (1991). A central role of perforin in cytolysis? Annu. Rev. Immunol. 9, 129–157.

Poenie, M., Tsien, R. Y., & Schmitt-Verhulst, A.-M. (1987). Sequential activation and lethal hit measured by $[Ca^{2+}]_i$ in individual cytolytic T cells and targets. EMBO J. 6, 2223–2232.

Rosenberg, S. A. (1988). Immunotherapy of cancer using interleukin 2. Immunol. Today 9, 58–62.

Russell, J. H. (1983). Internal disintegration model of cytotoxic lymphocyte-induced target damage. Immunol. Rev. 72, 97–118.

Ryan, J. C., Niemi, E. C., Goldfien, R. D., Hiserodt, J. C., & Seaman, W. E. (1991). NKR-P1, an activating molecule on rat natural killer cells, stimulates phosphoinositide turnover and a rise in intracellular calcium. J. Immunol. 147, 3244–3250.

Saito, T., Pardoll, M. D., Fowlkes, J. B., & Ohno, H. (1990). A murine thymocyte clone expressing γδ T cell receptor mediates natural killer-like cytolytic function and TH1-like lymphokine production. Cell. Immunol. 131, 284–301.

Sauer, H., Pratsch, L., Tschopp, J., Bhakdi, S., & Peters, R. (1991). Functional size of complement and perforin pores compared by confocal laser scanning microscopy and fluorescence micropholysis. Biochim. Biophys. Acta 1063, 137–146.

Shi, L., Kam, C.-M., Powers, J. C., Aebersold, R., & Greenberg, A. H. (1992a). Purification of three cytotoxic lymphocyte granule serine proteases that induce apoptosis through distinct substrate and target cell interaction. J. Exp. Med. 176, 1521–1529.

Shi, L., Kraut, R. P., Aebersold, R., & Greenberg, A. H. (1992b). A natural killer cell granule protein that induces DNA fragmentation and apoptosis. J. Exp. Med. 175, 553.

Shinkai, Y., Takio, K., & Okumura, K. (1988a). Homology of perforin to the ninth component of complement (C9). Nature 334, 525–527.

Shinkai, Y., Ishikawa, H., Hattori, M., & Okumura, K. (1988b). Resistance of mouse cytolytic cells to pore-forming protein-mediated cytolysis. Eur. J. Immunol. 18, 29–33.

Shinkai, Y., Yoshida, M., Maeda, K., Kobata, T., Maruyama, K., Yodoi, J., Yagita, H., & Okumura, K. (1989). Molecular cloning and chromosomal assignment of a human perforin (PFP) gene. Immunogenetics 30, 452–457.

Shiver, J. W., & Henkart, P. A. (1991). A noncytotoxic mast cell tumor line exhibits potent IgE-dependent cytotoxicity after transfection with the cytolysin/perforin gene. Cell 64, 1175–1181.

Smyth, M. J., Ortaldo, J. R., Shinkai, Y.-I., Yagita, H., Nakata, M., Okumura, K., & Young, H. A. (1990a). Interleukin-2 induction of pore-forming protein gene expression in human peripheral blood CD8[+] T cells. J. Exp. Med. 171, 1269–1281.

Smyth, M. J., Ortaldo, J. R., Bere, W., Yagita, H., Okumura, K., & Young, H. A. (1990b). IL-2 and IL-6 synergize to augment the pore-forming protein gene expression and cytotoxic potential of human peripheral blood T cells. J. Immunol. 145, 1159–1166.

Smyth, M. J., Norihisa, Y., Gerard, J. R., Young, H. A., & Ortaldo, J. R. (1991). Il-7 regulation of cytotoxic lymphocytes: Pore-forming protein gene expression, interferon-gamma production, and cytotoxicity of human peripheral blood lymphocyte subsets. Cell. Immunol. 138, 390–403.

Sprent, J., & Webb, S. R. (1987). Function and specificity of T cell subsets in the mouse. Adv. Immunol. 41, 39–133.

Stanley, K. K. (1988). The molecular mechanism of complement C9 insertion and polymerization in biological membranes. Curr. Top. Microbiol. Immunol. 140, 49–65.

Sugita, Y., Nakano, Y., & Tomita, M. (1988). Isolation from human erythrocytes of a new membrane protein which inhibits the formation of complement transmembrane channels. J. Biochem. 104, 633–637.

Takeuchi, Y., Nishimura, T., Gao, X., Watanabe, K., Akatsuka, A., Shinkai, Y., Okumura, K., & Habu, S. (1992). Perforin is expressed in CTL populations generated in vivo. Immunol. Lett. 31, 183–188.

Tian, Q., Streuli, M., Saito, H., Schlossman, S. F., & Anderson, P. (1991). A polyadenylate binding protein localized to the granules of cytolytic lymphocytes induces DNA fragmentation in target cells. Cell 67, 629.

Trapani, J. A., Kwon, B. S., Kozak, C. A., Chintamaneni, C., Young, J. D.-E., & Dupont, B. (1990). Genomic organization of the pore-forming protein (perforin) gene and localization to chromosome 10; similarities to and differences from C9. J. Exp. Med. 171, 545–557.

Trinchieri, G., & Perussia, B. (1984). Human natural killer cells: Biologic and pathologic aspects. Lab. Invest. 50, 489–513.

Tschopp, J., Masson, D., & Stanley, K. K. (1986). Structural/functional similarity between proteins involved in complement- and cytotoxic T-lymphocyte-mediated cytolysis. Nature 322, 831–834.

Tschopp, J., & Nabholz, M. (1987). The role of cytoplasmic granule components in cytolytic lympho-cyte-mediated cytolysis. Ann. Inst. Pasteur/Immunol. 138, 290–296.

Tschopp, J., & Nabholz, M. (1990). Perforin-mediated target cell lysis by cytolytic T lymphocytes. Annu. Rev. Immunol. 8, 279–302.

Verret, C. R., Firmenich, A. A., Kranz, D. M., & Eisen, H. N. (1987). Resistance of cytotoxic T lymphocytes to the lytic effects of their toxic granules. J. Exp. Med. 166, 1536–1547.

Waver, T. C., & Unanue, E. R. (1990). The costimulatory function of antigen presenting cells. Immunol. Today 11, 49–54.

Yannelli, J. R., Sullivan, J. A., Mandell, G. L., & Engelhard, V. H. (1986). Reorientation and fusion of cytotoxic T lymphocyte granules after interaction with target cells as determined by high resolution cinemicrography. J. Immunol. 136, 377–382.

Yokoyama, W. M. (1993). Recognition structures on natural killer cells. Curr. Opinion in Immunol. 5, 67–73.

Youn, B.-S., Liu, C.-C., Kim, K.-K., Young, J. D.-E., Kwon, M. H., & Kwon, B. S. (1991). Structure of the mouse pore-forming protein (perforin) gene: Analysis of transcription initiation site, 5' flanking sequence, and alternative splicing of 5' untranslated regions. J. Exp. Med. 173, 813–822.

Young, J. D.-E., Young, T. M., Lu, L. P., Unkeless, J. C., & Cohn, Z. A. (1982). Characterization of a membrane pore-forming protein from Entamoeba histolytica. J. Exp. Med. 156, 1677–1690.

Young, J. D.-E., & Cohn, Z. A. (1986). Cell-mediated killing: A common mechanism? Cell 46, 641–642.

Young, J. D.-E., Hengartner, H., Podack, E. R., & Cohn, Z. A. (1986a). Purification and characterization of a cytolytic pore-forming protein from granules of cloned lymphocytes with natural killer activity. Cell 44, 849–859.

Young, J. D.-E., Leong, L. G., Liu, C.-C., Daminao, A., Wall, D. A., & Cohn, Z. A. (1986b). Isolation and characterization of a serine esterase from cytolytic T cell granules. Cell 47, 183–194.

Young, J. D.-E., Liu, C.-C., Leong, L. G., & Cohn, Z. A. (1986c). The pore-forming protein (Perforin) of cytolytic T lymphocytes is immunologically related to the components of membrane attack complex of complement through cysteine-rich domains. J. Exp. Med. 164, 2077–2082.

Young, J. D.-E., Cohn, Z. A., & Podack, E. R. (1986d). The ninth component of complement and the pore-forming protein (Perforin 1) from cytotoxic T cells: Structural, immunological, and functional similarities. Science 233, 184–190.

Young, J. D.-E., Damiano, A., DiNome, M. A., Leong, L. G., & Cohn, Z. A. (1987). Dissociation of membrane binding and lytic activities of the lymphocyte pore-forming protein (perforin). J. Exp. Med. 165, 1371–1382.

Young, J. D.-E., & Cohn, Z. A. (1988). How killer cells kill. Sci. Amer. 256, 38–44.

Young, J. D.-E., & Liu, C.-C. (1988). How do cytotoxic T lymphocytes avoid self-lysis? Immunol. Today 9, 14–15.

Young, J. D.-E. (1989). Killing of target cells by lymphocytes: A mechanistic view. Physiol. Rev. 69, 250–314.

Young, L. H. Y., Peterson, L. B., Wicker, L. S., Persechini, P. M., & Young, J. D.-E. (1989a). *In vivo* expression of perforin by CD8+ lymphocytes in autoimmune disease: Studies on spontaneous and adoptively transferred diabetes in nonobese diabetic mice. J. Immunol. 143, 3994–3999.

Young, L. H. Y., Klavinskis, L. S., Oldstone, M. B. A., & Young, J. D.-E. (1989b). *In vivo* expression of perforin by CD8+ lymphocytes during an acute viral infection. J. Exp. Med. 169, 2159–2171.

Young, L. H. Y., Foster, C. S., & Young, J. D.-E. (1990a). *In vivo* expression of perforin by natural killer cells during a viral infection. Am. J. Pathol. 136, 1021–1030.

Young, L. H. Y., Joag, S. V., Zheng, L. M., Lee, C.-P., & Young, J. D.-E. (1990b). Perforin-mediated myocardial damage in acute myocarditis. Lancet 336, 1019–1021.

Zalman, L. S., Brothers, M. A., Chiu, F. J., & Müller-Eberhard, H. J. (1986a). Mechanism of cytotoxicity of human large granular lymphocytes: Relationship of the cytotoxic lymphocyte protein to the ninth component (C9) of human complement. Proc. Natl. Acad. Sci. USA 83, 5262–5266.

Zalman, L. S., Wood, L. M., & Müller-Eberhard, H. J. (1986b). Isolation of a human erythrocyte membrane protein capable of inhibiting expression of homologous complement transmembrane channels. Proc. Natl. Acad. Sci. USA 83, 6975–6979.

Zalman, L. S., Brothers, M. A., & Müller-Eberhard, H. J. (1988). Self-protection of cytotoxic lymphocytes: A soluble form of homologous restriction factor in cytoplasmic granules. Proc. Natl. Acad. Sci. USA 85, 4827–4831.

Zheng, L. M., Liu, C.-C., Ojcius, D. M., & Young, J. D.-E. (1991a). Expression of lymphocyte perforin in the mouse uterus during pregnancy. J. Cell Sci. 99, 317–323.

Zheng, L. M., Zychlinsky, A., Liu, C.-C., Ojcius, D. M., & Young, J. D.-E. (1991b). Extracellular ATP as a trigger for apoptosis or programmed cell death. J. Cell Biol. 112, 279–288.

Zychlinsky, A., Joag, S. V., Liu, C.-C., & Young, J. D.-E. (1990). Cytotoxic mechanisms of murine lymphokine-activated killer cells: Functional and biochemical characterization of homogeneous populations of spleen LAK cells. Cell. Immunol. 126, 377–390.

Zychlinsky, A., Zheng, L. M., Liu, C.-C., & Young, J. D.-E. (1991). Cytotoxic lymphocytes can induce both necrosis and apoptosis. J. Immunol. 146, 393–400.

THE ASIALOGLYCOPROTEIN RECEPTOR

Christian Fuhrer and Martin Spiess

Biomembranes
Volume 4, pages 175–199.
Copyright © 1996 by JAI Press Inc.
All rights of reproduction in any form reserved.
ISBN: 1-55938-661-4.

I. INTRODUCTION

An important function of hepatocytes is the selective removal of many different substances from the circulation for intracellular processing or degradation. Specific surface receptors promote the binding of serum macromolecules and their subsequent internalization by receptor-mediated endocytosis. Among the best characterized endocytic transporters is the hepatic asialoglycoprotein (ASGP) receptor, also called galactose (or galactose/N-acetyl galactosamine) receptor or hepatic lectin. The ASGP receptor was discovered in studies on the metabolism of serum glycoproteins in mammals (reviewed by Ashwell and Morell, 1974; Ashwell and Harford, 1982). After enzymatic desialylation, glycoproteins carrying N-linked oligosaccharides are rapidly cleared from the circulation, and accumulate and are degraded in parenchymal cells of the liver. Recognition of ASGPs depends on the normally penultimate galactose residues which are exposed upon desialylation. Interestingly, birds, whose serum proteins are largely unsialylated, have an analogous hepatic receptor specific for agalactoglycoproteins, that is, glycoproteins with terminal N-acetyl glucosamine residues (Lunney and Ashwell, 1976; Drickamer, 1981).

Expression of functional ASGP receptors is virtually unique to hepatocytes (Hubbard and Stukenbrok, 1979; Hubbard et al., 1979). Liver macrophages contain an unrelated galactose-specific receptor with a preference for large ligand particles (Schlepper-Schäfer et al., 1986; Kempka et al., 1990). Peritoneal macrophages, however, express an ASGP receptor that is homologous to, but distinct from, the hepatic ASGP receptor (Ii et al., 1990). This macrophage ASGP receptor may be involved in the binding of tumoricidal macrophages to tumor cells (Oda et al., 1989). Most of the information covered in this review originates from studies of the hepatic ASGP receptor, and it will be clearly stated when data is derived from the macrophage system.

The ASGP receptor is a typical transport receptor. Circulating ASGPs are concentrated on the cell surface by high-affinity binding to the receptor. Ligand–receptor complexes are clustered into clathrin-coated domains of the plasma membrane. The coated pits rapidly invaginate and pinch off as coated vesicles. Upon uncoating and fusion of the vesicles with endosomes, the complexes are dissociated in the acidic environment. The receptors are recycled to the cell surface, while ligand molecules are delivered to lysosomes by fluid-phase transport.

A likely physiological role of the hepatic ASGP receptor is to provide the degradative pathway in serum glycoprotein homeostasis. Consistent with this notion is a correlation (of potential diagnostic value) between increased levels of circulating ASGPs and reduced ASGP receptor activity in several liver diseases (cirrhosis, hepatocarcinomas, and various types of hepatitis; reviewed by Sawamura and Shiozaki, 1991). However, this mechanism is limited to glycoproteins with exposed tri- or tetra-antennary oligosaccharides, as required for efficient binding to the receptor. It is conceivable that there are also other specific ligands that need to be targeted to the liver for purposes other than mere degradation. There is some evidence that the ASGP receptor may contribute to the hepatic uptake of immunoglobulin A (Daniels et al., 1989) and chylomicron remnants (Windler et al., 1991). It has also been speculated that the natural ligands may include yet unidentified galactose-exposing glycoproteins (potentially involved in cell–cell or cell–matrix interactions) whose free concentrations need to be kept low (Weigel, 1992).

While all physiological ligands of the ASGP receptor may not be known at present, the receptor is being used to target exogenous molecules specifically to hepatocytes for diagnostic and therapeutic purposes. Reporter molecules are coupled to galactosylated carrier substances and injected into the circulation to distinguish between normal liver tissue and damaged liver tissue with reduced ASGP receptor levels. Technetium-99m coupled to ASGPs or galactosylated albumin has been used for scintigraphic imaging (e.g., Makdisi and Versland, 1991; Kudo et al., 1993), and arabinogalactan-coated ultrasmall superparamagnetic iron oxide for magnetic resonance imaging (e.g., Reimer et al., 1991). Similarly, pharmacological drugs have been coupled to ASGP receptor ligands or packaged into derivatized liposomes to develop liver-specific delivery systems (e.g., Ishihara et al., 1991; Jansen et al., 1993). DNA complexed to polylysine-coupled ASGPs has successfully been used for targeted gene transfer and gene therapy (Wu and Wu, 1991; Wu et al., 1991; Wilson et al., 1992).

In the two decades since its discovery, the intracellular itinerary of the ASGP receptor and its ligands has been characterized in great detail. The receptor proteins have been purified, and the corresponding cDNAs and one of the genes have been cloned and sequenced. *In vitro* mutagenesis and expression of wild-type and mutant receptor proteins in heterologous systems is beginning to reveal some of the signals required for the correct routing and functioning of the receptor. The aim of this chapter is to summarize what is currently known about the ASGP receptor with emphasis on the structural and molecular aspects.

II. INTRACELLULAR ROUTING OF RECEPTOR AND LIGAND

The ASGP receptor has been studied most extensively in perfused rat liver, isolated rat hepatocytes, and HepG2 cells, a human hepatoma cell line expressing the native receptor. Parenchymal hepatocytes contain 100,000–500,000 ASGP binding sites per cell. Surface receptors are randomly distributed over the basolateral plasma membrane domain facing the capillaries and are enriched in clathrin-coated pits, but they are essentially absent from the apical membrane facing the bile canaliculi (Wall and Hubbard, 1981; Matsuura et al., 1982). The minimum cycle time for the ASGP receptor in HepG2 cells at high ligand concentration is approximately 7 min. Ligand binding occurs within approximately 1 min and internalization with a mean time of 2 min; an additional 4 min are required for ligand dissociation and reappearance of receptor at the cell surface (Schwartz et al., 1982). With a mean lifetime of 30 h, a single receptor can, therefore, internalize up to 250 ligand molecules.

The intracellular pathways of receptor and ligand (illustrated in Figure 1) have been delineated in detail using electron microscopy (Geuze et al., 1983a,b, 1984b; Mueller and Hubbard, 1986; Stoorvogel et al., 1989, and others referenced by Breitfeld et al., 1985). ASGPs and their receptor have been found in coated pits, coated and uncoated vesicles, and peripheral, early endosomes together with other transport receptors, such as those for mannose-6-phosphate (M6P), transferrin and polymeric immunoglobulins (poly-Ig). Early endosomes are important sorting organelles at the crossroads of several transport routes: to and from the plasma membrane and the *trans*-Golgi, via late endosomes to lysosomes, and transcytosis. The endosomal compartment that contains both ASGP receptor and ligand has a tubulo-vesicular morphology. Entry of ligand–receptor complexes seems to occur from tubular regions. In rat liver cells subjected for 60 min to infusion of asialo-

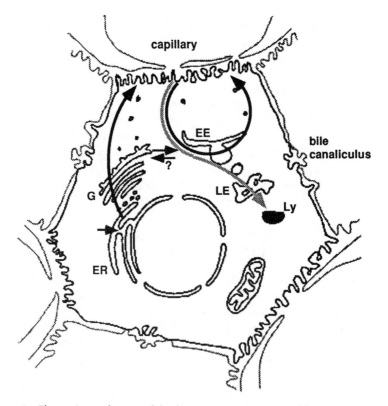

Figure 1. The major pathways of the hepatic ASGP receptor (black arrows) and its ligands (gray arrow). (ER, endoplasmic reticulum; G, Golgi apparatus; EE, early endosomes; LE, late endosomes; Ly, lysosomes.)

orosomucoid (ASOR), ligand was abundant within the vesicular portions, whereas receptors were concentrated in connected tubular extensions (Geuze et al., 1983b). This compartment was, therefore, also named "compartment for the uncoupling of receptors and ligands" (CURL). Segregation of receptors with different final destinations also takes place in this compartment. The transcytotic poly-Ig receptor is enriched in tubular regions separate from domains in which the ASGP receptor is co-distributed (Geuze et al., 1984b). Recently, the cation-dependent M6P receptor and the ASGP receptor were shown to be concentrated in different subpopulations of endosomal tubules and vesicles (Klumperman et al., 1993).

Since ligand binding is restricted to a pH range of 6.5–8, release of ligand from the receptor appears to be due to the acidic conditions in early endosomes (pH $\leq$ 6). Studies by Weigel and coworkers suggest an additional ATP-dependent receptor inactivation/reactivation mechanism for ligand release and binding (McAbee and Weigel, 1988; Medh and Weigel, 1991). The majority of internalized ligand, 50–75%, is retained intracellularly. By subcellular fractionation of perfused rat liver, distinct endosomal compartments have been identified which successively receive internalized ligand. After 2.5 min at 37 °C, [^{125}I]iodinated ASOR was detectable in receptor-positive, early endosomes. After 15 min, the ligand was found predominantly in receptor-negative, late endosomes (Mueller and Hubbard, 1986). After longer times, ligand reached lysosomes and was degraded. A considerable fraction of internalized ligand (up to 50%), however, is not released and is returned to the cell surface in a process called diacytosis (Regoeczi et al., 1982; Weigel and Oka, 1984).

Like other transport receptors (e.g., those for LDL and transferrin), the ASGP receptor is endocytosed and recycled constitutively (Geffen et al., 1989; Fuhrer et al., 1991). Therefore, a considerable fraction of the receptor (50–75%) is always intracellular, mainly in endosomes, irrespective of the presence or absence of ligand. Constitutive internalization proceeds at a rate of approximately 6%/min. Upon binding of ligand to the receptor, however, the rate is increased at least two-fold (Schwartz et al., 1982) which is reflected in a rapid and reversible decrease in the number of surface receptors (Ciechanover et al., 1983; Schwartz et al., 1984). By quantitative immunoelectron microscopy in HepG2 cells, a decrease of the surface receptor pool from 34% to 20% of total ASGP receptor and an increase of the endosomal pool from 37% to 50% was observed after incubation with ASOR (Zijderhand-Bleekemolen et al., 1987). A considerable fraction (~20%) was also detected in the Golgi apparatus of both HepG2 cells and hepatocytes (Geuze et al., 1984a; Zijderhand-Bleekemolen et al., 1987). It did not disappear upon inhibition of protein synthesis. However, during a 30 min incubation at 37 °C in the presence of external protease, all ASGP receptor molecules are degraded, indicating that they are recycling to the plasma membrane (Geffen et al., 1989). The function of the Golgi pool of ASGP receptors is not clear. The distribution of the ASGP receptor is reminiscent of that of the M6P receptors, except that the latter are most concentrated in late rather than early endosomes.

A number of pharmacological agents also affect the intracellular distribution of the ASGP receptor. Lysosomotropic agents (such as ammonium chloride, chloroquine, and monensin) cause a dose-dependent and reversible inhibition of receptor recycling in HepG2 cells by neutralizing the pH of acidic organelles (Tolleshaug and Berg, 1979; Tycko et al., 1983; Schwartz et al., 1984). As a result, receptors gradually disappear from the cell surface and accumulate in early endosomes.

Like virtually all endocytic receptors, the ASGP receptor is phosphorylated by protein kinase C at serine residues within its cytoplasmic portion (Schwartz, 1984b; Takahashi et al., 1985; Backer and King, 1991; Geffen et al., 1991). Activation of protein kinase C by phorbol esters results in hyperphosphorylation and in a concomitant reduction of surface receptors by 40–50% in HepG2 cells (down-regulation; Fallon and Schwartz, 1986). The rate of internalization and the total number of binding sites were found not to be affected, suggesting a role for receptor phosphorylation/dephosphorylation in the control of recycling to the cell surface (Fallon and Schwartz, 1987). In addition, the protein kinase inhibitor, staurosporine, but not the protein kinase C-specific inhibitor, H7, inhibits internalization of ASGPs and transferrin (Fallon and Danaher, 1992). Site-directed mutagenesis studies (discussed in detail below), however, indicate that receptor redistribution is independent of receptor phosphorylation at serine residues and is rather an indirect effect of protein kinases acting more generally on membrane traffic (Geffen and Spiess, 1992).

III. MOLECULAR STRUCTURE OF THE ASGP RECEPTOR

A. Protein Structure

ASGP receptor polypeptides are integral membrane proteins which, upon solubilization with nonionic detergents, can be purified by affinity chromatography using immobilized ASGPs or galactose. Hepatic ASGP receptors have been isolated from rabbit (Kawasaki and Ashwell, 1976), human (Baenziger and Maynard, 1980), rat (Drickamer et al., 1984), and mouse liver (Hong et al., 1988). The receptor preparations contained two (rat and mouse three) distinct, but immunologically related proteins in the range of 40–60 kD, of which one was roughly three times as abundant as the other(s). The ASGP receptor proteins are themselves glycosylated by the addition of two or three N-linked oligosaccharides.

cDNA sequences from rat (Holland et al., 1984; McPhaul and Berg, 1987; Halberg et al., 1987), human (Spiess and Lodish, 1985; Spiess et al., 1985), and mouse receptors (Sanford and Doyle, 1990; Takezawa et al., 1993) have been cloned and demonstrate the existence of two homologous genes. The two minor rat polypeptides originate from the same gene and differ only in their carbohydrate structures (Halberg et al., 1987). The more abundant ASGP receptor protein in each system is generically called HL-1 (hepatic lectin 1), and the minor form(s), HL-2.

In the literature, the rat receptors are usually referred to as RHL-1 and RHL-2/3, and the human receptors as H1 and H2.

The rat macrophage ASGP receptor (initially called ASGP binding protein) was affinity-purified as a single galactose binding polypeptide of 42 kD (Kawasaki et al., 1986; Kelm and Schauer, 1988) which is immunologically cross-reactive with the hepatic receptor. Corresponding cDNAs have been cloned from peritoneal macrophages of rat (Ii et al., 1990) and mouse (Sato et al., 1992). They are the product of a third related ASGP receptor gene.

The deduced protein sequences of the known liver and macrophage ASGP receptor proteins are highly homologous almost over the entire length, with at least 40% identical residues for any two sequences. HL-1 sequences of different species are more closely related to each other (80–90% identity) than to HL-2 (52–57% identity). HL-2 sequences are likewise more homologous to each other. With only ~65% identical residues between human and rodent HL-2, they appear to be less conserved than HL-1 sequences.

The overall structure of the ASGP receptor proteins as deduced from the sequence data and from biochemical analyses is summarized in Figure 2. The polypeptide chains consist of approximately 300 amino acids and span the membrane in a type II orientation: a single transmembrane domain of approximately 20 hydrophobic residues separates a small amino-terminal portion protruding into the cytoplasm (35–58 residues) from a large carboxy-terminal domain (~230–250 residues) exposed on the exoplasmic side of the membrane (Chiaccia and Drickamer, 1984).

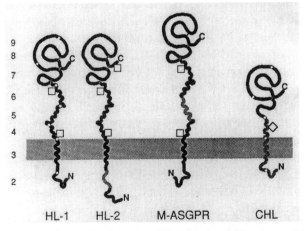

Figure 2. General structure of the subunits of the hepatic ASGP receptor (HL-1, HL-2), the macrophage ASGP receptor (M-ASGPR), and the chicken hepatic lectin (CHL). Black lines indicate segments with sequence homology, squares represent N-glycosylation sites, and white dots show exon boundaries as determined for RHL-1 and CHL (Leung et al., 1985; Bezouska et al., 1991). Exons for HL-1 are numbered on the left.

The carboxy-terminal half of the polypeptide constitutes the carbohydrate-recognition domain (CRD) that is sufficient for galactose binding (Hsueh et al., 1986).

Four different splice variants of H2 mRNA have been discovered. They differ in the presence or absence of an 18-codon insertion in the cytoplasmic domain and/or of a 5-codon insertion at the exoplasmic end of the transmembrane domain (Spiess and Lodish, 1985; Paietta et al., 1992b). Both these insertions are absent in H1. Using insert-specific antisera, the predominant species of mature H2 was shown to contain only the cytoplasmic insertion (Bischoff and Lodish, 1987; Bischoff et al., 1988; Paietta et al., 1992b). This species is also entirely colinear with the cloned cDNA sequences of rat and mouse HL-2. The significance of the 18-amino acid insertion for receptor function is unknown. The short exoplasmic insertion in some splice variants, however, causes complete retention and degradation of H2 polypeptides in the endoplasmic reticulum (Lederkremer and Lodish, 1991; Wikström and Lodish, 1991, 1992; Paietta et al., 1992a).

The general organization of the receptor protein is reflected in the structure of the gene as analyzed for RHL-1 (Leung et al., 1985). While exon 1 covers 5′ untranslated sequences, most of the cytoplasmic domain is encoded by exon 2. The 18-codon insertion in HL-2 coincides exactly with the position of intron 2. Exon 3 corresponds to the membrane-spanning signal–anchor sequence. Exons 4–6 constitute the glycosylated peptide segment that extends to the CRD, which is encoded by exons 7–9.

The connecting segment of exons 4–6 is composed of heptad repeats (Beavil et al., 1992), a sequence with a characteristic distribution of hydrophobic residues with a periodicity of seven. Hydrophobic residues predominate in the positions a and d of the heptads, which is due to the α-helical repeat of exactly 3.5 residues per turn in superhelically twisted coiled-coils, as exemplified by tropomyosin and influenza virus hemagglutinin, which form dimeric and trimeric coiled-coils, respectively. There are a few skip residues inserted in this segment of the ASGP receptor proteins, exactly at the two intron positions. The existence of a heptad repeat segment suggests a dimeric or a trimeric subunit structure of the receptor (or multiples thereof) with a coiled-coil stalk.

The macrophage ASGP receptor is more closely related to HL-1 (with ~55% identity) than to HL-2 (with ~40%); the CRD is clearly better conserved (65–78% identity to HL-1; ~54–59% to HL-2) than the amino-terminal rest of the molecule (40–45% identity to HL-1; ~24–29% to HL-2). The macrophage receptor also lacks the cytoplasmic insertion found in HL-2 sequences. The most striking difference to the hepatic receptors is an additional segment of 24 amino acids in the exoplasmic domain. It is located exactly at the boundary of exons 4 and 5 in the RHL-1 gene and appears to be an additional exon. It carries on the heptad periodicity of the flanking exons and thus appears to lengthen a potential coiled-coil stalk structure.

B. C-Type Lectins

The ASGP receptor proteins are members of a rapidly growing family of proteins which possess homologous CRDs (Drickamer, 1988; Bezouska et al., 1991; Drickamer and Taylor, 1993). The hallmark of these lectin domains is the absolute requirement for Ca^{2+} for carbohydrate binding. They have been classified as C-type animal lectins, as opposed to the S-type lectins which depend on free thiol groups (Drickamer, 1988) and to other lectins that belong to neither group (e.g., the M6P receptors). More than 30 proteins with C-type CRDs are currently known. The CRDs contain 14 invariant or 18 strongly conserved residues within a total of ~130 amino acids. Among the invariant amino acids are four cysteines which were shown to form disulfide bridges in several proteins. In a subclass of CRDs, which includes the ASGP receptors, an additional pair of cysteines near the amino-terminal end of the domain is conserved, forming an additional disulfide bond (as was shown, e.g., for tetranectin; Fuhlendorff et al., 1987).

The C-type lectins are "mosaic proteins" composed of several functional units, one of them the CRD, that have been combined in evolution by "exon shuffling" from different ancestral genes. Based on the general structure and on sequence homology, several subfamilies can be distinguished (Bezouska et al., 1991). Among the C-type lectins are secretory proteins containing glycosaminoglycan-attachment domains (group I, proteoglycans) or collagen-like domains (group III, humoral defense); others are type II membrane proteins (endocytic receptors, groups II and V) or type I membrane proteins with EGF-like repeats (group IV, cell adhesion molecules) or with multiple CRDs (group VI, macrophage mannose receptor). In some cases, the CRD constitutes essentially the entire protein except for a cleaved signal sequence (group VII).

Group II includes the ASGP receptor proteins, the chicken hepatic lectin (Drickamer, 1981), the lymphocyte IgE receptor (Lüdin et al., 1987), the Kupffer cell receptor (Hoyle and Hill, 1988), the B-cell differentiation antigen Lyb-2 (Nakayama et al., 1989), and the placental mannose receptor (Curtis et al., 1992). While sequence homology among these proteins is restricted to the CRD, they are very similarly organized. They are composed of a relatively short amino-terminal cytoplasmic domain (23–95 residues), a single transmembrane segment, a stalk segment of different lengths (26–288 residues) characterized by heptad repeats, and a CRD at the carboxy-terminus.

The CRD of one member of the C-type lectin family, the rat mannose-binding protein A (group III), has recently been crystallized and its structure determined (Weis et al., 1991a,b). The polypeptide adopts a previously undescribed fold with two Ca^{2+} ions pinning together several loop regions near the "top" of the molecule. The two ions are linked by two invariant acidic amino acids which are likely to contribute to the loss of ligand binding at mildly acidic pH by reducing the affinity for Ca^{2+} as they are protonated.

In the crystal structure of the CRD-ligand complex, the 3- and 4-hydroxyl groups of terminal mannose residues coordinate to one of the Ca^{2+} ions and are hydrogen-bonded to two pairs of glutamic acid and asparagine residues (Weis et al., 1992). In galactose-specific CRDs, one of these pairs is invariably replaced by glutamine and aspartic acid to allow the formation of hydrogen bonds to the axial 3- and 4-hydroxyl groups of galactose. By site-directed mutagenesis of the two critical residues, Glu and Asn, in the mannose-binding protein to Gln and Asp, the binding activity could indeed be changed from a preference for mannose to one for galactose (Drickamer, 1992).

C.   Oligomeric Organization

ASGP receptor proteins specifically recognize terminal nonreducing galactose and N-acetyl-galactosamine residues (reviewed by Schwartz, 1984a). Binding of individual residues of this type is of low affinity with dissociation constants in the order of 10^{-3} M. In contrast, galactose-terminal bi-, tri-, and tetra-antennary oligosaccharides exhibit dissociation constants of $\sim 10^{-6}$, $\sim 5 \times 10^{-9}$, and $\sim 10^{-9}$ M, respectively (Lee et al., 1983). This indicates that at least three galactoses may simultaneously interact with as many binding sites closely arranged within the span of a ligand oligosaccharide. In detergent solution, the receptor behaves as a complex of approximately 260 kD (gel filtration and sedimentation equilibrium analysis; Kawasaki and Ashwell, 1976; Andersen et al., 1982), which sets an upper limit for the subunit composition of the receptor complex at a hexamer.

Several lines of evidence show that the hepatic receptor is a hetero-oligomer composed of both HL-1 and HL-2. The cDNAs of both HL-1 and HL-2 must be expressed simultaneously to generate high-affinity ASGP-binding sites in trans-fected cells (McPhaul and Berg, 1986; Shia and Lodish, 1989). Both receptor species can be simultaneously immobilized (Henis et al., 1990), induced to inter-nalize (Bischoff et al., 1988), or co-immunoprecipitated (Sawyer et al., 1988), by antibodies specific for either of the receptor proteins. Chemical cross-linking of HepG2 cells produces covalent homo-dimers and -trimers of H1, as well as hetero-trimers of H1 and H2 (Bischoff et al., 1988).

The structural requirements for ligand molecules have been characterized exten-sively using natural and synthetic oligosaccharides (Lee et al., 1983, 1984; Hardy et al., 1985; Townsend et al., 1986). Among the triantennary ligands, the naturally occurring oligosaccharide shown in Figure 3a was found to have the highest affinity. Based on this structure, galactose-binding sites would be predicted to be positioned at the corners of a triangle with sides of 1.5-, 2.2-, and 2.5-nm (Lee et al., 1984). Molecules with intergalactose distances shorter than these lengths are invariably poor ligands. Among the other ligands, those with the most flexible structures have the highest affinities. Rice et al. (1990) used derivatives of a triantennary glycopep-tide, in which a photolyzable reagent was coupled to carbon-6 of either galactose-6, -6', or -8. Upon binding to rat hepatocytes and photolysis, the ligands derivatized

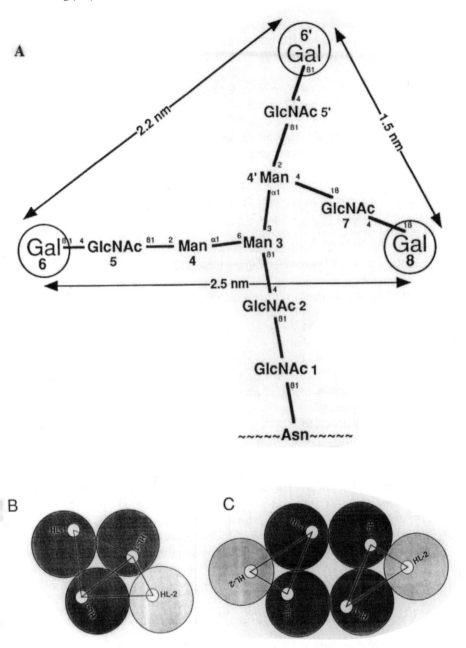

Figure 3. Ligand structure vs. oligomeric organization of the hepatic ASGP receptor. (**A**) Natural triantennary oligosaccharide. (**B**) Minimal receptor oligomer proposed by Lodish (1991). (**C**) Hexameric hetero-oligomer proposed by Weigel (1992).

at galactose-6 or -6' specifically labeled RHL-1, whereas the ligand derivatized at galactose-8 labeled only RHL-2/3. The second, minor subunits may serve to create an asymmetric arrangement of galactose-binding sites, as is illustrated in Figures 3b and 3c.

In contrast to the hepatic receptor, the macrophage ASGP receptor appears to consist of a single polypeptide species which upon expression in heterologous cells produces a functional ligand-binding receptor (Ozaki et al., 1992). The molecular basis for this difference in subunit requirement is not clear. Upon gel filtration in Triton X-100, the macrophage receptor behaves as a complex of ~410 kD, consistent with hexamers or octamers.

Generally, secretory and membrane proteins fold and oligomerize in the ER, and there are mechanisms to prevent the exit of unfolded or misfolded proteins (reviewed by Hurtley and Helenius, 1989; Pelham, 1989). In HepG2 cells, it has been shown that newly made H1 subunits do not exit the ER for at least 30 min before transport to the Golgi apparatus occurs over the next 1–2 h (Lodish et al., 1992). The ER form of H1, but not the Golgi-processed form, is sensitive to reduction by 1 mM dithiothreitol, indicating that it is not completely folded. However, H1 can be cross-linked to dimers and trimers already within 15 min of subunit synthesis, that is, before acquisition of the correctly folded conformation. Calcium is required for the intramolecular folding of H1, since the calcium ionophore, A23187, and thapsigargin (an inhibitor of the Ca^{2+} ATPase) block folding of H1 and inhibit its exit from the ER, while oligomerization is not affected (Lodish et al., 1992). High concentrations of Ca^{2+} might be necessary for the folding of the CRD; alternatively Ca^{2+} ions could have an indirect effect by activating one or more of the Ca^{2+}-binding chaperonins of the ER.

In transfected cells, H1 forms oligomers and is transported to the cell surface also in the absence of the second subunit (Shia and Lodish, 1989). In contrast, only ~30% of H2, when expressed in the absence of H1, reaches the cell surface (Lederkremer and Lodish, 1991). The fraction of H2 that exits the ER is significantly increased when co-expressed with H1, which is indicative of hetero-oligomer formation. A splice variant of H2 (called H2a) with a five-codon insertion at the exoplasmic end of the transmembrane domain is completely retained in the ER. Retained subunits are degraded in the ER (Amara et al., 1989; Wikström and Lodish, 1991, 1992, 1993). Recent studies suggest that two degradation pathways exist, one of which involves the initial generation of a 35 kD fragment corresponding to the extracytoplasmic portion of the polypeptide (Yuk and Lodish, 1993).

IV. SORTING DETERMINANTS

The function of an endocytic receptor is based on several sorting events: insertion into the ER, specific delivery to the appropriate plasma membrane domain in polarized cells, clustering into clathrin-coated pits during endocytosis, and segregation into recycling vesicles in early endosomes. By site-directed mutagenesis and expression in transfected cells, the structural requirements for the sorting of receptors are beginning to be elucidated.

A. Insertion into the Endoplasmic Reticulum

ASGP receptor proteins are targeted to the ER in a co-translational process involving signal recognition particle and signal recognition particle receptor. Using H1 and RHL-1 as a model, the transmembrane segment was shown to function as an internal uncleaved signal–anchor sequence that is necessary and sufficient for ER insertion with a type II (N-cytoplasmic/C-exoplasmic) orientation (Holland and Drickamer, 1986; Spiess and Lodish, 1986). The characteristics of the internal signal are similar to those of cleaved signals except that the apolar domain is ~20 residues in length to span the lipid bilayer (Spiess and Handschin, 1987). There is even a cryptic signal cleavage site at the carboxy-terminal end of the signal–anchor which is activated upon truncation of the amino-terminal domain (Schmid and Spiess, 1988).

B. Endocytosis

Using a ligand-independent assay, subunit H1 of the ASGP receptor was found to be internalized and recycled in the absence of H2 with kinetics very similar to those of the hetero-oligomeric complex (Geffen et al., 1989). Similar findings were reported for RHL-1 (Braiterman et al., 1989). The major receptor subunit thus contains all the signals required for endocytosis and recycling. In different receptor systems, the determinants for clathrin-dependent endocytosis have been identified in the cytoplasmic receptor domains as short sequences containing at least one aromatic residue, typically a tyrosine, with the potential to assume a turn conformation (reviewed by Trowbridge, 1991; Hopkins, 1992). The cytoplasmic sequences of H1 and H2, and of the rat macrophage ASGP receptor are shown in Figure 4. The cytoplasmic domain of H1 fused to the transmembrane and exoplasmic portions of aminopeptidase N, a resident plasma membrane protein, induces endocytosis of the fusion protein (Cescato, unpublished data). The only aromatic residue of the cytoplasmic domain of H1, Tyr-5, is critical for rapid endocytosis. Upon mutation to alanine, the internalization rate is reduced by a factor of approximately 4, from ~6%/min to ~1.5%/min and the surface pool of H1 is concomitantly increased from 50% to 85% (Fuhrer et al., 1991). The residual internalization of the mutant, which is significantly higher than that of a resident protein like hemagglutinin or glycophorin (Ktistakis et al., 1990), still occurs via the clathrin-dependent pathway. Deletion of residues 4–11 (including Tyr-5) does not further reduce internalization, suggesting that weak determinants elsewhere in the protein contribute to endocytosis (Geffen et al., 1993). Upon deletion of residues 12–33, internalization is still quite efficient, indicating that the tyrosine signal acts independently of its position relative to the membrane.

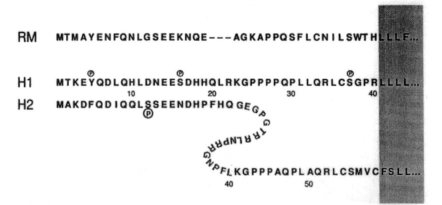

Figure 4. Aligned sequences of the cytoplasmic domains of the hepatic subunits H1 and H2, and of the rat macrophage ASGP receptor (RM). Known phosphorylation sites are indicated.

The critical tyrosine-5 of H1 was reported to be phosphorylated (Fallon, 1990). However, tyrosine phosphorylation is not essential for endocytosis, since mutation of Tyr-5 to phenylalanine yielded almost wild-type rates of internalization (Geffen et al., 1993). Tyr-5 is conserved in all known HL-1 sequences, as well as in the sequences of the macrophage ASGP receptor. Mutation of Tyr-5 of the rat macrophage receptor similarly resulted in a fourfold decrease of internalization (Ozaki et al., 1993). In all known HL-2 sequences, however, a phenylalanine is found at position 5. HL-1 and HL-2 conform equally well with the consensus motif *aromatic residue (mostly Tyr)–X–X–large hydrophobic residue* for endocytosis signals derived from sequence comparison and mutational analysis of the transferrin receptor, the cation-independent M6P receptor, and lysosomal acid phosphatase (Collawn et al., 1990, 1991; Canfield et al., 1991; Jadot et al., 1992; Lehmann et al., 1992). Yet, when expressed separately in fibroblasts, subunit H2 is internalized poorly (~1.2%/min), and mutation of Phe-5 does not further reduce endocytosis (Fuhrer et al., 1994). Consistent with this result, ligand uptake by receptors composed of H1 lacking Tyr-5 and wild-type H2 is inefficient, indicating that H2 cannot compensate for an internalization defect in H1. Endocytosis of the hetero-oligomeric ASGP receptor thus is exclusively driven by subunit H1 (Fuhrer et al., 1994).

The ASGP receptor is phosphorylated at cytoplasmic serine residues (Schwartz, 1984b; Takahashi et al., 1985). By site-directed mutagenesis, the major site of phosphorylation in the human receptor was identified to be Ser-12 of H2; weak phosphorylation was also detected at Ser-16 and Ser-37 of H1 (Geffen et al., 1991). Mutation of all cytoplasmic serines to either alanine or glycine did not affect internalization of H1 nor ligand uptake by hetero-oligomeric receptors. Therefore, neither endocytosis nor recycling seem to depend on serine residues or their phosphorylation. Mutant receptors lacking cytoplasmic serines were even found to redistribute to intracellular compartments upon stimulation of protein kinase C by phorbol esters in COS-7 cells (Geffen and Spiess, 1992). This indicates that phorbol ester-induced down-regulation is independent of receptor phosphorylation, a behavior also observed for the transferrin receptor (Davis and Meisner, 1987; Rothenberger et al., 1987; McGraw et al., 1988). The function of phosphorylation of endocytic transport receptors is not clear at present.

C. Sorting to the Basolateral Plasma Membrane

In polarized epithelial cells, mechanisms exist for the specific delivery of proteins to either the apical or the basolateral plasma membrane domain (reviewed by Mostov et al., 1992; Rodriguez-Boulan and Powell, 1992). The pathways from the Golgi apparatus to the cell surface differ between cell types. In hepatocytes, all membrane proteins studied so far are first transported to the basolateral domain from where apical proteins are transcytosed to the apical surface (Bartles and Hubbard, 1988). In Madin–Darby canine kidney (MDCK) cells, there are direct pathways from the Golgi apparatus to either surface domain (Simons and Wandinger-Ness, 1990). When expressed in MDCK cells, the hepatic ASGP receptor is specifically sorted to the basolateral cell surface (Wessels et al., 1989; Graeve et al., 1990; Fuhrer et al., 1994), suggesting that either basolateral transport occurs by default or that the receptor contains basolateral sorting information that is correctly recognized in MDCK cells. The latter hypothesis was confirmed by the finding that mutations in the cytoplasmic domain affected polarized surface delivery of subunit H1 in MDCK cells (Geffen et al., 1993). Deletion of amino acids 4–11 and mutation of Tyr-5 to Ala, that is, the two mutations that strongly reduce internalization, abolish specific basolateral targeting and result in nonpolarized surface expression (~50% on either surface). Deletion of residues 12–33, mutation of Tyr-5 to Phe, or of Ser-16 and Ser-37 to Gly or Ala, have no effect on basolateral specificity. The kinetics of appearance of newly synthesized wild-type and mutant H1 at the apical and basolateral surfaces indicate that these proteins are sorted intracellularly and are transported directly to the respective domains (and not via transcytosis). These results show that residues critical for efficient endocytosis at the plasma membrane are also important for specific basolateral surface transport in the Golgi apparatus (Geffen et al., 1993).

When expressed alone in MDCK cells, H2 is completely degraded intracellularly. However, co-expression of H2 (or the Phe-5-to-Ala mutant of H2) with the Tyr-5-to-Ala mutant of H1 results in specific basolateral expression of the ligand-binding receptor complex (Fuhrer et al., 1994). This indicates that subunit H2 contributes basolateral sorting information that is independent of an endocytosis signal. Consistent with this, RHL-2 was shown to be basolaterally expressed in MDCK cells in the absence of RHL-1 (Graeve et al., 1990). Unlike for endocytosis, polarized transport of the ASGP receptor to the basolateral domain of MDCK cells may thus involve two signals, only one of which is active for internalization. To what extent basolateral polarity of the ASGP receptor in hepatocytes is signal mediated is as yet unknown.

Correlations between signals for endocytosis and basolateral delivery have also been reported for influenza virus hemagglutinin HA-Y543, the nerve growth factor receptor, the Fc receptor, and lysosomal glycoprotein Igp-120, (Brewer and Roth, 1991; Hunziker et al., 1991; Le Bivic et al., 1991). In the case of lysosomal acid phosphatase, the two sorting signals appear to be overlapping, but distinct (Prill et al., 1993). In the poly-Ig receptor, vesicular stomatitis virus G protein, and the LDL receptor, however, the main basolateral sorting determinant is different from the internalization signal (Casanova et al., 1991; Matter et al., 1992; Aroeti et al., 1993; Thomas et al., 1993). A consensus among endocytosis-independent signals has not yet emerged.

D. Sorting Mechanisms

Endocytosis signals appear to be recognized in the plasma membrane by clathrin-associated proteins (APs, also called adaptor proteins; reviewed by Pearse and Robinson, 1990; Robinson, 1992). Two types of adaptor complexes, AP-1 and AP-2, have been characterized that are specific for clathrin-coated membranes of the *trans*-Golgi apparatus and the plasma membrane, respectively. An interaction between adaptor proteins and endocytic proteins was observed *in vitro* by binding of adaptor complexes to immobilized tail peptides of the LDL receptor, the M6P receptor, and LAP (Pearse, 1988; Glickman et al., 1989; Sosa et al., 1993). Binding was dependent on the tyrosine residues that are critical for endocytosis and, except for the M6P receptor sequence, for AP-2 complexes. In a ligand blot assay, purified human ASGP receptor was shown to bind to the β-adaptin of coated vesicle proteins and purified AP-2 complexes that had been fractionated by SDS-gel electrophoresis and transferred to nitrocellulose (Beltzer and Spiess, 1991). This interaction could be competed by an excess of dihydrofolate reductase fusion proteins containing the cytoplasmic domain of H1 or of hemagglutinin HA-Y543, but not by a fusion protein with the wild-type hemagglutinin tail. The receptor was shown to bind to the amino-terminal ~60 kD trypsin fragment of β-adaptin. The notion that AP-2 complexes recognize receptor tails is also confirmed by the finding that AP-2 complexes could be co-immunoprecipitated with the EGF receptor upon induction of internalization by ligand (Sorkin and Carpenter, 1993).

The components that recognize basolateral signals in the *trans*-Golgi apparatus are not known yet. At present, we can distinguish between basolateral determinants that are related to endocytosis signals and those that are not, which may operate by different mechanisms and target proteins to the basolateral surface by different routes. It is not known whether the ASGP receptor is transported to the surface directly in exocytic vesicles or via endosomes like the M6P receptors. In the latter case, AP-1 adaptor complexes might be involved in recognition of the tyrosine signal.

V. PERSPECTIVES

Analysis of the hepatic ASGP receptor has been facilitated by the relatively high expression levels in hepatocytes and by the ease of ligand preparation. Cell biological studies have made it one of the best characterized model systems for receptor-mediated endocytosis and have contributed to the identification and the understanding of endosomal compartments and of intracellular membrane traffic in general. Composed of two homologous subunits, the receptor turned out to be more complicated than other systems on a molecular level. The exact subunit stoichiometry of the hetero-oligomeric receptor complex remains one of the foremost structural questions to be answered. While there are plausible explanations for the requirement for a second non-identical subunit for ligand binding by the hepatic receptor, it is an unsolved puzzle that the homo-oligomeric macrophage ASGP receptor apparently can bind the same ligand molecules with similar affinities. Better knowledge of receptor structure (e.g., interactions between the subunits, flexibility of the individual galactose-binding sites relative to each other) will be essential to answer this question.

Of more general interest will be the further characterization of sorting signals in the ASGP receptor proteins and the identification of the corresponding recognition molecules and sorting machineries. Imminent questions are whether the receptor is transported from the *trans*-Golgi apparatus to the cell surface via endosomes or directly via dedicated basolateral transport vesicles, and to what extent basolateral sorting signals are relevant for surface transport in hepatocytes. Virtually nothing is known about the determinants that target internalized receptors into the recycling pathway to the cell surface or (like the EGF receptor) towards lysosomes. From what we know about the sorting signals, they all appear to be degenerate on the level of primary sequence. Progress is likely to come from the mosaic of data obtained in many different receptor systems, among them that of the ASGP receptors.

ACKNOWLEDGMENTS

We thank Dr. J. Wahlberg for helpful discussions on the manuscript. Our own research was supported by grants from the Swiss National Science Foundation and by the Incentive Award of the Helmut Horten Foundation.

REFERENCES

Amara, J. F., Lederkremer, G., & Lodish, H. F. (1989). Intracellular degradation of unassembled asialoglycoprotein receptor subunits—A pre-Golgi, nonlysosomal endoproteolytic cleavage. J. Cell Biol. 109, 3315–3324.

Andersen, T. T., Freytag, J. W., & Hill, R. L. (1982). Physical studies of the rabbit hepatic galactoside-binding protein. J. Biol. Chem. 257, 8036–8041.

Aroeti, B., Kosen, P. A., Kuntz, I. D., Cohen, F. E., & Mostov, K. E. (1993). Mutational and secondary structural analysis of the basolateral sorting signal of the polymeric immunoglobulin receptor. J. Cell Biol. 123, 1149–1160.

Ashwell, G., & Harford, J. (1982). Carbohydrate-specific receptors of the liver. Annu. Rev. Biochem. 51, 531–554.

Ashwell, G., & Morell, A. G. (1974). The role of surface carbohydrates in the hepatic recognition and transport of circulating glycoproteins. Adv. Enzymol. 41, 99–128.

Backer, J. M., & King, G. L. (1991). Regulation of receptor-mediated endocytosis by phorbol esters. Biochem. Pharmacol. 41, 1267–1277.

Baenziger, J. U., & Maynard, Y. (1980). Human hepatic lectin. Physicochemical properties and specificity. J. Biol. Chem. 255, 4607–4613.

Bartles, J. R., & Hubbard, A. L. (1988). Plasma membrane protein sorting in epithelial cells: Do secretory pathways hold the key? Trends Biochem. Sci. 13, 181–184.

Beavil, A. J., Edmeades, R. L., Gould, H. J., & Sutton, B. J. (1992). Alpha-helical coiled-coil stalks in the low-affinity receptor for IgE (FcepsilonRII/CD23) and related C-type Llectins. Proc. Natl. Acad. Sci. USA 89, 753–757.

Beltzer, J. P., & Spiess, M. (1991). In vitro binding of the asialoglycoprotein receptor to the beta adaptin of plasma membrane coated vesicles. EMBO J. 10, 3735–3742.

Bezouska, K., Crichlow, G. V., Rose, J. M., Taylor, M. E., & Drickamer, K. (1991). Evolutionary conservation of intron position in a subfamily of genes encoding carbohydrate-recognition domains. J. Biol. Chem. 266, 11604–11609.

Bischoff, J., Libresco, S., Shia, M. A., & Lodish, H. F. (1988). The H1 and H2 polypeptides associate to form the asialoglycoprotein receptor in human hepatoma cells. J. Cell Biol. 106, 1067–1074.

Bischoff, J., & Lodish, H. F. (1987). Two asialoglycoprotein receptor polypeptides in human hepatoma cells. J. Biol. Chem. 262, 11825–11832.

Braiterman, L. T., Chance, S. C., Porter, W. R., Lee, Y. C., Townsend, R. R., & Hubbard, A. L. (1989). The major subunit of the rat asialoglycoprotein receptor can function alone as a receptor. J. Biol. Chem. 264, 1682–1688.

Breitfeld, P. P., Simmons, C. F., Strous, G. J. A. M., Geuze, H. J., & Schwartz, A. L. (1985). Cell biology of the asialoglycoprotein receptor system: A model of receptor-mediated endocytosis. Int. Rev. Cytology 97, 47–95.

Brewer, C. B., & Roth, M. G. (1991). A single amino acid change in the cytoplasmic domain alters the polarized delivery of influenza virus hemagglutinin. J. Cell Biol. 114, 413–421.

Canfield, W. M., Johnson, K. F., Ye, R. D., Gregory, W., & Kornfeld, S. (1991). Localization of the signal for rapid internalization of the bovine cation-independent mannose 6-phosphate/insulin-like growth factor-ii receptor to amino acids 24–29 of the cytoplasmic tail. J. Biol. Chem. 266, 5682–5688.

Casanova, J. E., Apodaca, G., & Mostov, K. E. (1991). An autonomous signal for basolateral sorting in the cytoplasmic domain of the polymeric immunoglobulin receptor. Cell 66, 65–75.

Chiacchia, K. B., & Drickamer, K. (1984). Direct evidence for the transmembrane orientation of the hepatic glycoprotein receptors. J. Biol. Chem. 259, 15440–15446.

Ciechanover, A., Schwartz, A. L., & Lodish, H. F. (1983). The asialoglycoprotein receptor internalizes and recycles independently of the transferrin and insulin receptors. Cell 32, 267–275.

Collawn, J. F., Kuhn, L. A., Liu, L., Tainer, J. A., & Trowbridge, I. S. (1991). Transplanted LDL and mannose-6-phosphate receptor internalization signals promote high-efficiency endocytosis of the transferrin receptor. EMBO J. 10, 3247–3253.

Collawn, J. F., Stangel, M., Kuhn, L. A., Esekogwu, V., Jing, S. Q., Trowbridge, I. S., & Tainer, J. A. (1990). Transferrin receptor internalization sequence YXRF implicates a tight turn as the structural recognition motif for endocytosis. Cell 63, 1061–1072.

Curtis, B. M., Scharnowske, S., & Watson, A. J. (1992). Sequence and expression of a membrane-associated C-type lectin that exhibits CD4-independent binding of human immunodeficiency virus envelope glycoprotein gp120. Proc. Natl. Acad. Sci. USA 89, 8356–8360.

Daniels, C. K., Schmucker, D. L., & Jones, A. L. (1989). Hepatic asialoglycoprotein receptor-mediated binding of human polymeric immunoglobulin A. Hepatology 9, 229–234.

Davis, R. J., & Meisner, H. (1987). Regulation of transferrin receptor cycling by protein kinase C is independent of receptor phosphorylation at serine 24 in Swiss 3T3 fibroblasts. J. Biol. Chem. 262, 16041–16047.

Drickamer, K. (1981). Complete amino acid sequence of a membrane receptor for glycoproteins. J. Biol. Chem. 256, 5827–5839.

Drickamer, K. (1988). Two distinct classes of carbohydrate-recognition domains in animal lectins. J. Biol. Chem. 263, 9557–9560.

Drickamer, K. (1992). Engineering galactose-binding activity into a C-type mannose-binding protein. Nature 360, 183–186.

Drickamer, K., & Taylor, M. E. (1993). Biology of animal lectins. Ann. Rev. Cell Biol. 9, 237–264.

Drickamer, K., Mamon, J. F., Binns, G., & Leung, J. O. (1984). Primary structure of the rat liver asialoglycoprotein receptor. Structural evidence for multiple polypeptide species. J. Biol. Chem. 259, 770–778.

Fallon, R. J. (1990). Tyrosine phosphorylation of the asialoglycoprotein receptor. J. Biol. Chem. 265, 3401–3406.

Fallon, R. J., & Danaher, M. (1992). The effect of staurosporine, a protein kinase inhibitor, on asialoglycoprotein receptor endocytosis. Exp. Cell Res. 203, 420–426.

Fallon, R. J., & Schwartz, A. L. (1986). Regulation by phorbol esters of asialoglycoprotein and transferrin receptor distribution and ligand affinity in a hepatoma cell line. J. Biol. Chem. 261, 15081–15089.

Fallon, R. J., & Schwartz, A. L. (1987). Mechanism of the phorbol ester-mediated redistribution of asialoglycoprotein receptor: Selective effects on receptor recycling pathways in Hep G2 cells. Mol. Pharmacol. 32, 348–355.

Fuhlendorff, J., Clemmensen, I., & Magnusson, S. (1987). Primary structure of tetranectin, a plasminogen kringle 4 binding plasma protein: Homology with asialoglycoprotein receptors and cartilage proteoglycan core protein. Biochemistry 26, 6757–6764.

Fuhrer, C., Geffen, I., Huggel, K., & Spiess, M. (1994). The two subunits of the asialoglycoprotein receptor contain different sorting information. J. Biol. Chem. 269, 3277–3282.

Fuhrer, C., Geffen, I., & Spiess, M. (1991). Endocytosis of the ASGP receptor H1 is reduced by mutation of tyrosine-5 but still occurs via coated pits. J. Cell Biol. 114, 423–432.

Geffen, I., Fuhrer, C., Leitinger, B., Weiss, M., Huggel, K., Griffiths, G., & Spiess, M. (1993). Related signals for endocytosis and basolateral sorting of the asialoglycoprotein receptor. J. Biol. Chem. 268, 20772–20777.

Geffen, I., Fuhrer, C., & Spiess, M. (1991). Endocytosis by the asialoglycoprotein receptor is independent of cytoplasmic serine residues. Proc. Natl. Acad. Sci. USA 88, 8425–8429.

Geffen, I., & Spiess, M. (1992). Phorbol ester-induced redistribution of the ASGP receptor is independent of receptor phosphorylation. FEBS Lett. 305, 209–212.

Geffen, I., Wessels, H. P., Roth, J., Shia, M. A., & Spiess, M. (1989). Endocytosis and recycling of subunit H1 of the asialoglycoprotein receptor is independent of oligomerization with H2. EMBO J. 8, 2855–2862.

Geuze, H. J., Slot, J. W., Strous, G. J., Luzio, J. P., & Schwartz, A. L. (1984a). A cycloheximide-resistant pool of receptors for asialoglycoproteins and mannose 6-phosphate residues in the Golgi complex of hepatocytes. EMBO J. 3, 2677–2685.

Geuze, H. J., Slot, J. W., Strous, G. J., & Schwartz, A. L. (1983a). The pathway of the asialoglycoprotein-ligand during receptor-mediated endocytosis: A morphological study with colloidal gold/ligand in the human hepatoma cell line, Hep G2. Eur. J. Cell Biol. 32, 38–44.

Geuze, H. J., Slot, J. W., Strous, G. J. A. M., Lodish, H. F., & Schwartz, A. L. (1983b). Intracellular site of asialoglycoprotein receptor-ligand uncoupling. Double-label immunoelectronmicroscopy during receptor mediated endocytosis. Cell 32, 277–287.

Geuze, H. J., Slot, J. W., Strous, G. J. A. M., Peppard, J., von Figura, K., Hasilisk, A., & Schwartz, A. L. (1984b). Intracellular receptor sorting during endocytosis: Comparative immunoelectron microscopy of multiple receptors in rat liver. Cell 37, 195–204.

Glickman, J. N., Conibear, E., & Pearse, B. M. F. (1989). Specificity of binding of clathrin adaptors to signals on the mannose-6-phosphate insulin-like growth factor-ii receptor. EMBO J. 8, 1041–1047.

Graeve, L., Patzak, A., Drickamer, K., & Rodriguez-Boulan, E. (1990). Polarized expression of functional rat liver asialoglycoprotein receptor in transfected Madin-Darby canine kidney cells. J. Biol. Chem. 265, 1216–1224.

Halberg, D. F., Wager, R. E., Farrell, D. C., Hildreth, J., Quesenberry, M. S., Loeb, J. A., Holland, E. C., & Drickamer, K. (1987). Major and minor forms of the rat liver asialoglycoprotein receptor are independent galactose-binding proteins. Primary structure and glycosylation heterogeneity of minor receptor forms. J. Biol. Chem. 262, 9828–9838.

Hardy, M. R., Townsend, R. R., Parkhurst, S. M., & Lee, Y. C. (1985). Different modes of ligand binding to the hepatic galactose/N-acetylgalactosamine lectin on the surface of rabbit hepatocytes. Biochemistry 24, 22–28.

Henis, Y. I., Katzir, Z., Shia, M. A., & Lodish, H. F. (1990). Oligomeric structure of the human asialoglycoprotein receptor: Nature and stoichiometry of mutual complexes containing H1 and H2 polypeptides assessed by fluorescence photobleaching recovery. J. Cell Biol. 111, 1409–1418.

Holland, E. C., & Drickamer, K. (1986). Signal recognition particle mediates the insertion of a transmembrane protein which has a cytoplasmic NH2 terminus. J. Biol. Chem. 261, 1286–1292.

Holland, E. C., Leung, J. O., & Drickamer, K. (1984). Rat liver asialoglycoprotein receptor lacks a cleavable NH2-terminal signal sequence. Proc. Natl. Acad. Sci. USA 81, 7338–7342.

Hong, W., Le, A. V., & Doyle, D. (1988). Identification and characterization of a murine receptor for galactose-terminated glycoproteins. Hepatology 8, 553–558.

Hopkins, C. R. (1992). Selective membrane protein trafficking—vectorial flow and filter. Trends Biochem. Sci. 17, 27–32.

Hoyle, G. W., & Hill, R. L. (1988). Molecular cloning and sequencing of a cDNA for a carbohydrate binding receptor unique to rat Kupffer cells. J. Biol. Chem. 263, 7487–7492.

Hsueh, E. C., Holland, E. C., Carrera, G. M., & Drickamer, K. (1986). The rat liver asialoglycoprotein receptor polypeptide must be inserted into a microsome to achieve its active conformation. J. Biol. Chem. 261, 4940–4947.

Hubbard, A. L., & Stukenbrok, H. (1979). An electron microscope autoradiographic study of the carbohydrate recognition systems in rat liver. II. Intracellular fates of the ^{125}I-ligands. J. Cell Biol. 83, 56–81.

Hubbard, A. L., Wilson, G., Ashwell, G., & Stukenbrok, H. (1979). An electron microscope autoradiographic study of the carbohydrate recognition systems in rat liver. I. Distribution of ^{125}I-ligands among the liver cell types. J. Cell Biol. 83, 47–55.

Hunziker, W., Harter, C., Matter, K., & Mellman, I. (1991). Basolateral sorting in MDCK cells requires a distinct cytoplasmic domain determinant. Cell 66, 907–920.

Hurtley, S. M., & Helenius, A. (1989). Protein oligomerization in the endoplasmic reticulum. Annu. Rev. Cell Biol. 5, 277–307.

Ii, M., Kurata, H., Itoh, N., Yamashina, I., & Kawasaki, T. (1990). Molecular cloning and sequence analysis of cDNA encoding the macrophage lectin specific for galactose and N-acetylgalactosamine. J. Biol. Chem. 265, 11295–11298.

Ishihara, H., Hayashi, Y., Hara, T., Aramaki, Y., Tsuchiya, S., & Koike, K. (1991). Specific uptake of asialofetuin-tacked liposomes encapsulating interferon-gamma by human hepatoma cells and its inhibitory effect on hepatitis B virus replication. Biochem. Biophys. Res. Commun. 174, 839–845.

Jadot, M., Canfield, W. M., Gregory, W., & Kornfeld, S. (1992). Characterization of the signal for rapid internalization of the bovine mannose 6-phosphate/insulin-like growth factor-ii receptor. J. Biol. Chem. 267, 11069–11077.

Jansen, R. W., Kruijt, J. K., van-Berkel, T. J., & Meijer, D. K. (1993). Coupling of the antiviral drug ara-AMP to lactosaminated albumin leads to specific uptake in rat and human hepatocytes. Hepatology 18, 146–152.

Kawasaki, T., & Ashwell, G. (1976). Chemical and physical properties of an hepatic membrane protein that specifically binds asialoglycoproteins. J. Biol. Chem. 251, 1296–1302.

Kawasaki, T., Ii, M., Kozutsumi, Y., & Yamashina, I. (1986). Isolation and characterization of a receptor lectin specific for galactose/N-acetylgalactosamine from macrophages. Carbohydr. Res. 151, 197–206.

Kelm, S., & Schauer, R. (1988). The galactose-recognizing system of rat peritoneal macrophages; Identification and characterization of the receptor molecule. Biol. Chem. Hoppe-Seyler 369, 693–704.

Kempka, G., Roos, P. H., & Kolb-Bachofen, V. (1990). A membrane-associated form of C-reactive protein is the galactose-specific particle receptor on rat liver macrophages. J. Immunol. 144, 1004–1009.

Klumperman, J., Hille, A., Veenendaal, T., Oorschot, V., Stoorvogel, W., von Figura, K., & Geuze, H. J. (1993). Differences in the endosomal distributions of the 2 mannose 6-phosphate receptors. J. Cell Biol. 121, 997–1010.

Ktistakis, N. T., Thomas, D., & Roth, M. G. (1990). Characteristics of the tyrosine recognition signal for internalization of transmembrane surface glycoproteins. J. Cell Biol. 111, 1393–1407.

Kudo, M., Todo, A., Ikekubo, K., Yamamoto, K., Vera, D. R., & Stadalnik, R. C. (1993). Quantitative assessment of hepatocellular function through *in vivo* radioreceptor imaging with technetium 99m galactosyl human serum albumin. Hepatology 17, 814–819.

Le Bivic, A., Sambuy, Y., Patzak, A., Patil, N., Chao, M., & Rodriguezboulan, E. (1991). An internal deletion in the cytoplasmic tail reverses the apical localization of human NGF receptor in transfected MDCK cells. J. Cell Biol. 115, 607–618.

Lederkremer, G. Z., & Lodish, H. F. (1991). An alternatively spliced miniexon alters the subcellular fate of the human asialoglycoprotein receptor H2 subunit—Endoplasmic reticulum retention and degradation or cell surface expression. J. Biol. Chem. 266, 1237–1244.

Lee, R. T., Lin, P., & Lee, Y. C. (1984). New synthetic cluster ligands for galactose/N-acetylgalactosamine-specific lectin of mammalian liver. Biochemistry 23, 4255–4261.

Lee, Y. C., Townsend, R. R., Hardy, M. R., Lonngren, J., Arnarp, J., Haraldsson, M., & Lonn, H. (1983). Binding of synthetic oligosaccharides to the hepatic Gal/GalNAc lectin. Dependence on fine structural features. J. Biol. Chem. 258, 199–202.

Lehmann, L. E., Eberle, W., Krull, S., Prill, V., Schmidt, B., Sander, C., Vonfigura, K., & Peters, C. (1992). The internalization signal in the cytoplasmic tail of lysosomal acid phosphatase consists of the hexapeptide PGYRHV. EMBO J. 11, 4391–4399.

Leung, J. O., Holland, E. C., & Drickamer, K. (1985). Characterization of the gene encoding the major rat liver asialoglycoprotein receptor. J. Biol. Chem. 260, 12523–12527.

Lodish, H. F. (1991). Recognition of complex oligosaccharides by the multi-subunit asialoglycoprotein receptor. Trends Biochem. Sci. 16, 374–377.

Lodish, H. F., Kong, N., & Wikstrom, L. (1992). Calcium is required for folding of newly made subunits of the asialoglycoprotein receptor within the endoplasmic reticulum. J. Biol. Chem. 267, 12753–12760.

Lüdin, C., Hofstetter, H., Sarfati, M., Levy, C. A., Suter, U., Alaimo, D., Kilchherr, E., Frost, H., & Delespesse, G. (1987). Cloning and expression of the cDNA coding for a human lymphocyte IgE receptor. EMBO J. 6, 109–114.

Lunney, J., & Ashwell, G. (1976). A hepatic receptor of avian origin capable of binding specifically modified glycoproteins. Proc. Natl. Acad. Sci. USA 73, 341–343.

Makdisi, W. F., & Versland, M. R. (1991). Asialoglycoproteins as radiodiagnostic agents for detection of hepatic masses and evaluation of liver function. Targeted Diagn. Ther. 4, 151–162.

Matsuura, S., Nakada, H., Sawamura, T., & Tashiro, Y. (1982). Distribution of an asialoglycoprotein receptor on rat hepatocyte cell surface. J. Cell Biol. 95, 864–875.

Matter, K., Hunziker, W., & Mellman, I. (1992). Basolateral sorting of LDL receptor in MDCK cells—the cytoplasmic domain contains 2 tyrosine-dependent targeting determinants. Cell 71, 741–753.

McAbee, D. D., & Weigel, P. H. (1988). ATP-dependent inactivation and reactivation of constitutively recycling galactosyl receptors in isolated rat hepatocytes. Biochemistry 27, 2061–2069.

McGraw, T. E., Dunn, K. W., & Maxfield, F. R. (1988). Phorbol ester treatment increases the exocytic rate of transferrin receptor recycling pathway independent of serine-24 phosphorylation. J. Cell Biol. 106, 1061–1066.

McPhaul, M., & Berg, P. (1986). Formation of functional asialoglycoprotein receptor after transfection with cDNAs encoding the receptor proteins. Proc. Natl. Acad. Sci. USA 83, 8863–8867.

McPhaul, M., & Berg, P. (1987). Identification and characterization of cDNA clones encoding two homologous proteins that are part of the asialoglycoprotein receptor. Mol. Cell. Biol. 7, 1841–1847.

Medh, J. D., & Weigel, P. H. (1991). Reconstitution of galactosyl receptor inactivation in permeabilized rat hepatocytes is ATP-dependent. J. Biol. Chem. 266, 8771–8778.

Mostov, K., Apodaca, G., Aroeti, B., & Okamoto, C. (1992). Plasma membrane protein sorting in polarized epithelial cells. J. Cell Biol. 116, 577–583.

Mueller, S. C., & Hubbard, A. L. (1986). Receptor-mediated endocytosis of asialoglycoproteins by rat hepatocytes: Receptor-positive and receptor-negative endosomes. J. Cell Biol. 102, 932–942.

Nakayama, E., von Hoegen, I., & Parnes, J. R. (1989). Sequence of the Lyb-2 B-cell differentiation antigen defines a gene superfamily of receptors with inverted membrane orientation. Proc. Natl. Acad. Sci. USA 86, 1352–1356.

Oda, S., Sato, M., Toyoshima, S., & Osawa, T. (1989). Binding of activated macrophages to tumor cells through a macrophage lectin and its role in macrophage tumoricidal activity. J. Biochem. 105, 1040–1043.

Ozaki, K., Ii, M., Itoh, N., & Kawasaki, T. (1992). Expression of a functional asialoglycoprotein receptor through transfection of a cloned cDNA that encodes a macrophage lectin. J. Biol. Chem. 267, 9229–9235.

Ozaki, K., Itoh, N., & Kawasaki, T. (1993). Role of tyrosine-5 in the cytoplasmic tail of the macrophage asialoglycoprotein receptor in the rapid internalization of ligands. J. Biochem. Tokyo 113, 271–276.

Paietta, E., Stockert, R. J., & Racevskis, J. (1992a). Alternatively spliced variants of the human hepatic asialoglycoprotein receptor, H2, differ in cellular trafficking and regulation of phosphorylation. J. Biol. Chem. 267, 11078–11084.

Paietta, E., Stockert, R. J., & Racevskis, J. (1992b). Differences in the abundance of variably spliced transcripts for the second asialoglycoprotein receptor polypeptide, H2, in normal and transformed human liver. Hepatology 15, 395–402.

Pearse, B. M. F. (1988). Receptors compete for adaptors found in plasma membrane coated pits. EMBO J. 7, 3331–3336.

Pearse, B. M. F., & Robinson, M. S. (1990). Clathrin, adaptors, and sorting. Annu. Rev. Cell Biol. 6, 151–171.

Pelham, H. (1989). The selectivity of secretion—protein sorting in the endoplasmic reticulum. Biochem. Soc. Trans. 17, 795–802.

Prill, V., Lehmann, L., Vonfigura, K., & Peters, C. (1993). The cytoplasmic tail of lysosomal acid phosphatase contains overlapping but distinct signals for basolateral sorting and rapid internalization in polarized MDCK cells. EMBO J. 12, 2181–2193.

Regoeczi, E., Chindemi, P. A., Debanne, M. T., & Hatton, W. C. (1982). Dual nature of the hepatic lectin pathway for human asialotransferrin type 3 in the rat. J. Biol. Chem. 257, 5431–5436.

Reimer, P., Weissleder, R., Lee, A. S., Buettner, S., Wittenberg, J., & Brady, T. J. (1991). Asialoglycoprotein receptor function in benign liver disease: Evaluation with MR imaging. Radiology 178, 769–774.

Rice, K. G., Weisz, O. A., Barthel, T., Lee, R. T., & Lee, Y. C. (1990). Defined geometry of binding between triantennary glycopeptide and the asialoglycoprotein receptor of rat hepatocytes. J. Biol. Chem. 265, 18429–18434.

Robinson, M. S. (1992). Adaptins. Trends Cell Biol. 2, 293–297.

Rodriguez-Boulan, E., & Powell, S. K. (1992). Polarity of epithelial and neuronal cells. Annu. Rev. Cell Biol. 8, 395–427.

Rothenberger, S., Iacopetta, B. J., & Kühn, L. C. (1987). Endocytosis of the transferrin receptor requires the cytoplasmic domain but not its phosphorylation site. Cell 49, 423–431.

Sanford, J. P., & Doyle, D. (1990). Mouse asialoglycoprotein receptor cDNA sequence: Conservation of receptor genes during mammalian evolution. Biochim. Biophys. Acta 1087, 259–261.

Sato, M., Kawakami, K., Osawa, T., & Toyoshima, S. (1992). Molecular cloning and expression of cDNA encoding a galactose/N-acetylgalactosamine-specific lectin on mouse tumoricidal macrophages. J. Biochem. Tokyo 111, 331–336.

Sawamura, T., & Shiozaki, Y. (1991). Mechanism and clinical relevance of elevated levels of circulating asialoglycoproteins. Targeted Diagn. Ther. 4, 215–234.

Sawyer, J. T., Sanford, J. P., & Doyle, D. (1988). Identification of a complex of the three forms of the rat liver asialoglycoprotein receptor. J. Biol. Chem. 263, 10534–10538.

Schlepper-Schäfer, J., Hülsmann, D., Djovkar, A., Meyer, H. E., Herbertz, L., Kolb, H., & Kolb-Bachofen, V. (1986). Endocytosis via galactose receptors *in vivo*. Ligand size directs uptake by hepatocytes and/or liver macrophages. Exp. Cell Res. 165, 494–506.

Schmid, S. R., & Spiess, M. (1988). Deletion of the amino-terminal domain of asialoglycoprotein receptor H1 allows cleavage of the internal signal sequence. J. Biol. Chem. 263, 16886–16891.

Schwartz, A. L. (1984a). The hepatic asialoglycoprotein receptor. CRC Crit. Rev. Biochem. 16, 207–233.

Schwartz, A. L. (1984b). Phosphorylation of the human asialoglycoprotein receptor. Biochem. J. 223, 481–486.

Schwartz, A. L., Bolognesi, A., & Fridovich, S. E. (1984). Recycling of the asialoglycoprotein receptor and the effect of lysosomotropic agents in hepatoma cells. J. Cell Biol. 98, 732–738.

Schwartz, A. L., Fridovich, S. E., & Lodish, H. F. (1982). Kinetics of internalization and recycling of the asialoglycoprotein receptor in a hepatoma cell line. J. Biol. Chem. 257, 4230–4237.

Shia, M. A., & Lodish, H. F. (1989). The two subunits of the human asialoglycoprotein receptor have different fates when expressed alone in fibroblasts. Proc. Natl. Acad. Sci. USA 86, 1158–1162.

Simons, K., & Wandinger-Ness, A. (1990). Polarized sorting in epithelia. Cell 62, 207–210.

Sorkin, A., & Carpenter, G. (1993). Interaction of activated EGF receptors with coated pit adaptins. Science 261, 612–615.

Sosa, M. A., Schmidt, B., von Figura, K., & Hille-Rehfeld, A. (1993). *In vitro* binding of plasma membrane-coated vesicle adaptors to the cytoplasmic domain of lysosomal acid phosphatase. J. Biol. Chem. 268, 12537–12543.

Spiess, M., & Handschin, C. (1987). Deletion analysis of the internal signal-anchor domain of the human asialoglycoprotein receptor H1. EMBO J. 6, 2683–2691.

Spiess, M., & Lodish, H. F. (1985). Sequence of a second human asialoglycoprotein receptor: Conservation of two receptor genes during evolution. Proc. Natl. Acad. Sci. USA 82, 6465–6469.

Spiess, M., & Lodish, H. F. (1986). An internal signal sequence: The asialoglycoprotein receptor membrane anchor. Cell 44, 177–185.

Spiess, M., Schwartz, A. L., & Lodish, H. F. (1985). Sequence of human asialoglycoprotein receptor cDNA. An internal signal sequence for membrane insertion. J. Biol. Chem. 260, 1979–1982.

Stoorvogel, W., Geuze, H. J., Griffith, J. M., Schwartz, A. L., & Strous, G. J. (1989). Relations between the intracellular pathways of the receptors for transferrin, asialoglycoprotein, and mannose 6-phosphate in human hepatoma cells. J. Cell Biol. 108, 2137–2148.

Takahashi, T., Nakada, H., Okumura, T., Sawamura, T., & Tashiro, Y. (1985). Phosphorylation of the rat hepatocyte asialoglycoprotein receptor. Biochem. Biophys. Res. Commun. 126, 1054–1060.

Takezawa, R., Shinzawa, K., Watanabe, Y., & Akaike, T. (1993). Determination of mouse major asialoglycoprotein receptor cDNA sequence. Biochim. Biophys. Acta 1172, 220–222.

Thomas, D. C., Brewer, C. B., & Roth, M. G. (1993). Vesicular stomatitis virus glycoprotein contains a dominant cytoplasmic basolateral sorting signal critically dependent upon a tyrosine. J. Biol. Chem. 268, 3313–3320.

Tolleshaug, H., & Berg, T. (1979). Chloroquine reduces the number of asialoglycoproteins receptors in the hepatocyte plasma membrane. Biochem. Pharmacol. 28, 2919–2922.

Townsend, R. R., Hardy, M. R., Wong, T. C., & Lee, Y. C. (1986). Binding of N-linked bovine fetuin glycopeptides to isolated rabbit hepatocytes: Gal/GalNAc hepatic lectin discrimination between Gal beta(1,4)GlcNAc and Gal beta(1,3)GlcNAc in a triantennary structure. Biochemistry 25, 5716–5725.

Trowbridge, I. S. (1991). Endocytosis and signals for internalization. Curr. Op. Cell Biol. 3, 634–641.

Tycko, B., Keith, C. H., & Maxfield, F. R. (1983). Rapid acidification of endocytic vesicles containing asialoglycoprotein in cells of a human hepatoma cell line. J. Cell Biol. 255, 5971–5978.

Wall, D. A., & Hubbard, A. (1981). Galactose-specific recognition system of mammalian liver: Receptor distribution on the hepatocyte cell surface. J. Cell Biol. 90, 687–696.

Weigel, P. H. (1992). Mechanisms and control of glycoconjugate turnover. In: Glycoconjugates: Composition, Structure, and Function (Allen, H. J., & Kisailus, E. C., Eds.). Marcel Dekker, Inc., New York, pp. 421–497.

Weigel, P. H., & Oka, J. A. (1984). Recycling of the hepatic asialoglycoprotein receptor in isolated rat hepatocytes. Receptor-ligand complexes in an intracellular slowly dissociating pool return to the cell surface prior to dissociation. J. Biol. Chem. 259, 1150–1154.

Weis, W. I., Crichlow, G. V., Murthy, H., Hendrickson, W. A., & Drickamer, K. (1991a). Physical characterization and crystallization of the carbohydrate-recognition domain of a mannose-binding protein from rat. J. Biol. Chem. 266, 20678–20686.

Weis, W. I., Drickamer, K., & Hendrickson, W. A. (1992). Structure of a C-type mannose-binding protein complexed with an oligosaccharide. Nature 360, 127–134.

Weis, W. I., Kahn, R., Fourme, R., Drickamer, K., & Hendrickson, W. A. (1991b). Structure of the calcium-dependent lectin domain from a rat mannose-binding protein determined by MAD phasing. Science 254, 1608–1615.

Wessels, H. P., Geffen, I., & Spiess, M. (1989). A hepatocyte-specific basolateral membrane protein is targeted to the same domain when expressed in Madin-Darby canine kidney cells. J. Biol. Chem. 264, 17–20.

Wikström, L., & Lodish, H. F. (1991). Nonlysosomal, pre-golgi degradation of unassembled asialoglycoprotein receptor subunits—A TLCK-sensitive and TPCK-sensitive cleavage within the ER. J. Cell Biol. 113, 997–1007.

Wikström, L., & Lodish, H. F. (1992). Endoplasmic reticulum degradation of a subunit of the asialoglycoprotein receptor in vitro—Vesicular transport from endoplasmic reticulum is unnecessary. J. Biol. Chem. 267, 5–8.

Wikström, L., & Lodish, H. F. (1993). Unfolded H2b asialoglycoprotein receptor subunit polypeptides are selectively degraded within the endoplasmic reticulum. J. Biol. Chem. 268, 14412–14416.

Wilson, J. M., Grossman, M., Wu, C. H., Chowdhury, N. R., Wu, G. Y., & Chowdhury, J. R. (1992). Hepatocyte-directed gene transfer in vivo leads to transient improvement of hypercholesterolemia in low density lipoprotein receptor-deficient rabbits. J. Biol. Chem. 267, 963–967.

Windler, E., Greeve, J., Levkau, B., Kolbbachofen, V., Daerr, W., & Greten, H. (1991). The human asialoglycoprotein receptor is a possible binding site for low-density lipoproteins and chylomicron remnants. Biochem. J. 276, 79–87.

Wu, G. Y., Wilson, J. M., Shalaby, F., Grossman, M., Shafritz, D. A., & Wu, C. H. (1991). Receptor-mediated gene delivery *in vivo*. Partial correction of genetic analbuminemia in Nagase rats. J. Biol. Chem. 266, 14338–14342.

Wu, G. Y., & Wu, C. H. (1991). Targeted delivery and expression of foreign genes in hepatocytes. Targeted Diagn. Ther. 4, 127–149.

Yuk, M. H., & Lodish, H. F. (1993). Two pathways for the degradation of the H2 subunit of the asialoglycoprotein receptor in the endoplasmic reticulum. J. Cell Biol. 123, 1735–1749.

Zijderhand-Bleekemolen, J. E., Schwartz, A. L., Slot, J. W., Strous, G. J., & Geuze, H. J. (1987). Ligand- and weak base-induced redistribution of asialoglycoprotein receptors in hepatoma cells. J. Cell Biol. 104, 1647–1654.

THE LOW DENSITY
LIPOPROTEIN RECEPTOR

Adrian Ozinsky, Gerhard A. Coetzee, and
Deneys R. van der Westhuyzen

Biomembranes
Volume 4, pages 201–221.
Copyright © 1996 by JAI Press Inc.
All rights of reproduction in any form reserved.
ISBN: 1-55938-661-4.

I. INTRODUCTION

The low density lipoprotein (LDL) receptor is a cell surface membrane protein that removes cholesterol-containing lipoproteins from plasma (Brown and Goldstein, 1986). It is the prototype of a family of structurally related surface receptors (the LDL receptor gene family) that bind and internalize lipoproteins. Mutations that disrupt the LDL receptor gene impair the clearance of the cholesterol-rich intermediate density and low density lipoproteins. The disease that results, familial hypercholesterolemia, is characterized by the accumulation of cholesterol in the plasma and premature atherosclerosis (Goldstein and Brown, 1989). This review will emphasize the structural requirements for the LDL receptor's function as an endocytic receptor and its ability to follow a defined, though complex intracellular itinerary. We will also highlight those functions of the LDL receptor which as yet do not have a known molecular determinant. Other members of the LDL receptor gene family, the LDL receptor-related protein/α_2-macroglobulin receptor (abbreviated as LRP), the rabbit very low density lipoprotein (VLDL) receptor, and glycoprotein 330 (GP330), will also be discussed.

II. LDL RECEPTOR GENE AND PROTEIN STRUCTURE

The LDL receptor is translated as a polypeptide of 860 amino acids (Yamamoto et al., 1984). A signal sequence of 21 amino acids is cleaved off, leaving a functional protein of 839 amino acids. The mature protein contains 2 N-linked and about 18 O-linked oligosaccharides (Cummings et al., 1983). The molecular weight of the glycosylated receptor is about 115 kD, whereas the apparent molecular weight, determined by SDS-PAGE, is 160 kD (Tolleshaug et al., 1982). The protein spans the cell membrane once and is orientated with its N-terminus being extracellular (Schneider et al., 1983). In total, 768 amino acids are extracellular, 22 amino acids span the cell membrane, and 50 amino acids form the intracellular cytoplasmic tail (Russell et al., 1984; Yamamoto et al., 1984). Overall, the LDL receptor is a modular protein composed of distinct domains (Figure 1). The extracellular portion consists of the ligand binding domain, an epidermal growth factor (EGF) precursor homology domain, and an O-linked sugar domain (Sudhof et al., 1985). The structure of these domains is discussed below.

The LDL receptor gene is found on the distal short arm of chromosome 19 (p13.1–p13.3) (reviewed by Hobbs et al., 1990). The locus spans 45 kilobases including the known upstream promoter-elements which are found within 200 base pairs 5' of the methionine codon for translation initiation. This promoter consists of three imperfect repeats and two TATA boxes. Repeats 1 and 3 are binding sites for the transcription factor, Sp1, which constitutively promotes transcription. The activity of the LDL receptor promotor is regulated by the intracellular pool of unesterified cholesterol. This effect is mediated through repeat 2, which does not bind Sp1. Repeat 2 is a conditional-positive element; low intracellular sterols

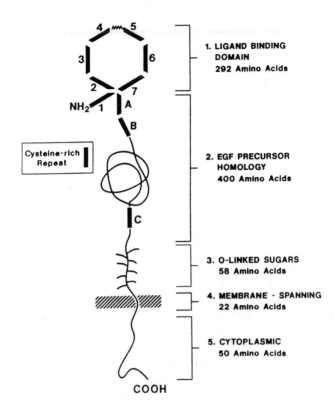

Figure 1. Model for the human LDL receptor. The domains of the LDL receptor are indicated. The ligand binding repeats are numbered 1–7, and the growth factor repeats of the EGF precursor homology domain are labeled A–C. (Figure used with permission from Esser et al., 1986.)

stimulate maximal positive transcriptional activity, but its positive effect is lost when intracellular sterols accumulate. A nuclear protein which binds repeat 2, sterol regulatory element binding protein, has recently been isolated (Wang et al., 1993; Yokoyama et al., 1993).

The division of the LDL receptor gene into exons generally corresponds to the organization of the protein into domains and subdomains (Figure 2). In many instances, single exons encode entire subdomains of the protein. This, together with the homology between several of the domains of the LDL receptor and the domains of other proteins, including the other members of the gene family, is suggestive of the shuffling of exons during the evolution of the LDL receptor (Südhof et al., 1985).

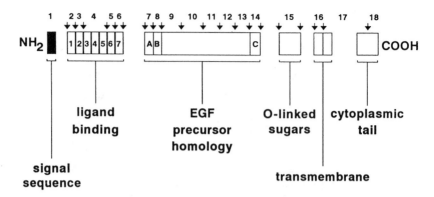

Figure 2. Exon organization and protein domains in the human LDL receptor. The six domains of the protein are labeled in the lower portion. The seven cysteine-rich, 40-amino acid repeats in the binding domain are numbered 1–7. The three growth factor repeats in the EGF precursor homology domain are lettered A–C. The positions at which introns interrupt the coding region are indicated by arrows. Exon numbers are shown between the arrows. (Figure based on Südhof et al., 1985.)

The LDL receptor gene is composed of 18 exons. Exon 1 encodes a signal sequence which is cleaved from the protein in the endoplasmic reticulum. Exons 2–6 encode the ligand binding domain which consists of 292 amino acids. It is divided into seven homologous, though imperfect, repeats of 40 amino acids, which are aligned head-to-tail and which are homologous to the C9 component of complement. Each repeat contains six highly conserved cysteine residues. These cysteines are thought to be fully disulphide bonded within their given repeat. The COOH-terminal of each repeat has a conserved cluster of negatively charged amino acids which are necessary for ligand binding.

The domain with homology to the epidermal growth factor (EGF) precursor is encoded by exons 7–14. This domain extends from amino acids 293–692. It contains three growth factor repeats (A–C) which are found in four copies in the precursor for EGF. The A and B repeats of the LDL receptor are separated from the C repeat by five copies of a 40–60 amino acid repeat, each of which contains a conserved YWTD (tyrosine-tryptophan-threonine-aspartate) motif. The EGF precursor homology domain is needed for the binding of LDL on the cell surface, as well as for the dissociation of ligand from the receptor in endosomes.

The O-linked sugar domain is encoded by exon 15. This 57 amino acid domain (amino acids 693–750) contains a cluster of 18 serine or threonine residues which are potential sites for the addition of O-linked oligosaccharide chains. The function of this domain has not been clearly defined. Deletion of the domain by site-directed mutagenesis, did not affect the synthesis, transport, binding, recycling, nor degradation of the receptor when transfected into hamster fibroblasts (Davis et al., 1986a). Familial hypercholesterolemic (FH) patients homozygous for a naturally-occurring deletion of the entire exon 15, express a phenotype milder than classic FH (Koivisto et al., 1993). Their LDL receptors retain residual function, resulting in a moderate impairment of LDL catabolism, and a serum cholesterol concentration intermediate between normal and FH levels. The manner in which the cell biology of the LDL receptor is disturbed in these patients has not been determined. It has been speculated that the extensive O-linked glycosylation of this domain may hold the receptor in an extended conformation, so as to facilitate ligand binding, but this has not been shown directly (Goldstein et al., 1985).

The membrane-spanning domain, which anchors the receptor in the membrane of the various compartments of the cell, consists of a hydrophobic 22 amino acid sequence which is encoded by exon 16 and the 5' part of exon 17. The 50 amino acid cytoplasmic tail of the LDL receptor (amino acids 790–839) is encoded by the 3' end of exon 17 and exon 18. This domain is well conserved between species, and contains signals for targeting the receptor to the various plasma membrane domains of polarized cells, clustering in coated pits and internalization, as well as the self-association of receptors into multimeric structures.

The oligomerization of LDL receptors has been demonstrated by the cross-linking of up to 20% of the monomers into dimeric and trimeric structures (Van Driel et al., 1987). Higher order multimers were not detected. The self-association of LDL receptors is thought to be mediated by noncovalent interactions between part of their cytoplasmic tails, from amino acids 812–829. The role of oligomerization has not been clarified but may be important in the binding of certain ligands (see following).

III.   THE LDL RECEPTOR GENE FAMILY

Members of the LDL receptor gene family have related structures (Figure 3). These proteins use the same domains found in the LDL receptor as building blocks, assembling them in different numbers into different structures. The LDL receptor gene family comprises a growing number of members from different species (Schneider and Nimpf, 1993). These proteins are far less-well characterized than the LDL receptor itself, but they begin to reveal how a different pattern of assembly modifies a protein's function. The structures of three members, the VLDL receptor, LRP, and GP330 are discussed in relation to the LDL receptor.

A.   The VLDL Receptor

The 846 amino acid rabbit VLDL receptor shows remarkable protein sequence and domain homology with the LDL receptor (Takahashi et al., 1992). This receptor is able to bind and endocytose apoE-containing lipoproteins (including VLDL) with high affinity. However, unlike the LDL receptor, it is not able to bind the apoB-containing ligand, LDL. mRNA for the VLDL receptor is expressed abundantly in

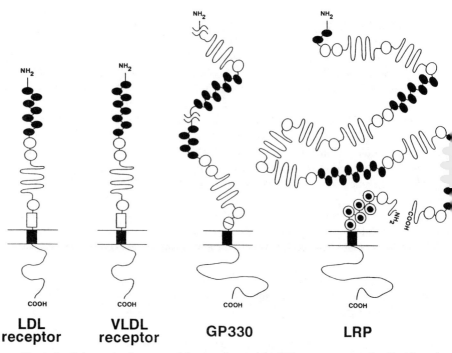

Figure 3. Schematic diagrams of the members of the LDL receptor gene family. Ligand binding repeat (●); growth factor repeats (○) of the EGF precursor homology domain with intervening YWTD-repeat spacer (~); incomplete growth factor repeats (⊖); unidentified regions of GP330 (≈); 0-linked sugar domain (□); EGF repeats (◉); transmembrane domain (■). (Figure modified with permission from Hobbs et al., 1990.)

various extra-hepatic tissues, but not in the liver, leading to the proposal that this receptor could mediate the extra-hepatic clearance of triglyceride-rich lipoproteins such as VLDL. The key structural feature of the VLDL receptor that could account for its inability to bind LDL is in its binding domain, where it has eight binding repeats compared to the seven repeats of the LDL receptor. The VLDL receptor repeats retain the spacing of the cysteine residues and the presence of a carboxy-terminal cluster of acidic amino acids that are characteristic of the LDL receptor's binding repeats. The binding of ligand to the LDL receptor and the VLDL receptor is discussed below under "ligand binding."

B. LDL Receptor-related Protein

LRP is the largest cell-surface protein yet identified and is equivalent to the α_2-macroglobulin receptor (Herz et al., 1988). This 600 kD protein of 4525 amino acids, binds lipoproteins that have been enriched with apoprotein E (apoE), and thus has been proposed to be the receptor that removes chylomicron remnants from the circulation. LRP is also able to bind a diverse group of other ligands, including protease complexes, toxins, and other proteins of uncertain function (reviewed by Herz, 1993). LRP has an extracellular domain of 4,400 amino acids, which grossly corresponds to four copies of the extracellular domain of the LDL receptor, a predicted single transmembrane span of 25 amino acids, and a 100 amino acid cytoplasmic tail. Whereas the LDL receptor contains a single binding domain made up of a cluster of seven binding repeats, LRP has 31 binding repeats, arranged in four clusters with 2-, 8-, 10-, and 11-repeats. Adjacent to each cluster of binding repeats, is a domain with homology to the EGF precursor, though, in LRP, each domain is expanded when compared to that found in the LDL receptor. Whereas the LDL receptor has three growth factor repeats and a single set of five YWTD repeats, LRP has a total of 22 growth factor repeats and eight copies of the set of five YWTD repeats (see structure in Figure 3). LRP does not have an O-linked sugar domain. In its place, adjacent to the cell membrane, is a cassette of six repeats which are homologous to a second type of growth factor repeat found in the EGF precursor. These repeats differ in the spacing of their cysteine residues from the growth factor repeats of the EGF precursor homology domain found in both the LDL receptor and the LRP.

Although translated as a single peptide of 600 kD, LRP is cleaved in the Golgi apparatus into a 515 kD fragment, which contains the bulk of the predicted extracellular domain, and an 85 kD fragment, which contains the single transmembrane domain and the cytoplasmic tail (Herz et al., 1990). These cleaved peptides remain noncovalently associated, and thus the extracellular domain remains membrane bound. This proteolysis occurs in LRP at a proteolytic consensus sequence that is not found in the LDL receptor. The binding of ligands to LRP is discussed below under "ligand binding."

C. Glycoprotein 330

GP330 is a membrane glycoprotein whose physiological function remains to be determined. It is expressed in several types of epithelial cells, including the proximal tubule of the kidney, where it is localized in the clathrin coated pits on the apical cell membrane (Kerjaschki and Farquhar, 1983; Kounnas et al., 1992). GP330 functions as an endocytic receptor (Moestrup et al., 1993a) and has the ability to bind many of the ligands of LRP *in vitro*. The binding of ligand to GP330 is discussed under "ligand binding." The structure of GP330 has not been completely determined. Partial cDNA clones have been isolated from rat kidney (Raychowdhury et al., 1989) and a human homologue has been identified (Kanalas and Makker, 1990). The deduced protein has an extracellular domain containing more than 13 ligand binding repeats homologous to those of the LDL receptor (Figure 3). GP330 also contains four complete growth factor repeats, together with two incomplete repeats. There is a transmembrane span of 29 amino acids and a cytoplasmic tail of 188 amino acids which includes two NPXY internalization signals.

IV. ITINERARY OF THE LDL RECEPTOR

The LDL receptor is a migrant membrane protein which traverses many membrane compartments during its complex, though defined, intracellular itinerary (Goldstein and Brown, 1985). This section provides an outline of the route travelled by the LDL receptor and of the characteristics of ligand binding. Emphasis is placed on the structural features of the LDL receptor which determine events at different locations in the cell.

A. Synthesis of LDL Receptors

Synthesis of the LDL receptor is initiated with a 21 amino acid signal sequence which is later cleaved from the receptor, presumably in the ER following membrane insertion of the polypeptide. A precursor form of the LDL receptor with an apparent molecular weight of 120 kD can be identified in the ER. This precursor is glycosylated with core O-linked sugar chains and the high mannose N-linked chains characteristic of an ER protein (Cummings et al., 1983). The receptor is transported to the cell surface via the Golgi apparatus where the oligosaccharide chains are processed. This processing alters the apparent molecular weight of the receptor precursor from 120 kD to the apparent 160 kD of the mature receptor. Partially glycosylated intermediates are not seen, probably due to the small pool sizes in the various compartments of the secretory pathway, resulting from the rapid transit rate from the ER to the surface ($t_{1/2}$ = 15 minutes).

B. Targeting of LDL Receptors

The LDL receptor moves passively and constitutively though the secretory pathway (Goldstein et al., 1985). It does not contain a retention signal characteristic of proteins resident in intracellular locations (Pelham, 1991). Targeting signals within the cytoplasmic tail confer specific itineraries on certain migrant proteins and this also applies to the LDL receptor. This targeting includes the sorting of membrane proteins to the different surfaces of polarized cells. An understanding of the expression of the LDL receptor in polarized cells is particularly important since hepatocytes perform a central role in the clearance of cholesterol from the plasma. The sinusoidal surface of the hepatocyte corresponds to the basolateral surface, from which cholesterol-containing lipoproteins are removed from the circulation by the LDL receptor. The bile cannelicular surface corresponds to the apical surface (Mostov et al., 1992). Direct experiments on hepatocytes are confounded by their low level of expression of LDL receptors and their difficulty to culture (Pathak et al., 1990; Yokode et al., 1992). Targeting events have thus been studied in Madin–Darby canine kidney (MDCK) cells, which are a common model system for polarized cells (Simons and Fuller, 1985; Mostov et al., 1992), and in the hepato-cytes of mice containing a transgene causing a high level of expression of the human LDL receptor.

The signals for the targeting of the LDL receptor reside in its cytoplasmic tail. In polarized MDCK cells (Hunziker et al., 1991) and hepatocytes (Yokode et al., 1992), these signals direct the LDL receptor to the basolateral surface. Extensive mutagenesis studies in MDCK cells have shown that there are two basolateral sorting signals in the tail of the LDL receptor (Figure 4; Matter et al., 1992). The position and nature of these targeting signals have not been precisely demarcated. The proximal signal, which includes amino acids 802–816, overlaps with the amino acids necessary for endocytosis (amino acids 804–807; see below). Truncation of the tail with a stop signal at amino acid 812, therefore, allows normal endocytosis, but disrupts basolateral targeting. The distal targeting signal (amino acids 821–839) includes the last eight amino acids of the tail. This potentially explains why this region of the LDL receptor, which is encoded by the last exon of the gene, is highly conserved in diverse species (Mehta et al., 1991). The two signals are able to function independently of each other, with each being able individually to direct targeting. However, they appear to have differing sorting capacities, with the proximal determinant being unable to effectively sort LDL receptors during in-creased levels of expression, indicating a low sorting capacity. In transgenic mice, however, the surface distribution of the LDL receptor was not altered by differing levels of expression of the LDL receptor transgene (Pathak et al., 1990). The mechanisms underlying this targeting and the difference between the two signals are not known, nor are the reasons for the presence of multiple targeting signals. Both targeting signals also sort LDL receptors in endosomes, to distinguish the recycling pathway from the transcytosis pathway (Matter et al., 1993; also see "recycling of LDL receptors"). The distal signal is active in the basolateral en-dosomes, while the proximal signal is active in the apical endosomes. The converse experiments have not been performed, thus it has not been determined whether these activities are unique to these sites, or whether both signals are active in all endosomes. Another possibility is that the two signals display tissue specificity.

Membrane	basolateral targeting proximal determinant		basolateral targeting distal determinant		
KNWRLKNINSINFDNPVYQKTTEDEVHICHNQDGYSYPSRQMVSLEDDVA
7	8	8	8	8	8
9	0 *endocytosis* 1	2	3	3
0	0	0	0	0	9

Figure 4. Localization of the signals for internalization and targeting in the cytoplas-mic tail of the LDL receptor. The sequence of the LDL receptor cytoplasmic tail is given in single letter code. The position of the signal for internalization through clathrin-coated pits is indicated, as are the positions for the proximal and distal signals for basolateral targeting in polarized cells.

Indeed, the adaptor proteins which are necessary for the recognition of targeting signals in various cell pathways have different tissue distributions as well as different intracellular distributions (Pearse and Robinson, 1990).

The steady-state distribution of LDL receptors on the surface of hepatocytes is different to that of fibroblasts, which are not polarized. In fibroblasts, about 70% of LDL receptors are found clustered in coated pits, through which efficient receptor-mediated endocytosis occurs (Anderson et al., 1982). These clustered LDL receptors are derived from a dispersed population of receptors which appear on the cell surface during receptor recycling (Sanan et al., 1989). In the hepatocytes of transgenic mice, LDL receptors on the basolateral surface are spread diffusely, with few receptors seen within coated pits (Pathak et al., 1990). These LDL receptors, nevertheless, mediate the efficient internalization of LDL. No LDL receptors could be detected on the apical surface of hepatocytes in the transgenic animals. This distribution is, therefore, consistent with the known basolateral targeting signals in the LDL receptor, as identified in MDCK cells. Evidently, basolaterally-targeted LDL receptors were not redirected to the apical surface by the transcytosis pathway followed by apical proteins in hepatocytes. The distribution of LDL receptors was different in the kidney epithelial cells of transgenic mice. Here the LDL receptors were found preferentially on the apical surface and in the apical endocytic pathway. The mechanism underlying these sorting differences in the various epithelial cells are not known. In the case of the polymeric immunoglobulin receptor, phosphory-lation of a serine residue in its tail causes apical expression by inactivating basolateral targeting (Casanova et al., 1990). However, this mechanism does not apply for the LDL receptor; substitution of the serine residue (amino acid 833) in the cytoplasmic tail did not affect its basolateral distribution (Yokode et al., 1992), even though this residue can be phosphorylated (Kishimoto et al., 1987).

The targeting signal sequences of the LDL receptor are not well conserved in the other members of the LDL receptor gene family, though this assessment is compli-cated by disparate sequences being able to achieve the structure of a targeting signal. A targeting signal has not been identified in the tail of GP330, though it is confined to the apical pole of kidney cells (Kerjaschki and Farquhar, 1983; Kounnas et al., 1992). The cell biology of the VLDL receptor in polarized cells has not been studied and the receptor may not be targeted. Indeed, the tissues expressing the most VLDL receptor mRNA namely, heart, muscle, and adipose tissue, are not polarized, whereas the liver has barely detectable mRNA levels (Takahashi et al., 1992). In the case of LRP, its distribution in polarized cells has not been reported.

C. Ligand Binding

Binding to the LDL Receptor

The LDL receptor binds lipoproteins containing apolipoprotein B-100 (apoB) or apolipoprotein E (apoE; Goldstein and Brown, 1977). These apolipoproteins are found on several lipoproteins, including LDL, VLDL, βVLDL, IDL, chylomicron remnants, and HDL$_c$, all of which are ligands for the LDL receptor. The binding of the apoB-containing ligand, LDL, and apoE-containing ligand, βVLDL, is described, demonstrating how a receptor utilizes multiple repeats within a single binding domain to recognize more than one ligand.

The binding between the LDLR and apoB and apoE occurs via ionic interactions between acidic residues of the LDLR and regions rich in basic amino acids in both ligands. This is also true for the binding of ligands to LRP and GP330 (Moestrup et al., 1993b). The ligand binding domain of the LDL receptor consists of seven binding repeats. Mutation studies have shown that these repeats have a modular structure, with different combinations of the repeats being required for the binding of the different ligands to the LDL receptor (Esser et al., 1988; Russell et al., 1989). In these studies, where the binding repeats were individually deleted, deletion of repeat 5 reduced apoE binding by 60% whereas any other repeat could be removed without affecting binding. The simplest interpretation of these findings is that apoE only binds to repeat 5. However, this result does not exclude other more complex models, which include a role for the other repeats in apoE binding. Repeat 5 has the unique presence of 4 acidic amino acids compared to a triplet cluster of acidic amino acids found in the other repeats (serine-aspartate-glutamate-glutamate vs. serine-aspartate-glutamate) which are important for ligand binding (see following). Interestingly, the binding of apoE-containing ligands occurs with a 10-fold higher affinity than apoB-containing ligands. This is thought, in part, to be due to the many apoE molecules of a single lipoprotein particle being able to interact simultaneously, with several LDL receptors of an oligomeric partnership, with each receptor providing a single apoE binding site. The binding of ApoB (LDL) tolerated only the deletion of repeat 1, indicating that repeats 2–7 were all required for apoB binding. In addition to the ligand binding domain, repeat A of the EGF precursor homology domain is also required for the binding of apoB (Esser et al., 1988); its deletion reduces LDL binding by 75%. However, it is not required for the binding of LDL to receptors which are immobilized on nitrocellulose (ligand blot). Thus repeat A appears to perform a permissive role in allowing LDL binding at the cell surface. The receptor binding domain of apoB is composed of two clusters of basic amino acids; one of the clusters is homologous to the receptor binding domain of apoE. In apoB (4536 amino acids), a region totalling about 800 amino acids which flanks this site, is necessary for its integrity (Milne et al., 1989). Due to its large size, especially of its receptor binding domain, apoB is probably relatively rigid on the surface of the lipoprotein particle. This might explain its poor tolerance to mutations in the LDL receptor's binding repeats. By contrast apoE, is a far smaller, 299 amino acid protein (Lalazar et al., 1988), probably more able to adapt to distortions introduced into the receptor outside of repeat 5, possibly due to its mobility on the surface of the lipoprotein particle (induced fit model).

The modular character of each binding repeat is emphasized by the finding that point mutations within different binding repeats cause the same disruption of binding as deletion of the entire corresponding repeat (Esser et al., 1988; Russell et al., 1989). This indicates that the effects of point mutations extend only within their own repeat. The numerous mutations within the ligand binding domain that have been described (reviewed by Hobbs et al., 1993) all conform to these general principles for the role of the different repeats in the binding of ligand.

Binding to Members of the LDL Receptor Gene Family

The ligand binding repeats of the other members of the LDL receptor gene family retain the same overt features thought to be important for ligand binding in the LDL receptor. They all retain about 50% amino acid homology between their own repeats, which is comparable to the homology between the repeats of the LDL receptor's binding domain, as well as between the repeats of the different members of the gene family. All the members of the gene family are able to bind apoE-containing ligands, though none are able to bind the apoB ligand, LDL (Takahashi et al., 1992; Willnow et al., 1992). The structural features that confer this binding specificity have not been defined. It is intriguing that the VLDL receptor, with eight binding repeats, is unable to bind LDL (Takahashi et al., 1992). LDL binding is also impaired in other instances where the receptor has more than seven binding repeats. In LRP, it is also a cluster of 8 binding repeats that has been shown to bind ligands, yet LRP is unable to bind LDL (Moestrup et al., 1993b). In a naturally-occurring mutant LDL receptor, which has a duplication of the entire ligand binding domain, LDL binding is also impaired (a low binding capacity, albeit with a normal binding affinity; Lehrman et al., 1987). In all these instances, the impaired LDL binding could result from an unidentified, subtle alteration in the amino acid sequence compared to the native LDL receptor, rather than from the increased number of binding repeats.

The binding of receptor associated protein (RAP) to the members of the LDL receptor gene family is also intriguing. This 39 kD protein is able to compete with the binding of all the various ligands to LRP (reviewed by Herz, 1993), as well as being able to bind to GP330 (Kounnas et al., 1992). Several of these ligands do not compete with each other, indicating that LRP has more than one ligand binding site, though each is able to bind RAP. Despite its apparent broad binding specificity, RAP is unable to bind to the LDL receptor (Herz et al., 1991). Indeed, RAP limits the binding of apoE-enriched βVLDL (an apoE ligand) to LRP, but does not inhibit the binding of the same ligand to the LDL receptor. These findings are difficult to interpret because competition between two ligands does not necessarily mean that they are binding to the identical site (Herz, 1993).

Role of Proteoglycans and Lipase in Ligand Binding to Cells

The binding of lipoproteins to cells may not be initiated by an interaction with a specific surface receptor, such as the LDL receptor. The lipoproteins may first become cell-associated, by binding to heparin and dermatan sulphate proteoglycans

(HSPG), prior to interacting with the LDL receptor (Saxena et al., 1993). Although the binding of lipoproteins directly to HSPGs may be poor, at least two factors are known to increase this binding by up to 80-fold (Mulder et al., 1993). First, lipoprotein lipase (LPL) acts as an intermediary in associating lipoproteins (LDL, LP(a), and VLDL) with HSPGs (Rumsey et al., 1992; Saxena et al., 1993; Williams et al., 1992; Chappell et al., 1993; Mulder et al., 1993). Second, apoE-enriched VLDL and chylomicron remnants also associate with HSPGs (Ji et al., 1993). The subsequent endocytosis of the lipoproteins does not occur via the HSPGs, but is thought to be mediated by lipoprotein receptors. The role of the LDL receptor in these processes is controversial. Fibroblasts from FH patients, which lack functional LDL receptors, were reported to internalize LDL associated with HSPGs (Rumsey et al., 1992). In another study (Williams et al., 1992), down-regulation of LDL receptor activity did not reduce the internalization of HSPG bound lipoproteins. In contrast, however, FH fibroblasts were reported to exhibit less than 10% of HSPG-mediated LDL uptake of normal fibroblasts (Mulder et al., 1993), thereby indicating a major role for the LDL receptor in this process. This interpretation was supported by the observation that the internalization of LDL bound to HSPGs was suppressed in parallel to down-regulation of LDL receptor activity (Mulder et al., 1993).

While it is an appealing hypothesis that the HSPGs might act as an abundant source of low affinity sites to initially concentrate ligand at the cell surface, the physiological demonstration of these findings is less clear. The level of LPL used to demonstrate enhanced binding of lipoproteins to HSPGs is 2 orders of magnitude higher than the concentration of LPL in the circulation (Goldberg et al., 1986; Williams et al., 1992; Mulder et al., 1993). The LPL concentration, though, may be far greater in confined compartments such as the Space of Disse, the site of clearance of lipoproteins by the liver (Williams et al., 1992). In such a localized environment, the interaction of lipoproteins with HSPGs would be influenced not only by the availability of LPL, but also by the availability of apoE which prevents lipoprotein binding to LPL (Saxena et al., 1993). Interpretation may be further complicated as apoE secreted from the hepatocyte may associate with lipoprotein remnant particles (apoE-enriched) which would enable them to bind directly to HSPGs (Ji et al., 1993). ApoE secreted by hepatocytes into this local environment thus might alter the type of lipoprotein associated with the HSPGs.

D. Internalization of LDL Receptors

The internalization of ligand by the LDL receptor is the prototype for the paradigm of receptor-mediated endocytosis occurring through clathrin-coated pits (Brown et al., 1983; Goldstein et al., 1985). The first identification of a signal causing internalization of a surface protein was identified in the LDL receptor (Lehrman et al., 1985; Davis et al., 1986b). The internalization signal was localized to the first 22 amino acid (amino acids 790–812) of the 50 amino acid tail (Davis et al., 1987a). Mutagenesis studies revealed that the internalization signal for the LDL receptor includes NPXY (amino acids 804–807) where X could be any amino acid (Davis et al., 1987b; Chen et al., 1990). This sequence is conserved in the LDL receptor tail across six species and is found in the cytoplasmic tail of several other proteins known to be internalized via coated pits, though this is not the only amino acid sequence able to mediate internalization through coated pits. These internalization signals are all assumed to adopt a similar conformation (Collawn et al., 1990) and structural studies on a nona-peptide containing the NPVY internalization signal of the LDL receptor show that it forms a reverse β-turn (Bansal and Gierach, 1991). Peptides derived from LDL receptors that are defective in internalization do not form reverse turns. The aromatic amino acid (tyrosine 807) is crucial to the structural integrity of the turn, whereas there is a lax requirement for the amino acid at position 806. The region on the COOH-terminal side of the signal does not affect the structure of the turn (Bansal and Gierach, 1991). The importance of the turn was further highlighted by the correlation of the propensity of the nona-peptide to assume a reverse turn and the efficiency of internalization. The endocytosis signal of lysosomal acid phosphatase, which has no sequence similarity with the signal of the LDL receptor, also adopts a reverse turn conformation, emphasizing the diversity of sequences which may adopt a reverse turn (Eberle et al., 1991). Thus the internalization signal is thought to be expressed in the structural context of a tight reverse turn conformation, which is presented to its putative cytoplasmic receptor.

   NPXY internalization signals are conserved in the cytoplasmic tails of the other members of the LDL receptor gene family, all of which have been shown to be endocytic receptors. LRP and GP330 have two NPXY internalization signals in their tails. Duplication of internalization signals in the transferrin receptor increases its rate of internalization (Collawn et al., 1993), though this has not been assessed in the LDL receptor gene family.

   What are the cellular components that recognize the internalization signal? A strong candidate is the adaptor protein complex that is localized to the plasma membrane (termed hydroxyapatite [HA] type II; reviewed by Pearse and Robinson, 1990). Adaptors bind to the cytoplasmic tail of different receptors including the LDL receptor and are also required for receptor inclusion into coated pits (Pearse, 1988; Smythe et al., 1992). These complexes (molecular weight 250–300 kD) are composed of two adaptin subunits (α-, β-adaptins) of 100 kD each, together with one 50 kD protein and another 20 kD protein. The interaction between the adaptor and the LDL receptor tail can be prevented by competition by other receptors which are internalized via coated pits, indicating the common role of a single protein complex in recognizing diverse internalization signals. Importantly, a receptor which is not internalized via coated pits, the hemaglutinin receptor, does not compete with the LDL receptor for the binding of adaptors (Lazarovits and Roth, 1988; Pearse, 1988). The actual binding site for adaptors on the LDL receptor has not been studied. On other receptors, adaptors bind to the aromatic internalization signal (Pearse and Robinson, 1990). Adaptors also bind clathrin and promote the assembly of clathrin triskelions. Thus, they have a central role in receptor mediated endocytosis, both by localizing specific receptors within coated pits and by promoting the formation of the clathrin coat.

E. Recycling of LDL Receptors

The itinerary of the LDL receptor, in the endosomal pathway, was initially characterized in fibroblasts (reviewed by Goldstein and Brown, 1985). After internalization, the route followed by the LDL receptor parts from that of the ligand, which is delivered toward the lysosomes. The LDL receptor recycles back to the cell surface to constitutively mediate further rounds of endocytosis (in the presence or absence of ligand). The average time for the round trip of one endocytic cycle of the LDL receptor, has been estimated to be about 12 min in fibroblasts (Brown et al., 1983). Given a receptor half-life of 12 h in fibroblasts (Casciola et al., 1988; and see below), each LDL receptor therefore performs on average, about 60 rounds of endocytosis. Dissociation of ligand from LDL receptors occurs due to the acidification of the endosome by proton pumps (Brown et al., 1983). A change in the conformation of the LDL receptor occurs, probably dependent on repeats A and B of the EGF precursor homology domain (Davis et al., 1987b). Deletion of these repeats prevents ligand–receptor dissociation and causes receptors to become trapped within the cell (Davis et al., 1987b; Van der Westhuyzen et al., 1991). This also occurs where the acidification of endosomes is prevented by incubation with NH_4Cl (Grant et al., 1990). Under these conditions, recycling is impaired only in the presence of ligand, indicating that the NH_4Cl affects receptor–ligand dissociation, without a general effect on the recycling process. Interestingly, the recycling of LDL receptors is completely inhibited while they are only partially occupied by LDL. This is consistant with a model where LDL receptors bind ligand and recycle as partners in an oligomeric assembly.

In polarized cells, the itinerary of LDL receptors is more complex than in fibroblasts. LDL receptors require sorting in endosomes either to recycle back to the surface from where they were internalized, or to cross the cell in a transcytosis pathway (Li et al., 1991; Matter et al., 1993). In MDCK cells, LDL receptors follow different paths, depending on the surface from which they entered the cell. LDL receptors in basolateral endosomes recycle to the basolateral surface, while delivering ligand to the lysosomes. These receptors do not enter the transcytosis pathway. By contrast, LDL receptors in endosomes derived from the apical surface, are directed to transcytose ligand to the basolateral surface. Thus the LDL receptors, from either source of endosomes, are sorted to the basolateral surface, though via different pathways. These basolateral targeting signals are the same as those which direct the sorting of newly-synthesized LDL receptors within the *trans*-Golgi network, indicating that a common sorting mechanism probably recognizes the signals in both sites (Matter et al., 1993). Disruption of the targeting signals causes missorting in endosomes, with both the basolateral and the apical LDL receptors being targeted to the apical surface.

F. Turnover of LDL Receptors

The site and the mechanism of the degradation of normal LDL receptors has not been identified. In fibroblasts, LDL receptors are turned over with a half-life of about 12 h (Casciola et al., 1988). This rate is not changed by conditions that alter the number of receptors present within the cell, indicating that the number of receptors is solely regulated by the rate of synthesis, and not through degradation. Lysosomes, responsible for the degradation of the LDL ligand, appear not to be the primary site of receptor turnover, since their activity can be inhibited without affecting the half-life of the LDL receptor (Casciola et al., 1989). The presence of ligand, though, has no effect on the stability of normal LDL receptors (Casciola et al., 1988). These results indicate that the receptor and the ligand follow independent routes to their sites of degradation (Casciola et al., 1988). The initial degradation events of the LDL receptor may occur on the cell surface, as mutations which prevent receptor internalization do not alter their rate of degradation (Casciola et al., 1989).

The stability of pre-existing LDL receptors is prolonged when new protein synthesis is inhibited by cycloheximide (Casciola et al., 1988). This suggests that the turnover of the LDL receptor requires a short-lived protein, though its function is not known. In the case of the normal LDL receptor expressed in cells defective in glycosylation ability, rapid receptor degradation occurs with the release of a large soluble receptor fragment into the medium (Kozarsky et al., 1988).

Certain mutant LDL receptors are degraded at an enhanced rate, sometimes due to their being unable to follow the normal LDL receptor itinerary. However, the features which dictate the sites at which LDL receptors are rendered unstable, are not always clear. Thus certain LDL receptors are retained within the ER, probably due to an impairment in their folding (Hobbs et al., 1990). Some of these mutant LDL receptors are degraded from this site, but others are stable while being retained in the ER. LDL receptors with certain other mutations escape from the ER and reach the cell surface, where they exhibit their unstable phenotype (Fourie et al., 1988). In all these instances, the mechanisms of degradation are not known. LDL receptors with mutations in the EGF precursor homology domain are prevented from recycling, as they are unable to dissociate from ligand (Davis et al., 1987; van der Westhuyzen et al., 1991). These intracellularly trapped LDL receptors are rendered unstable in the presence of ligand, indicating that it is their inability to dissociate from ligand which targets them to degradation. Normal LDL receptors can also be prevented from dissociating from ligand by incubation with NH_4Cl which prevents the acidification of endosomes. This too enhances the degradation of LDL receptors in a ligand-dependent manner (Grant et al., 1990). NH_4Cl inhibits the function of lysosomes, indicating that they are not involved in this enhanced LDL receptor degradation.

V. CONCLUSIONS

The LDL receptor has performed a leading role in the dramatic revelation of receptor-mediated endocytosis as an important cellular process. We have reviewed the studies addressing the known structure/function relationships of the LDL receptor, which show that it contains a collage of structural motifs that specify its functions as an endocytic receptor. The structural determinants have been identified for ligand binding and uptake, and also receptor targeting and recycling. The LDL receptor's various domains are used in different combinations in the assembly of a number of other proteins which, together with the LDL receptor, constitute the LDL receptor gene family. Structural comparisons within the family reveal how modifications of the binding domain have led to the acquisition of new ligand binding specificities. Knowledge of the structures of the domains of the LDL receptor enable predictions regarding the probable functions of similar domains in the other members of the gene family. Characterization of their function will, in turn, expand our insight into the structural requirements for receptor-mediated endocytosis.

REFERENCES

Anderson, R. G. W., Brown, M. S., Beisiegel, U., & Goldstein, J. L. (1982). Surface distribution and recycling of the LDL receptor as visualized by anti-receptor antibodies. J. Cell Biol. 93, 523–531.

Bansal, A., & Gierach, L. M. (1991). The NPXY internalization signal of the LDL receptor adopts a reverse-turn conformation. Cell 67, 1195–1201.

Brown, M. S., Anderson, R. G. W., & Goldstein, J. L. (1983). Recycling receptors: The round-trip itinerary of migrant membrane proteins. Cell 663–667.

Brown, M. S., & Goldstein, J. L. (1986). A receptor-mediated pathway for cholesterol homeostasis. Science 232, 34–47.

Casanova, J. E., Breitfeld, P. P., Ross, S. A., & Mostov, K. E. (1990). Phosphorylation of the polymeric immunoglobulin receptor is required for its efficient transcytosis. Science 248, 742–746.

Casciola, L. A. F., Van der Westhuyzen, D. R., Gevers, W., & Coetzee, G. A. (1988). Low density lipoprotein receptor degradation is influenced by a mediator protein(s) with a rapid turnover rate, but is unaffected by receptor up- or down-regulation. J. Lipid Res. 29, 1481–1489.

Casciola, L. A. F., Grant, K. I., Gevers, W., Coetzee, G. A., & Van der Westhuyzen, D. R. (1989). Low-density-lipoprotein receptors in human fibroblasts are not degraded in lysosomes. Biochem. J. 262, 681–683.

Chappel, D. A., Fry, G. L., Waknitz, M. A., Muhonen, L. E., Pladet, M. W., Iverius, P-H., & Strickland, D. K. (1993). Lipoprotein lipase induces catabolism of normal triglyceride-rich lipoproteins via the low density lipoprotein receptor-related protein/α_2-macroglobulin receptor *in vitro*. J. Biol. Chem. 268, 14168–14175.

Chen, W-J., Goldstein, J. L., & Brown, M. S. (1990). NPXY, a sequence often found in cytoplasmic tails, is required for coated pit-mediated internalization of the low density lipoprotein receptor. J. Biol. Chem. 265, 3116–3123.

Collawn, J. F., Stangel, M., Kuhn, L. A., Esekogwu, V., Jing, S., Trowbridge, I. S., & Tainer, J. A. (1990). Transferrin receptor internalization sequence YXRF implicates a tight turn as the structural recognition motif for endocytosis. Cell 63, 1061–1072.

Collawn, J. F., Lai, A., Domingo, D., Fitch, M., Hatton, S., & Trowbridge, I. S. (1993). YTRF is the conserved internalization signal of the transferrin receptor, and a second YTRF signal at position 31-34 enhances endocytosis. J. Biol. Chem. 268, 21686–21692.

Cummings, R. D., Kornfeld, S., Schneider, W. J., Hobgood, K. K., Tolleshaug, H., Brown, M. S., & Goldstein, J. L. (1983). Biosynthesis of N- and O-linked oligosaccharides of the low density lipoprotein receptor. J. Biol. Chem. 258, 15261–15273.

Davis, C. G., Elhammer, A., Russell, D. W., Schneider, W. J., Kornfeld, S., Brown, M. S., & Goldstein, J. L. (1986a). Deletion of clustered O-linked carbohydrates does not impair function of low density lipoprotein receptor in transfected fibroblasts. J. Biol. Chem. 261, 2828–2838.

Davis, C. G., Lehrman, M. A., Russell, D. W., Anderson, R. G. W., Brown, M. S., & Goldstein, J. L. (1986b). The J.D. mutation in familial hypercholesterolemia: Amino acid substitution in cytoplasmic domain impedes internalization of LDL receptors. Cell 45, 15–24.

Davis, C. G., Van Driel, I., Russell, D. W., Brown, M. S., & Goldstein, J. L. (1987a). The low density lipoprotein receptor. Identification of amino acids in cytoplasmic domain required for rapid endocytosis. J. Biol. Chem. 262, 4075–4082.

Davis, C. G., Goldstein, J. L., Sudhof, T. C., Anderson, R. G. W., Russell, D. W., & Brown, M. S. (1987b). Acid-dependent ligand dissociation and recycling of LDL receptor mediated by growth factor homology region. Nature 326, 760–765.

Eberle, W., Sander, C., Klaus, W., Schmidt, B., von Figura, K., & Peters, C. (1991). The essential tyrosine of the internalization signal in lysosomal acid phosphatase is part of a β turn. Cell 67, 1203–1209.

Esser, V., Limbird, L. E., Brown, M. S., Goldstein, J. L., & Russell, D. W. (1988). Mutational analysis of the ligand binding domain of the low density lipoprotein receptor. J. Biol. Chem. 263, 13282–13290.

Fourie, A. M., Coetzee, G. A., Gevers, W., & van der Westhuyzen, D. R. (1988). Two mutant low-density-lipoprotein receptors in Afrikaners slowly processed to surface forms exhibiting rapid degradation or functional heterogeneity. Biochem. J. 255, 411–415.

Goldberg, I. J., Kandel, J. J., Blum, C. B., & Ginsberg, H. N. (1986). Association of plasma lipoproteins with postheparin plasma lipase activities. J. Clin. Invest. 78, 1523–1528.

Goldstein, J. L., & Brown, M. S. (1977). The low-density lipoprotein pathway and its relation to atherosclerosis. Ann. Rev. Biochem. 46, 897–930.

Goldstein, J. L., Brown, M. S., Anderson, R. G. W., Russell, D. W., & Schneider, W. J. (1985). Receptor-mediated endocytosis: Concepts emerging from the LDL receptor system. Ann. Rev. Cell. Biol. 1, 1–39.

Goldstein, J. L., & Brown, M. S. (1989). Familial Hypercholesterolaemia. In: *The Metabolic Basis of Inherited Disease* (Scriver, C. R., Beaudet, A. L., Sly, W. S., & Valle, D., Eds.). McGraw-Hill, New York, pp. 1215–1250.

Grant, K. I., Casciola, L. A. F., Coetzee, G. A., Sanan, D. A., Gevers, W., & Van der Westhuyzen, D. R. (1990). Ammonium chloride causes reversible inhibition of low density lipoprotein receptor recycling and accelerates receptor degradation. J. Biol. Chem. 265, 4041–4047.

Herz, J., Hamann, U., Rogne, S., Myklebost, O., Gausepohl, H., & Stanley, K. K. (1988). Surface location and high affinity for calcium of a 500-kd liver membrane protein closely related to the LDL-receptor suggest a physiological role as lipoprotein receptor. EMBO J. 7, 4119–4127.

Herz, J., Kowal, R. C., Goldstein, J. L., & Brown, M. S. (1990). Proteolytic processing of the 600 kd low density lipoprotein receptor-related protein (LRP) occurs in a *trans*-Golgi compartment. EMBO J. 9, 1769–1776.

Herz, J., Goldstein, J. L., Strickland, D. K., Ho, Y. K., & Brown, M. S. (1991). 39-kDa protein modulates binding of ligands to low density lipoprotein receptor-related protein α2-macroglobulin receptor. J. Biol. Chem. 266, 21232–21238.

Herz, J. (1993). The LDL-receptor-related protein—portrait of a multifunctional receptor. Curr. Opin. Lipidol. 4, 107–113.

Hobbs, H. H., Brown, M. S., & Goldstein, J. L. (1993). Molecular genetics of the LDL receptor gene in familial hypercholesterolemia. Human Mutation 1, 445–466.

Hobbs, H. H., Russell, D. W., Brown, M. S., & Goldstein, J. L. (1990). The LDL receptor locus in familial hypercholesterolemia: Mutational analysis of a membrane protein. Annu. Rev. Genet. 24, 133–170.

Hunziker, W., Harter, C., Matter, K., & Mellman, I. (1991). Basolateral sorting in MDCK cells requires a distinct cytoplasmic domain determinant. Cell 66, 907–920.

Ji, Z-S., Brecht, W. J., Miranda, R. D., Hussain, H. M., Innerarity, T. L., & Mahley, R. W. (1993). Role of heparan sulfate proteoglycans in the binding and uptake of apolipoprotein E-enriched remnant lipoproteins by cultured cells. J. Biol. Chem 268, 10160–10167.

Kanalas, J. J., & Makker, S. P. (1990). Isolation of a 330-kDa glycoprotein from human kidney similar to the Heymann nephritis autoantigen (GP330). J. Am. Soc. Nephrol. 1, 792–798.

Kerjaschki, D., & Farquhar, M. G. (1983). Immunocytochemical localization of the Heymann nephritis antigen (GP330) in glomerular epithelial cells of normal Lewis rats. J. Exp. Med. 147, 667–686.

Kishimoto, A., Brown, M. S., Slaughter, C. A., & Goldstein, J. L. (1987). Phosphorylation of serine 833 in cytoplasmic domain of the LDL receptor by a high molecular weight enzyme resembling casein kinase II. J. Biol. Chem. 262, 1344–1351.

Koivisto, P. V. I., Koivisto, U-M., Kovanen, P. T., Gylling, H., Miettinen, T. A., & Kontula, K. (1993). Deletion of exon 15 of the LDL receptor gene is associated with a mild form of familial hypercholesterolemia FH-Espoo. Arterioscler. Thromb. 13, 1680–1688.

Kounnas, M. Z., Scott Argraves, W., & Strickland, D. K. (1992). The 39-kDa receptor-associated protein interacts with two members of the low density lipoprotein receptor family, α2-macroglobulin receptor and glycoprotein 330. J. Biol. Chem. 267, 21162–21166.

Kozarsky, K., Kingsley, D., & Krieger, M. (1988). Use of a mutant cell line to study the kinetics and function of O-linked glycosylation of low density lipoprotein receptors. Proc. Natl. Acad. Sci. USA 85, 4335–4339.

Lalazar, A., Weisgraber, K. H., Rall, S. C., Giladi, H., Innerarity, T. L., Levanon, A. Z., Boyles, J. K., Amit, B., Gorecki, M., Mahley, R. W., & Vogel, T. (1988). Site-specific mutagenesis of human apolipoprotein E. Receptor binding activity of variants with single amino acid substitutions. J. Biol. Chem. 263, 3542–3545.

Lazarovits J., & Roth, M. (1988). A single amino acid change in the cytoplasmic domain allows the influenza virus hemagglutinin to be endocytosed through coated pits. Cell 53, 743–752.

Lehrman, M. A., Goldstein, J. L., Brown, M. S., Russell, D. W., & Schneider, W. L. (1985). Internalization-defective LDL receptors produced by genes with nonsense and frameshift mutations that truncate the cytoplasmic domain. Cell 41, 735–743.

Lehrman, M. A., Goldstein, J. L., Russell, D. W., & Brown, M. S. (1987). Duplication of seven exons in LDL receptor gene caused by Alu-Alu recombination in a subject with familial hypercholesterolemia. Cell 48, 827–835.

Li, C., Stifani, S., Schneider, W. J., & Poznansky, M. J. (1991). Low density lipoprotein receptors on epithelial cell (Madin-Darby canine kidney) monolayers. J. Biol. Chem. 266, 9263–9270.

Matter, K., Hunziker, W., & Mellman, I. (1992). Basolateral sorting of LDL receptor in MDCK cells: The cytoplasmic domain contains two tyrosine-dependent targeting determinants. Cell 71, 741–753.

Matter, K., Whitney, J. A., Yamamoto, E. M., & Mellman, I. (1993). Common signals control low density lipoprotein receptor sorting in endosomes and the Golgi complex of MDCK cells. Cell 74, 1053–1064.

Mehta, K. D., Chen, W-J., Goldstein, M. S., & Brown, M. S. (1991). The low density lipoprotein receptor in *Xenopus laevis*. J. Biol. Chem. 266, 10406–10414.

Milne, R., Theolis, R., Maurice, R., Pease, R. J., Weech, P. K., Rassart, E., Fruchart, J-C., Scott, J., & Marcel, Y. L. (1989). The use of monoclonal antibodies to localize the low density lipoprotein receptor-binding domain of apolipoprotein B. J. Biol. Chem. 264, 19754–19760.

Moestrup, S. K., Nielsen, S., Andreasen, P., Jorgensen, K. E., Nykjaer, A., Roigaard, H., Gliemann, J., & Christensen, E. I. (1993a). Epithelial glycoprotein-330 mediates endocytosis of plasminogen activator—plasminogen activator inhibitor type-1 complexes. J. Biol. Chem. 268, 16564–16570.

Moestrup, S. K., Holtet, T. L., Etzerodt, M., Thogersen, H. C., Nykjaer, A., Andreasen, P. A., Rasmussen, H. H., Sottrup-Jensen, L., & Gliemann, J. (1993b). α_2-Macroglobulin-proteinase complexes, plasminogen activator inhibitor type-1-plasminogen activator complexes, and receptor-associated protein bind to a region of the α_2-macroglobulin receptor containing a cluster of eight comple-ment-type repeats. J. Biol. Chem. 268, 13691–13696.

Mostov, K., Apodaca, B., Aroeti, B., & Okamoto, C. (1992). Plasma membrane protein sorting in polarized epithelial cells. Cell 116, 577–583.

Mulder, M., Lombardi, P., Jansen, H., Van Berkel, T. J. C., Frants, R. R., & Havekas, L. M. (1993). Low density lipoprotein receptor internalizes low density and very low density lipoproteins that are bound to heparan sulfate proteoglycans via lipoprotein lipase. J. Biol. Chem. 268, 9369–9375.

Pathak, R. K., Yokode, M., Hammer, R. E., Hofmann, S. L., Brown, M. S., Goldstein, J. L., & Anderson, R. G. W. (1990). Tissue-specific sorting of the human LDL receptor in polarized epithelia of transgenic mice. J. Cell Biol. 111, 347–359.

Pearse, B. M. F. (1988). Receptors compete for adaptors found in plasma membrane coated pits. EMBO J. 7, 3331–3336.

Pearse, B. M. F., & Robinson, M. S. (1990). Clathrin, adaptors and sorting. Annu. Rev. Cell Biol. 6, 151–171.

Pelham, H. R. (1991). Recycling of proteins between the endoplasmic reticulum and Golgi complex. Curr. Opin. Cell Biol. 3, 585–591.

Raychowdhury, R., Niles, J. L., McCluskey, R. T., & Smith, J. A. (1989). Autoimmune target in Heymann nephritis is a glycoprotein with homology to the LDL receptor. Science 244, 1163–1165.

Rumsey, S. C., Obunike, J. C., Arad, Y., Deckelbaum, R. J., & Goldberg, I. J. (1992). Lipoprotein lipase-mediated uptake and degradation of low density lipoproteins by fibroblasts and macro-phages. J. Clin. Invest. 90, 1504–1512.

Russell, D. W., Schneider, W. J., Yamamoto, T., Luskey, K. L., Brown, M. S., & Goldstein, J. L. (1984). Domain map of the LDL receptor: Sequence homology with the epidermal growth factor precursor. Cell 37, 577–585.

Russell, D. W., Brown, M. S., & Goldstein, J. L. (1989). Different combinations of cysteine-rich repeats mediate binding of low density lipoprotein receptor to two different proteins. J. Biol. Chem. 264, 21682–21688.

Sanan, D. A., van der Westhuyzen, D. R., Gevers, W., & Coetzee, G. A. (1989). Early appearance of dispersed low density lipoprotein receptors on the fibroblast surface during recycling. Eur. J. Cell Biol. 48, 327–336.

Saxena, U., Ferguson, E., & Bisgaier, C. L. (1993). Apolipoprotein E modulates low density lipoprotein retention by lipoprotein lipase anchored to the subendothelial matrix. J. Biol. Chem. 268, 14812–14819.

Schneider, W. J., Slaughter, C. J., Goldstein, J. L., Anderson, R. G. W., Caproe, D. J., & Brown, M. S. (1983). Use of anti-peptide antibodies to demonstrate external orientation of NH_2 terminus of LDL receptor in the plasma membrane of fibroblasts. J. Cell Biol. 97, 1635–1640.

Schneider, W. J., & Nimpf, J. (1993). Lipoprotein receptors: Old relatives and new arrivals. Curr. Opin. Lipidol. 4, 205–209.

Simons, K., & Fuller, S. D. (1985). Cell surface polarity in epithelia. Annu. Rev. Cell Biol. 1, 234–288.

Smythe, E., Carter, L. L., & Schmid, S. L. (1992). Cytosol- and clathrin-dependent stimulation of endocytosis *in vitro* by purified adaptors. J. Cell Biol. 119, 1163–1171.

Südhof, T. C., Goldstein, J. L., Brown, M. S., & Russell, D. W. (1985). The LDL receptor gene: A mosaic of exons shared with different proteins. Science 228, 815–822.

Takahashi, S., Kawarabayasi, Y., Nakai, T., Sakai, J., & Yamamoto, T. (1992). The rabbit very low density lipoprotein receptor: A low density lipoprotein receptor-like protein with distinct ligand specificity. Proc. Natl. Acad. Sci. USA 89, 9252–9256.

Tolleshaug, H., Goldstein, J. L., Schneider, W. J., & Brown, M. S. (1982). Post-translational processing of the LDL receptor and its genetic disruption in familial hypercholesterolaemia. Cell 30, 715–724.

Van der Westhuyzen, D. R., Stein, M. L., Henderson, H. E., Marais, A. D., Fourie, A. M., & Coetzee, G. A. (1991). Deletion of two growth-factor repeats from the low-density-lipoprotein receptor accelerates its degradation. Biochem. J. 278, 677–682.

Van Driel, I. R., Davis, C. G., Goldstein, J. L., & Brown, M. S. (1987). Self-association of the low density lipoprotein receptor mediated by the cytoplasmic domain. J. Biol. Chem. 262, 16127–16134.

Wang, W., Briggs, M. R., Hua, X., Yokoyama, C., Goldstein, J. L., & Brown, M. S. (1993). Nuclear protein that binds sterol regulatory element of low density lipoprotein receptor promoter. J. Biol. Chem. 268, 14497–14504.

Williams, K. J., Fless, G. M., Petrie, K. A., Snyder, M. L., Brocia, R. W., & Swenson, T. L. (1992). Mechanisms by which lipoprotein lipase alters cellular metabolism of lipoprotein(a), low density lipoprotein, and nascent lipoproteins. Roles for low density lipoprotein receptors and heparan sulfate proteoglycans. J. Biol. Chem. 267, 13284–13292.

Willnow, T. E., Goldstein, J. L., Orth, K., Brown, M. S., & Herz, J. (1992). Low density lipoprotein receptor-related protein and GP330 bind similar ligands, including plasminogen activator-inhibitor complexes and lactoferrin, an inhibitor of chylomicron remnant clearance. J. Biol. Chem. 267, 26172–26180.

Yamamoto, T., Davis, C. G., Brown, M. S., Schneider, W. J., Casey, M. L., Goldstein, J. L., & Russell, D. W. (1984). The human LDL receptor: A cysteine-rich protein with multiple Alu sequences in its mRNA. Cell 39, 27–38.

Yokode, M., Pathak, R. K., Hammer, R. E., Brown, M. S., Goldstein, J. L., & Anderson, R. G. W. (1992). Cytoplasmic sequence required for basolateral targeting of LDL receptor in livers of transgenic mice. J. Cell Biol. 117, 39–46.

Yokoyama, C., Wang, X., Briggs, M. R., Admon, A., Wu, J., Hua, X., Goldstein, J. L., & Brown, M. S. (1993). SREBP-1, a basic-helix-loop-helix-leucine zipper protein that controls transcription of the low density lipoprotein receptor gene. Cell 75, 187–197.

MANNOSE-6-PHOSPHATE RECEPTORS

Regina Pohlmann

Biomembranes
Volume 4, pages 223–253.
ISBN: 1-55938-661-4.

I. INTRODUCTION

The role of mannose-6-phosphate receptors (MPRs) in the targeting of lysosomal proteins to lysosomes is firmly established. The MPRs represent a particular well-examined example for sorting mechanisms accompanying the vesicle-mediated transport in eucaryotic cells.

Lysosomes are subcellular organelles that are specialized for intracellular digestion. They contain a wide variety of acid hydrolases and unique membrane proteins. The biogenesis of lysosomes requires a continuous flow of constituents that are synthesized at the membrane-bound ribosomes of the endoplasmic reticulum (ER) and transported through the Golgi apparatus. Transport vesicles that deliver proteins to lysosomes bud from the *trans*-Golgi network (TGN). These vesicles must incorporate lysosomal proteins and exclude the many other proteins (secretory glycoproteins and membrane proteins) that are packaged into different transport vesicles for delivery elsewhere.

Mannose-6-phosphate (M6P) residues are exclusively added to the N-linked oligosaccharides of soluble lysosomal proteins. These are recognized by specific transmembrane proteins, the mannose-6-phosphate receptors (MPRs). Ligands bind to the MPRs at the near neutral pH of the TGN. Thus, the lysosomal proteins are concentrated in coated transport vesicles that fuse with a prelysosomal compartment (endosome) where the dissociation of the receptor–ligand complexes takes place due to the acid pH. The lysosomal proteins become incorporated into lysosomes via an as yet undetermined mechanism. The receptors recycle to the Golgi apparatus or to the plasma membrane from where extracellular M6P-ligands can be endocytosed. Cytoplasmic factors recognize signal structures in the cytoplasmic domains of the receptors and regulate the vesicular transport.

The routing of lysosomal membrane glycoproteins is independent of M6P-markers. The structural determinants that direct their intracellular trafficking are beginning to be defined and seem to be localized in the protein structure of their cytoplasmic tails (for review see Kornfeld and Mellman, 1989; Fukuda, 1991; Peters and von Figura, 1994). Besides these two pathways for targeting proteins to lysosomes, M6P-independent transport of soluble lysosomal proteins has been reported for some mammalian cells (van Dongh et al., 1984; Lemansky et al., 1985; Rijnboutt et al., 1990).

This review is focused on the structure, topogenic signals, and function of the MPRs. The biogenesis of lysosomes as well as the cyclic transport of the MPRs have been subject of several reviews (von Figura and Hasilik, 1986; Nolan and Sly, 1987; Pfeffer, 1988; Dahms et al., 1989; Kornfeld and Mellman, 1989; von Figura, 1991; Kornfeld, 1992; Hille-Renfeld, 1995).

II. SYNTHESIS OF THE MANNOSE-6-PHOSPHATE MARKER

Mannose-6-phosphate groups are added to soluble lysosomal glycoproteins. This takes place most probably in a pre-Golgi compartment or the *cis*-Golgi (Pohlmann et al., 1982; Pelham, 1988). The synthesis of M6P-groups is a two step procedure in which two enzymes are involved sequentially. First, N-acetylglucosamine-phosphotransferase transfers the N-acetylglucosamine-phosphate portion of UDP-GlcNAc onto the C6-hydroxyl group of mannose. The second enzyme, a phosphoglycosidase (N-acetylglucosamine-1-phosphodiester-α-N-acetylglucosaminidase) clips off the terminal N-acetylglucosamine and exposes the M6P-marker. The primary structures of the many lysosomal enzymes that have been cloned do not reveal any sequence identity. Heat-denatured, SDS-denatured, or trypsin-treated lysosomal enzymes do not function as substrates for the phosphotransferase. This suggests that the phosphotransferase recognizes a conformation-specific protein determinant present only in lysosomal hydrolases. For cathepsin D, it has been shown that multiple protein regions contribute to this recognition domain (Baranski et al., 1992; Cantor and Kornfeld, 1992; Cantor et al., 1992).

The failure of the synthesis of M6P markers results in the missorting of all lysosomal soluble hydrolases which is the basis of the rare disorder I cell disease. Due to a defect of the phosphotransferase, the soluble acid hydrolases cannot be targeted to lysosomes resulting in an accumulation of undigested substrates (inclusions). The soluble hydrolases are lost from the cell and excessive amounts of lysosomal proteins are found in the serum of I cell patients. The analysis of I cell disease has greatly contributed to the discovery of the M6P recognition marker-dependent transport of lysosomal hydrolases and their receptors.

III. MANNOSE-6-PHOSPHATE RECEPTORS: STRUCTURAL AND BIOCHEMICAL FEATURES

Two mannose-6-phosphate specific receptors are known which differ in size and function (see Table 1). Both are type I transmembrane glycoproteins and are

Table 1. Properties of the MPR

	MPR 46	IGF II/M6PR
Transport of newly synthesized lysosomal enzymes	+	+
pH-dependence of binding	pH 6–6.3	pH 5.7–6.5
Cation-dependence	+	–
Quaternary structure	2 (4)	≥ 1
Secretion of newly synthesized lysosomal enzymes	+	–
Endocytosis of external M6P-containing ligands	–	+
Binding of IGF II	–	+

involved in the transport of lysosomal enzymes. Using a lysosomal enzyme affinity matrix, the larger MPR was first isolated (Sahagian et al., 1981). The detection of a second, smaller receptor was facilitated by the existence of cells lines that lack the larger receptor, for example, P388D₁ mouse macrophages (Hoflack and Kornfeld, 1985). Separation of the two receptors was achieved by the selective binding of the larger MPR to an affinity matrix containing methyl-6-phosphomannosyl residues present in lysosomal enzymes from *Dictyostelium discoideum*.

A. MPR 46

The small MPR was described first in 1985 by Hoflack and Kornfeld. The mannose-6-phosphate receptor with an apparent molecular mass of 46,000 (MPR 46), is a cation-dependent MPR due to the dependence on divalent cations for *in vitro* ligand binding in species like mouse and cow (Hoflack and Kornfeld, 1985; Ma et al., 1991).

The cDNA for the MPR 46 has been cloned from several species including man (Pohlmann et al., 1987), cow (Dahms et al., 1987), and mouse (Köster et al., 1991; Ma et al., 1991; Ludwig et al., 1992), and partially from chicken (Matzner et al., 1996). The mRNA of the human MPR 46 codes for a protein of 277 amino acids which has a N-terminal signal sequence of 20- or 26-amino acids. The following amino acids are exposed at the plasma membrane or oriented to the vesicle lumen. MPR 46 contains five potential N-glycosylation sites, four of which bear oligosaccharides (Dahms et al., 1987; Wendland et al., 1991b; Figure 1). The carbohydrate contributes 40% to the apparent molecular mass of the MPR 46. The transmembrane domain contains 20 amino acids and is followed by 67 amino acids of the cytoplasmic domain. The sequence identity of bovine and mouse MPR 46 to the human receptor is 93–95% and the cytoplasmic domains of man, mouse, cow, and chicken are identical. The gene for the human MPR 46 has been localized to chromosome 12 (Pohlmann et al., 1987) and consists of seven exons (110–1573 base pairs in length), spanning more than 12 kilobases (Klier et al., 1991). Exon 1 codes for a 5' untranslated sequence. Exon 2 codes for the signal sequence and the initial portion of the luminal domain which extends to exon 5. The transmembrane domain of the receptor spans exons 5 and 6. The cytoplasmic domain is encoded by exons 6 and 7. The latter includes the 3' untranslated region.

A series of post-translational modifications occur within the ER that generate a binding competent receptor. This was demonstrated by analysis of the biosynthesis of the human MPR 46 in BHK-21 cells that over-express the protein and by *in vitro* translation studies (Hille et al., 1989, 1990). The modifications include (1) the pairing of intramolecular disulfide bridges (compare Figure 1; Wendland et al., 1991c), (2) the dimerization of the receptor by noncovalent association, and (3) changes in its conformation. In this way, the MPR 46 gets its binding competent conformation before reaching the Golgi complex where it binds its ligands. Processing of the high-mannose type to complex type oligosaccharides is not a require-

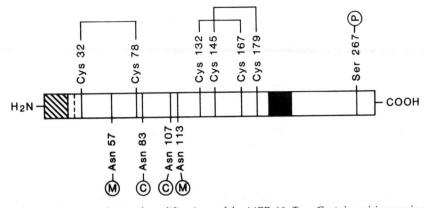

Figure 1. Post-translational modifications of the MPR 46. Top: Cystein pairing; serine phosphorylation. Bottom: positions of N-glycosylation with mannose-rich (M) and complex (C) oligosaccharides. The hatched area indicates the signal peptide, the filled box marks the transmembrane domain. The numbering of the amino acids starts with the initiator methionine.

ment for binding competence. Furthermore, N-glycosylation is not essential for generation of the binding conformation of the receptor. N-Glycosylation, however, increases the stability of the binding competent form after solubilization of the receptor (Wendland et al., 1991b). Two differentially glycosylated isoforms of the MPR 46 have been isolated from bovine testis (Li and Jourdian, 1991). One form (MPR-2B, 42,000D) is found in Leydig cells and bears terminal galactose residues on the outer branches of the complex chains. The other form (MPR-2A, 45,000D) has been detected in Sertoli cells and contains a linear poly-lactosamine chain. This form of the receptor has a lower affinity for mannose-6-phosphate containing ligands. This might result from the poly-lactosamine chain and the presence of terminal sialic acid residues on the outer branches of the complex chains. After treatment with neuraminidase and endo-β-galactosidase, the affinity of the MPR-2A form for M6P-ligands was comparable to the MPR-2B form. The mechanism by which poly-lactosamine and sialic acid residues affect the affinity is not apparent.

Phosphorylation of the human MPR 46 has been reported (Hemer et al., 1993). *In vitro*, casein kinase II phosphorylated serine 267 of the full length MPR 46 (identical to serine 56 in the cytoplasmic domain of the MPR 46). Recombinant mutants of the MPR 46 where serine 267 was exchanged for alanine demonstrated normal ligand binding and recycling after expression in BHK 21 cells. Recently, this phosphorylation was shown to determine the high affinity interaction of the AP-1 Golgi adapter proteins with membranes (Mauxian et al., 1996).

Detergent-solubilized MPR 46 exists in an equilibrium of monomeric, dimeric, and tetrameric forms which can be separated by sucrose density centrifugation or chromatography on a mannose-6-phosphate affinity matrix. The monomeric form of the human MPR 46 does not bind to the matrix, while dimeric and tetrameric forms bind readily to the affinity matrix (Waheed et al., 1990a, 1990b). Tetrameric receptors bind with the highest affinity due to their multivalency. However, the monomeric form of the bovine receptor did bind in the presence of Mn^{2+} (Li et al., 1990). This is probably due to differences in the affinity matrices used or might indicate a species-specific characteristic. The quaternary structure of the solubilized MPR 46 depends on receptor concentration, temperature, pH, and ligand concentration. Low receptor concentration, increasing temperature, pH $\leq$ 5.0, and the absence of ligands promote the dissociation of the receptor to dimeric and monomeric forms. The kinetics of receptor dissociation and association are fast enough to result in significant changes of the quaternary structure during trafficking of the receptor. This supports the attractive hypothesis that cycling of the receptor between a neutral compartment, where it binds its ligands and an acidic compartment, where the ligands dissociate, is accompanied by changes of its quaternary structure. In BHK-21 cells which over-express the human MPR 46, the monomeric, dimeric, and tetrameric forms were detected (Waheed et al., 1990a). However, it remains to be shown that the quaternary structure of the MPR 46 indeed changes along its pathway.

Ligand Binding

Tong et al. (1989b) showed by equilibrium dialysis that the bovine MPR 46 binds 1 mol of M6P/polypeptide chain (mannose-6-phosphate with a K_d of 8×10^{-6} M and pentamannose-6-phosphate with a K_d of 6×10^{-6} M). A divalent ligand, a high mannose oligosaccharide with two phosphomonoesters was bound with a K_d of 2×10^{-7} M. From studies with a variety of mannose-6-phosphate analogues, the authors concluded that the 6-phosphate and the 2-hydroxyl of the mannose-6-phosphate are the basis of the specificity of the receptors for this sugar phosphate. Binding occurs at a rather narrow pH optimum of pH 6–6.3.

The ligand binding capacity resides in the luminal portion of the MPR 46. Truncated receptors that lack the cytoplasmic and transmembrane domain are secreted, but retain their binding capacity (Dahms and Kornfeld, 1989; Wendland et al., 1989). Chemical modification of specific amino acids has revealed the involvement of histidine and arginine residues in the binding site. The loss of binding activity was not detected when the chemical modification was performed in the presence of M6P (Stein et al., 1987a). *In vitro* mutagenesis of the eight arginine and five histidine residues in the luminal portion of the MPR 46 demonstrated that the change of arginine 137 to lysine or glutamine and of histidine 131 to serine results in a loss of ligand binding, while the conformation, quaternary structure, and glycosylation of the receptor are not affected. This implies that arginine 137 and histidine 131 are essential for the binding activity of MPR 46 (Wendland et al., 1991a).

B. IGF II/M6PR (MPR 300)

The large MPR, M_r 300,000 mannose-6-phosphate receptor (MPR 300), was first isolated by Sahagian et al. (1981). It binds M6P-ligands independently of cations. In mammalian species, the receptor binds also the insulin-like growth factor II (IGF II) and hence is referred to as IGF II/M6PR (Morgan et al., 1987; Kiess et al., 1988; Tong et al., 1988; MacDonald et al., 1988). The cDNA for the IGF II/M6PR has been cloned from several species, including man (Morgan et al., 1987; Oshima et al., 1988), cow (Lobel et al., 1987, 1988), rat (MacDonald et al., 1988), and chicken (Zhoe et al., 1995). The cDNA for the human IGF II/M6PR encodes 2491 amino acids. This includes a signal sequence of 40 amino acids; luminal portion of 2264 amino acids; a transmembrane domain of 23 residues; and a cytoplasmic domain of 164 amino acids. The luminal domain is built by 15 repetitive sequences each about 147 amino acids in length, which have a sequence identity of 14–28% in the bovine receptor. The entire luminal domain of the bovine MPR 46 has a sequence identity of 14–37% with each of the repeats (Lobel et al., 1988). This shows that the two receptors are related and that MPR 46 may represent the evolutionary older receptor. The positions of the cysteine residues are highly conserved among the repeats and in the MPR 46 luminal domain (Wendland et al., 1991c). The thirteenth repeat contains an insertion of 43 base pairs that is a fibronectin type II repeat region, which in fibronectin is part of a collagen-binding domain; however, nothing is known about its function in the IGF II/M6PR. The gene for the IGF II/M6PR has been localized to chromosome 6 in man and to chromosome 17 in mouse (Laureys et al., 1988). The gene spans about 130 kilobase pairs in mouse (Stöger et al., 1993) and consists of 48 exons (Szebenyi and Rotwein, 1994; Figure 2). The distribution of the exons on the receptor domains is shown in Figure 3.

Several post-translational modifications are known for the IGF II/M6PR. The luminal domain has 19 potential N-glycosylation sites and at least two are utilized (Lobel et al., 1987, 1988). Therefore, the size of the mature receptor is likely to be between 275- and 300-kD. The receptor is known to be phosphorylated (Sahagian and Neufeld, 1983). The cytoplasmic domain contains consensus sequences for protein kinase C and casein kinases I and II (MacDonald et al., 1988). Phosphorylation on two serine residues of the cytoplasmic domain of the bovine IGF II/M6PR has been shown *in vivo* and *in vitro* (residues 2421 and 2492 of the full length protein; Meresse et al., 1990). Phosphorylation is believed to occur in the TGN by a casein II kinase-type kinase (Meresse and Hoflack, 1993). Similar conserved casein II kinase consensus sequences have been detected in the human IGF II/M6PR and phosphorylation has been shown for the corresponding serine residues (Rosorius et al., 1993a).

Palmitoylation of the IGF II/M6PR is another post-translational modification whose function is still to be identified (Westcott and Rome, 1988).

It is contested whether the biologically active form of the IGF II/M6PR is a monomer (Perdue et al., 1983; Stein et al., 1987b). Chemical cross-linking of

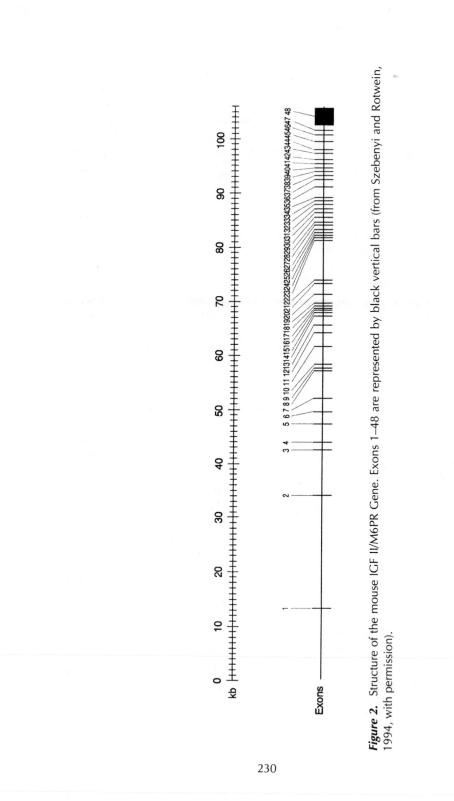

Figure 2. Structure of the mouse IGF II/M6PR Gene. Exons 1–48 are represented by black vertical bars (from Szebenyi and Rotwein, 1994, with permission).

230

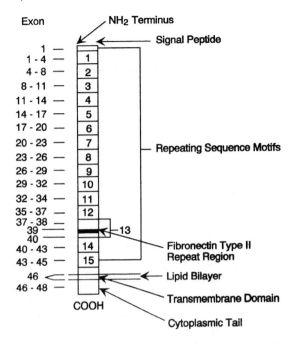

Figure 3. IGF II/M6PR protein and exon organization. The signal peptide, the 15 repeating sequence motifs, the transmembrane- and cytoplasmic-domain are indicated. The numbers at the left correspond to the exons coding for the respective protein area (from Szebenyi and Rotwein, 1994, with permission).

membranes with bifunctional reagents results in high molecular weight aggregates. However, this might be due to cross-linking to nonreceptor polypeptides (Stein et al., 1987b).

Ligand Binding

The IGF II/M6PR does not bind IGF II in chicken (Canfield and Kornfeld, 1989) nor in frog (Clairmont and Czech, 1989), while it exhibits high affinity binding in several mammalian species. IGF II is a nonglycosylated polypeptide and one mole of ligand is bound per receptor molecule ($K_d = 2 \times 10^{-10}$ M; Tong et al., 1988). Optimal IGF II binding occurs at pH 7.4 (cell surface), while M6P-ligands bind best at pH 6.3, the pH that is ascribed to the TGN. Binding of the two ligands, M6P and IGF II, can occur simultaneously (Roth et al., 1987; Braulke et al., 1988; Kiess et al., 1988; Tong et al., 1988; Waheed et al., 1988). The occupation of one site with its ligand interferes with binding of the other site (Roth et al., 1987; MacDonald et al., 1988; Kiess et al., 1989, 1990). Lysosomal enzymes, but not M6P interfere with IGF II binding and IGF II binding can inhibit lysosomal enzyme binding. The IGF

II binding connects the receptor to the GTP-binding protein G_{i-2} (Murayama et al., 1990). Stimulation of the GTPase of the $G_{i-2\alpha}$ protein is not mediated by M6P or M6P-ligands, but can be inhibited by M6P-binding. The biological meaning of IGF II-induced G_{i-2} activation and its inhibition by M6P containing ligands is not understood. The IGF II-binding site was localized to amino acids 1508–1566 in repeat 11 of the IGFII/M6PR (Schmidt et al., 1995).

The IGF II/M6PR binds diphosphorylated oligosaccharides with a much higher affinity than the MPR 46 (2×10^{-9} M and 2×10^{-7} M, respectively) (Tong et al., 1989a; Distler et al., 1991). Furthermore, only the IGF II/M6PR binds the methyl-6-phosphomannosyl diesters present in lysosomal enzymes from *Dictyostelium discoideum* (Sahagian et al., 1981). It has a broad pH-optimum for binding between pH 6–7.5 and binds independently of divalent cations (Sahagian et al., 1981; Hoflack et al., 1987; Tong et al., 1989a). Ligand binding studies indicate that the IGF II/M6PR binds two moles of M6P or one of phosphorylated disaccharide per monomer (Tong et al., 1989a; Distler et al., 1991). This indicates that only two of the 15 repetitive sequences contain a M6P-ligand binding site. Dahms et al. (1993) analyzed the M6P-binding function of domains 1–9 by site-directed mutagenesis. They showed that combined replacement of Arg 435 in domain 3 and Arg 1334 in domain 9 with lysine results in a dramatic loss of binding to an affinity matrix. No change in the glycosylation pattern nor in immunoreactivity could be detected. Arg 435 and Arg 1334 of the IGF II/M6PR correspond to the position of Arg 137 of the MPR 46 which has been shown to be essential for ligand binding. Whether Arg 435 and Arg 1334 are the only arginine residues of the IGF II/M6PR that participate in ligand binding has to be analyzed as well as their direct involvement in receptor–ligand interactions.

IV. TRANSPORT FUNCTION OF THE MPR

A. Subcellular Distribution

Both MPR recycle between the TGN and endosomes, as well as between endosomes and plasma membrane, and are primarily located in internal membranes. At the TGN and at the plasma membrane, both receptors are packaged into clathrin-coated vesicles for sorting. In double-labeling experiments with immunogold, Bleekemolen et al. (1988) found that both MPR occur in U937 monocytes in essentially the same intracellular compartments. About 12% of each receptor was present at the cell surface, 2% in the Golgi stacks, and about 25% in vacuoles resembling endosomal vacuoles. About 50% of each receptor was found in tubules, presumably associated with endosomes and TGN. The only exceptions were electron dense vesicles occurring in the *trans*-Golgi region and surrounding endosomes, which were slightly enriched in the IGF II/M6PR. Similar results were recently reported for HepG2 cells. In these cells, Klumperman et al. (1993) noted a lateral segregation of the two MPR in endosomes. Quantification of immunogold-labeling experiments indicated a concentration of the IGF II/M6PR in the central vacuole compared to associated electron-dense tubules and vesicles, while the MPR 46 is equally distributed between both structures. The lateral segregation of the two MPR may be connected to their differential sorting function. Double-immunolabeling of the two MPR with the asialoglycoprotein receptor, which is known to cycle between endosomes and the plasma membrane revealed only limited colocalization. This suggests that MPR localize to a distinct subpopulation of endosome-associated tubules and vacuoles. The endosome-associated tubules and vacuoles are probably destined to carry MPR from endosomes to the TGN. The relative enrichment of the MPR 46 in these structures may reflect that upon arrival in the endosomes, the MPR 46 is retrieved more rapidly than IGF II/M6PR from the maturing endosome. Similar results were obtained for BHK-21 cells that over-express the human MPR 46 and IGF II/M6PR.

B. Sorting and Endocytosis of Lysosomal Enzymes

The involvement of both MPR in the targeting of newly synthesized soluble lysosomal hydrolases is suggested by several observations. Some tumor cell lines are known to lack the IGF II/M6PR and contain the MPR 46 only (Gabel et al., 1983; Mainferme et al., 1985). These cells secrete high amounts of newly synthesized soluble lysosomal hydrolases (70%) and do not endocytose extracellular M6P-ligands. The 30% residual sorting in cells that do not express the IGF II/M6PR appears to be mediated by the MPR 46 and can be improved to up to 60% by over-expression of MPR 46 in these cells (Watanabe et al., 1990; Ma et al., 1991). Transfection with the IGF II/M6PR-cDNA completely corrects the sorting defect and qualifies the cells for uptake of extracellular M6P-ligands (Kyle et al., 1988; Lobel et al., 1989; Nolan et al., 1990). From MPR46 knock-out mice primary cell cultures have been established. A secretion of about 50% of newly synthesized Man6P-containing ligands was detected in cells that do not express the MPR 46 (Köster et al., 1993). Similar results were obtained by Ludwig et al. (1993). This argues for an essential function of MPR 46 in the targeting of newly synthesized enzymes. When double MPR-deficient and single MPR-deficient embryonic fibroblasts from knock-out mice were analyzed for lysosomal enzyme sorting the same species of Man6P-containing ligands were found in the secretions of double MPR-deficient fibroblasts as in the secretions of single MPR-deficient fibroblasts but at different ratios (Pohlmann et al., 1995).This indicates that neither MPR has an exclusive affinity for one or several lysosomal proteins. In an independent study Ludwig et al. (1994) detected differences in the pattern of the secreted phosphorylated proteins suggesting that the two receptors may interact *in vivo* with different subtypes of hydrolases. The missorting of soluble lysosomal proteins was much more pronounced in double MPR-deficient fibroblasts than in single MPR-deficient fibroblasts. Re-expression of either type of MPR in double MPR-deficient fibroblasts was not sufficient for the intracellular targeting of lysosomal enzymes.

Only a partial correction of the missorting was seen after re-expression of MPR 46. Even at MPR 46 levels that are five times higher than the wild type level more than one-third of the newly synthesized lysosomal proteins accumulates in the secretions. Two-fold over-expression of IGF II/M6PR completely corrects the missorting of lysosomal enzymes. However, at least one-fourth of the lysosomal enzymes are transported along a secretion-recapture pathway which is sensitive to M6P in the medium (Kasper et al., 1996).

The basal secretion of about 10% of lysosomal enzymes found in the extracellular medium of cells that express both MPR can at least in part be ascribed to the secretory function of the MPR 46 (Chao et al., 1990). This feature of the MPR 46 is easily detectable in BHK-21 cells and mouse L-cells that express normal levels of endogenous IGF II/M6PR and over-express the human MPR 46, it can also be demonstrated in nontransfected cells. The sorting of lysosomal enzymes is decreased in these cells and can be corrected to normal levels by subsequent transfection with the IGF II/M6PR. These results indicate that the two MPR cooperate in the binding of newly synthesized ligands. In contrast to ligands bound to the IGF II/M6PR, those bound to the MPR 46 may be transported to and released at a site from which they can exit either to the medium or to the lysosomes. This might be reflected in the differences in the subcellular distribution of the two MPR (Klumperman et al., 1993). The secretion property of the MPR 46 explains the observation that the sorting of lysosomal enzymes in IGF II/M6PR-deficient cells can only in part be corrected by transfection with the MPR 46 (Watanabe et al., 1990; Ma et al., 1991).

Endocytosis of M6P-ligands is mediated exclusively by the IGF II/M6PR (Stein et al., 1987c). The reason for the inability of the MPR 46 to bind ligands at the cell surface and to mediate their uptake does not depend on a failure in recycling (Stein et al., 1987b, 1987c; Duncan and Kornfeld, 1988; Ma et al., 1991). Watanabe et al. (1990) could demonstrate endocytosis of exogenous β-glucuronidase by the human MPR 46 in IGF II/M6PR-deficient mouse L-cells under specific conditions. These include high levels of over-expression of MPR 46 (50-fold), high ligand concentration, and an extracellular pH of 6.5. Even under these conditions, the uptake was only 1–2% of that mediated by physiological levels of the IGF II/M6PR.

The pathways for the delivery of soluble lysosomal proteins are summarized in Figure 4.

C. IGF II Binding and Transport by the IGF II/M6PR

The IGF II/M6PR has been mapped to the mouse Tme locus and was shown to be imprinted. It has a function in embryonic growth regulation. The maternal inheritance of an IGF II/M6PR null allele as well as homozygosity for the inactive allele is generally lethal at birth and mutants are about 30% larger. This indicates that maternal expression of the IGF II/M6PR is essential for late embryonic development and growth regulation. The phenotype is probably caused by an excess

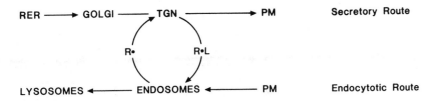

Figure 4. Pathways for the delivery of soluble lysosomal proteins. After their synthesis in the rough endoplasmic reticulum, lysosomal proteins follow the secretory pathway in the Golgi apparatus, where they are separated from this pathway by binding to the MPR. They are delivered to endosomes via clathrin-coated vesicles, dissociated from the receptors, and sorted to lysosomes. Extracellular ligands can be internalized by MPR located at the plasma membrane. The MPR cycle continuously between TGN, endosomes, and the plasma membrane. RER, rough endoplasmic reticulum; TGN, trans Golgi network; R, MPR; PM, plasma membrane; L, M6P-containing ligand).

of IGF II because the introduction of an IGF II null allele rescued the IGF II/M6PR mutant mice (Wang et al., 1994). IGF II that is bound to IGF II/M6PR at the cell surface is internalized and subjected to lysosomal degradation (Oka et al., 1985; Kiess et al., 1987). The IGF II/M6PR is thought to function as a "sink" for the IGF II and thus to control the effect of this growth factor during development (Haig and Graham, 1991). Binding of IGF II to the cell surface IGF II/M6PR can modulate the binding and uptake of lysosomal enzymes and the cell surface expression of this receptor (Kiess et al., 1988, 1989; Braulke et al., 1989, 1990a, b). In addition to binding to the IGF II/M6PR, the insulin-like growth factor II also binds to the IGF I- and insulin-receptors and activitates their tyrosine kinase activities. Some of the metabolic effects of IGF II, especially its proliferative effect, seem to be mediated by binding to the IGF I receptor (Rechler et al., 1980; Mottola and Czech, 1984; Conover et al., 1986; Kiess et al., 1987; Roth, 1988). These findings are supported by the study of Sakano et al. (1991), which demonstrated an effect on the stimulation of DNA-synthesis in BALB c3T3 cells and the glycogen synthesis in HepG2 cells only from IGF II mutants that bind to the IGF I receptor, but not to the IGF II receptor. However, numerous publications report on a variety of effects induced by IGF II binding to the IGF II/M6PR in many cell types. Among these are stimulation of Na^+/H^+ exchange and production of inositol trisphosphate, glycogen synthesis, stimulation of Ca^{2+} influx, DNA synthesis, and amino acid uptake (Mellas et al., 1986; Hari et al., 1987; Kojima et al., 1988; Nishimoto et al., 1987; Rogers and Hammerman, 1988, 1989). No IGF II-stimulated inositol trisphosphate production was detected in primary human hepatocytes, rat cardiac myocytes, or human fibroblasts suggesting that this effect is restricted to certain cell types (Thakker et al., 1989; Guse et al., 1992; Damke et al., 1993). It was reported that the production of inositol trisphosphate is increased by mannose-6-phosphate and is inhibited by pertussis toxin (Rogers and Hammerman, 1988,

1989). These observations promoted the search for the mechanism underlying the intracellular signal activation by the IGF II/M6PR. The coupling of the IGF II/M6PR to a G_{i-2} protein has been reported (Kojima et al., 1988; Nishimoto et al., 1989; Okamoto et al., 1990a, 1990b). Identity comparison between the cytoplasmic tails of human, rat, and bovine IGF II/M6PR revealed a conserved sequence of 14 amino acids (residue 2410–2423 of the human IGF II/M6PR) that shared the characteristics of mastoparan, a peptide toxin known to activate G-proteins directly (Higashijima et al., 1988; Okamoto et al., 1990a). This conserved sequence could activate the G_{i-2} protein in a mode of action similar to G-coupled receptors. In basolateral membranes from dog kidney cells, IGF II, transforming growth factor (TGF) β1-precursor and proliferin, two M6P-containing growth and differentiation factors, are reported to induce inositol trisphosphate production (Rogers et al., 1990). These results are in contrast to data obtained from IGF II/M6PR reconstituted in G_{i-2}-containing phospholipid vesicles (Murayama et al., 1990).

Taken together, the results indicate that the IGF II/M6PR can mediate intracellular signaling, but clearly the identification of the elements activated *in vivo* is necessary to understand the IGF II/M6PR-mediated signal transduction.

V. CYCLING OF THE MPR

A. Transport Signals

The MPRs recycle constitutively between TGN, endosomes, and plasma membrane, independent of ligand binding. Ligands are bound in the TGN and transported via endosomes to the lysosomes. The recycling of the MPR to the TGN is a characteristic of MPRs that is not observed for receptors like the LDL receptor or the asialoglycoprotein receptor that recycle between endosomes and plasma membrane. The recycling of MPRs to the TGN is much faster than that for the bulk of cell surface glycoproteins (Jin et al., 1989; Johnson et al., 1990). This indicates the presence of signals that mediate an active sorting of the MPRs into transport vesicles that fuse with distinct target membranes. The packaging of the receptors into transport vesicles is induced by binding to specific cytoplasmic tetrameric proteins, the adaptor protein complexes. Clathrin and adaptor proteins regulate the vesicular transport from TGN to endosomes and the receptor-mediated endocytosis. Two types of adaptor proteins, AP-1 and AP-2, have been isolated that are specific for either TGN membranes or plasma membranes (Pearse and Robinson, 1990). The cytoplasmic domains of the IGF II/M6PR (Pearse, 1988; Glickman et al., 1989) and MPR 46 (Sosa et al., 1993) bind, *in vitro*, to both adaptor types. The transport to the TGN has been reconstituted *in vitro* in the presence of ATP and cytoplasmic factors (Goda and Pfeffer, 1988).

The signals that mediate the endocytosis of the bovine IGF II/M6PR have been investigated in several studies. IGF II/M6PR that are mutated in the 163 amino acid cytoplasmic tail were expressed in IGF II/M6PR-deficient mouse L-cells, which then were analyzed for the sorting and endocytosis of lysosomal enzymes. The endocytosis signal has been localized to the inner half of the 163 amino acid cytoplasmic tail (Lobel et al., 1989), at amino acids 24–29 (Canfield et al., 1991). The internalization signal of IGF II/M6PR depends on a tyrosine residue in the sequence Tyr^{26}–Ser^{27}–Lys^{28}–Val^{29}. Substitution of Tyr^{26} for phenylalanine or tryptophan inactivates the signal (Jadot et al., 1992). The authors suggest that it represents a four-amino acid motif, with the essential elements being an aromatic residue in the amino-terminal position separated by two amino acids from a bulky hydrophobic residue in the carboxy-terminal position. Several studies have suggested that the internalization sequences have a propensity to form tight-turn structures. This has been shown for the transferrin receptor (Collawn et al., 1990, 1991), the LDL receptor (Bansal and Gierasch, 1991), and the lysosomal acid phosphatase (Eberle et al., 1991). All of these studies indicate that the critical tyrosine residue of the internalization signal is presented to the adaptor AP-2 protein in the context of a tight-turn motif. For the IGF II/M6PR, the internalization motif has to be analyzed further, particularly at the structural level.

Multiple signals appear to contribute to the internalization of the MPR 46. In the study of Johnson et al. (1990) on the bovine MPR 46, two separate sequences were analyzed by *in vitro* mutagenesis. One sequence contains the only tyrosine residue (Tyr^{45}) present in the 67-amino acid cytoplasmic domain and consists of four amino acids (Tyr–Arg–Gly–Val). The other, more effective, sequence includes two phenylalanine residues, Phe^{13} and Phe^{18}, within the 6-amino acid motif Phe–Pro–His–Leu–Ala–Phe.

The signals that mediate MPR recycling to the TGN are not as extensively analyzed as the internalization signals. Deletion of the outer 40 C-terminal amino acids of the cytoplasmic IGF II/M6PR domain partially impaired the cyclic transport between TGN and endosomes, while the cycling between plasma membrane and endosomes was unaffected (Lobel et al., 1989). The same partial effect could be demonstrated by the deletion of the four carboxy-terminal amino acids Leu–Leu–His–Val (Johnson et al., 1992a). The authors concluded from these results that probably more than one signal for efficient sorting at the TGN is present in the IGF II/M6PR cytoplasmic tail. In the MPR 46, deletion of a related carboxy-terminal sequence (His–Leu–Leu) similarly impaired the sorting of lysosomal enzymes (Johnson et al., 1992b). The internalization at the plasma membrane was not affected. Controversely, a study from Denzer et al. (1996) reports on the function of the two carboxy-terminal leucine residues 64 and 65 in the internalization of the MPR 46 from the plasmamembrane. Also, in the cytoplasmic domain of the insulin receptor several di-leucin motifs were found that can mediate internalization (Haft et al., 1994). Further evidence that a di-leucine motif functions as a Golgi complex sorting sequence was reported by Letourneur and Klausner (1992). The T-cell antigen receptor γ- and δ-chains contain a di-leucine and a tyrosine-containing motif required for their efficient sorting in the Golgi complex. The di-leucine containing motif appeared to also function as an internalization signal at the plasma membrane.

In the search for sorting signals in the cytoplasmic domains of proteins, site-directed mutagenesis of amino acid sequences is the general approach chosen. This method may be confounded by the introduction of conformational artifacts in the newly constructed proteins. A different method for the identification of sorting signals is based on the injection of cultured cells with F_{ab}-fragments directed against short peptide sequences (5 amino acids) of the cytoplasmic tail of a protein and subsequent immunocytochemical detection of the protein (Schulze-Garg et al., 1993). The antibodies are expected to compete with cytoplasmic transport factors and to interfere with their binding to the cytoplasmic domain of the transported protein. Injection of antibodies against peptide 43–47 of the cytoplasmic tail of MPR 46 into cultured human fibroblasts and BHK cells transfected with human MPR 46 resulted in loss of the normally observed perinuclear TGN location of MPR 46. Accumulation of the MPR 46 occurred in an endosomal compartment that did not co-localize with established endosomal markers like rab5, rab7, or endocytosed transferrin. This compartment was, however, in equilibrium with MPR 46 at the plasma membrane. The authors concluded that epitope 43–47 of the cytoplasmic tail of MPR46 contains information for return from endosomes to TGN and that this transport, or part of it, depends on specific signals. This suggests that receptor transport from endosomes to the Golgi complex is not a constitutive process (Green and Kelly, 1992). Peptide 43–47 included the tyrosine residue which has been reported to be an internalization signal in the bovine MPR 46 (Johnson et al., 1990). Taken together, these findings suggest that the tyrosine residue might be part of a multifunctional sorting signal in the cytoplasmic domain of the MPR 46.

A determinant that mediates the sorting of the MPR 46 in endosomes has been identified. In the MPR 46 cytoplasmic tail the amino acids 34–39 are required to avoid transport to dense lysosomes and subsequent proteolytic degradation (Rohrer et al., 1995). The transmembrane domain appears to contribute to this function. Cys^{34} and Cys^{30} in the MPR 46 cytoplasmic tail are amino acid residues that are reversibly palmitoylated via thioester linkage (Schweizer et al., 1996). The change of Cys^{34} but not of Cys^{30} to alanine results in missorting of cathepsin D and gradual accumulation of the receptor in dense lysosomes. This implies that the palmitoylated residue is essential for normal trafficking and lysosomal enzyme sorting function. Thus, Cys^{34} is involved in preventing the receptor from trafficking to lysosomes. A highly conserved CysCysArgArg sequence is found at position 15–18 of the IGF II/M6PR cytoplasmic tail. The palmitoylation of this receptor has been reported (Westcott and Rome, 1988) but it has yet to be determined whether the palmitoylation influences its trafficking and function in lysosomal enzyme sorting. In both MPRs, di-leucine motifs are found at the C-terminal ends which are flanked by casein kinase II phosphorylation sites that are phosphorylated. *In vitro*, casein kinase II phosphorylated serine 56 in the cytoplasmic domain of the human MPR 46 (Hemer et al., 1993). Serine phosphorylation was not essential for internalization of the receptor from the plasma membrane since recombinant serine mutants that were expressed in BHK 21 cells were internalized at the same rate as wild-type receptor (Hemer et al., 1993). Expression of serine mutants in mouse L-cells deficient for the IGF II/M6PR did not affect the sorting of newly synthesized cathepsin D to lysosomes (Johnson and Kornfeld, 1992b; Hemer et al., 1993). Serine phosphorylation of the IGF II/M6PR cytoplasmic tail is believed to occur in the TGN and/or after budding of clathrin-coated vesicles from the TGN and to be catalyzed by casein kinase II-like kinase. Such a post-translational modification occurs when the IGF II/M6PR exits from the TGN and represents a major, albeit transient modification (Meresse and Hoflack, 1993). Corresponding serine residues (serine 82 and 157) in the human IGF II/M6PR were phosphorylated by casein kinase II *in vitro* (Rosorius et al., 1993a). The recombinant human IGF II/M6PR cytoplasmic domain has been shown to interact with a 35kD cytosolic protein *in vitro*. The interaction depended on this phosphorylation (Rosorius et al., 1993b). The functional importance of the phosphorylation sites in the IGF II/M6PR trafficking remains controversial (Johnson and Kornfeld, 1992b; Chen et al., 1993).

While the involvement of the cytoplasmic domains of the MPR in intracellular recycling is well established, the extracellular and transmembrane domains might also determine the subcellular distribution of the MPR. A chimeric receptor in which the cytoplasmic domain of the IGF II/M6PR was fused to the extracytoplasmic and transmembrane domains of the EGF-receptor was detected predominantly at the cell surface (85%) (Dintzis and Pfeffer, 1990a). The complementary chimera that had the extracytoplasmic and transmembrane domain of the IGF II/M6PR fused to the cytoplasmic domain of the epidermal growth factor (EGF) receptor was detected in endosomal structures (Dintzis and Pfeffer, 1990b). The authors suggest that the luminal and transmembrane IGF II/M6PR domains might contain an "endosome retention signal" that retains the IGF II/M6PR in endosomal compartments. Alternatively, the EGF-receptor sequences might direct the rapid recycling to the cell surface. Experiments with chimeras containing lysozyme and the IGF II/M6PR cytoplasmic tail also suggest that the luminal domain of the IGF II/ M6PR contributes to the MPR subcellular distribution pattern (Conibear and Pearse, 1994). The MPRs have been shown to be essential components for the efficient translocation of the cytosolic AP-1 onto membranes of the TGN. Together with the ADP-ribosylation factor GTPase ARF-1 they are involved in the first step of clathrin coat assembly. The MPRs are critical components for the recruitment of AP-1 onto membranes. Two distinct determinants in the MPR 46 carboxy-terminal domain are required for the transport of lysosomal enzymes to lysosomes. A casein kinase II phosphorylation site is critical for the efficient interactions of AP-1 with its target membranes and the adjacent di-leucine motif appears to have important function for a post AP-1 binding step in the MPR 46 cycling pathway (Mauxion et al., 1996).

B.   Regulation of MPR Transport

Many reports indicate that transport of the IGF II/M6PR can be modulated by its ligands and by growth factors like IGF I, IGF II, PDGF, and EGF and hormones like insulin. Early studies on the subcellular distribution of the IGF II/M6PR indicated that M6P-ligands might influence the transport of their receptors. In the absence of M6P-ligands (I-cells), IGF II/M6PR were found to accumulate in the Golgi apparatus region (Brown and Farquhar, 1984; Brown et al., 1984). This suggested that a single step in the transport, the departure of IGF II/M6PR from the Golgi apparatus, might be triggered by ligand binding. The recycling of the MPR was reported to be independent of ligand cargo (Oka and Czech, 1986; Braulke et al., 1987; Pfeffer, 1988).

A well-studied effect is the influence of insulin on rat adipocytes. Besides the insulin-sensitive glucose transporter translocation, a mobilization of IGF II/M6PR from internal membranes to the cell surface has been reported (Cushman and Wardzala, 1980; Oka et al., 1984; Wardzala et al., 1984). The two proteins are known to reside in different intracellular compartments (Zorzano et al., 1989). The increase of IGF II/M6PR at the plasma membrane was coupled to a decrease in serine/threonine phosphorylation of the cytoplasmic domain of the translocated IGF II/M6PR by casein kinase II (Corvera et al., 1988a, 1988b). While in clathrin-coated regions a more highly phosphorylated subpopulation of IGF II/M6PR was found, the less phosphorylated receptors seemed to be excluded from clathrin-coated regions of the cell membrane and thus, from endocytosis. This might contribute to the increase of the IGF II/M6PR level at the cell surface.

Mannose-6-phosphate and IGF II doubled the cell surface expression of IGF II/M6PR in human skin fibroblasts (Braulke et al., 1989). The affinity of the receptors was unaffected, the effect was reversible, and due to receptor mobilization from internal membranes. IGF I and EGF similarly stimulated the redistribution of receptors. Combinations of growth factors and M6P increased the cell surface expression of the receptor in an additive manner, indicating independent intracellular signaling mechanisms. As the internalization rate of the IGF II/M6PR was not changed due to the increase of IGF II/M6PR at the cell surface, subsequently more ligands were endocytosed (Braulke et al., 1990a). However, the increased endocytosis could be demonstrated only for M6P-ligands, but not for IGF II which was hampered by the high background of IGF II binding proteins at the surface of fibroblasts. The increased uptake of M6P-ligands might control the concentration of extracellular M6P-ligands, which include not only lysosomal enzymes, but also proliferin and transforming growth factor β1-precursor which are thought to be involved in growth and differentiation processes (Lee and Nathans, 1988; Kovacina et al., 1989). The effects of M6P and IGF II on the distribution and function of IGF II/M6PR appear to be cell type-dependent. Mannose-6-phosphate, but not IGF I and IGF II, induced a redistribution of IGF II/M6PR in HepG2 cells, but in rat C6 glial and BRL-cells neither effector induced receptor translocation, but they did alter affinity (Kiess et al., 1989; Braulke et al., 1990a).

A simultaneous redistribution of both MPR and transferrin receptors by insulin-like growth factors has been reported for human fibroblasts, while the surface expression of receptors for low density lipoprotein and epidermal growth factor remained unaltered under these conditions (Damke et al., 1992). The increased surface expression of IGF II/M6PR and MPR 46 was accompanied by an increased uptake of receptor ligands. The number of transferrin receptors did not correlate with iron-uptake, although neither the rate nor the extent of transferrin internalization was changed. These results indicate that the redistribution of endocytic receptors shows selectivity and that the uptake of ligand may become uncoupled from the surface expression of the receptor via distinct mechanisms.

The translocation of receptors from intracellular membranes does not diminish the number of intracellular receptors. The overwhelming majority (90–95%) of the IGF II/M6PR are localized intracellularly and only one communication reported on impaired sorting of a low-affinity M6P-ligand (Cathepsin L) after PDGF-treatment of NIH 3T3 fibroblasts (Prence et al., 1990). The compartment from which the receptors are translocated has not yet been characterized and it remains to be determined whether the translocated receptors derive from membranes of the secretory and/or endocytotic route or from a different pool.

VI. DIFFERENTIAL EXPRESSION OF MPR

With the exception of a few tumor-derived cell lines that lack the IGF II/M6PR (Gabel et al., 1983; Mainferme et al., 1985), all mammalian cell lines and tissues examined contain both receptors. However, the concentration and the relative ratios of the two receptors vary among several human cell lines and tissues by about one order of magnitude (Wenk et al., 1991).

The expression of the IGF II/M6PR has been studied by *in situ* hybridization during the pre- and post-implantation period in rat and mouse (Tollefsen et al., 1989; Senior et al., 1990; Harvey and Kaye, 1991; Matzner et al., 1992). IGF II/M6PR transcripts were detected in the cardiovascular system and developing muscle, but only low levels were detected in the liver. For a number of tissues, high expression of IGF II/M6PR transcripts correlated with a high expression of IGF II transcripts (Ohlsson et al., 1989; Tollefsen et al., 1989; Senior et al., 1990). The coordinately regulated expression has suggested a role of the receptor in a paracrine/autocrine function of IGF II. There were, however, exceptions and the most notable is liver. The liver parenchyma, as one of the major sites of IGF II expression during embryonic development (Beck et al., 1987; Stylianopoulou et al., 1988), was found to express only low levels of IGF II/M6PR suggesting an endocrine function for the secretion of IGF II (Senior et al., 1990; Matzner et al., 1992). In the rat, the IGF II/M6PR levels were found to be high in fetal tissues and to decline in the early postnatal period (Sklar et al., 1989). This suggests that the IGF II/M6PR plays an important role in fetal growth and development of rodents.

MPR 46 expression was studied in the embryonic mouse and showed a pattern that was spatially and temporally different from IGF II/M6PR during mouse embryogenesis (Matzner et al., 1992). Tissues that are active in hematopoiesis (liver, bone marrow) and T cell differentiation (thymus) were the first that expressed high levels of MPR 46 RNA. There is, however, no insight in the specific function of MPR 46 in these cells. At later stages of embryogenic development, a wide variety of tissues expressed both receptors, but the expression pattern was almost non-overlapping. This points to specific functions of the two mannose-6-phosphate receptors during mouse embryogenesis. In lower vertebrates like the chicken and frog the IGF II/M6PR does not bind IGF II (Clairmont and Czech, 1989; Canfield and Kornfeld, 1989; Yang et al., 1991). *In situ* hybridization data from chicken show an expression pattern that differs from that of the mouse and rat embryo (Matzner et al., 1996). In chicken the IGF II/M6PR expression in the embryo is not higher than in the adult animal. The pattern of MPR46 expression in the chicken embryo resembles that of IGF II/M6PR expression. This suggests that the IGF II/M6PR expression in the mouse embryo depends on IGF II.

A few studies report on the regulated expression of the IGF II/M6PR. An inverse correlation was detected between the levels of IGF II/M6PR and the lysosomal enzyme, procathepsin D, that was found to be overexpressed in human breast cancer cells. The levels of IGF II/M6PR mRNA and protein were down-regulated in the presence of estradiol, while the gene expression of the lysosomal enzyme, cathepsin D, and of IGF II were increased (Cavailles et al., 1988; Yee et al., 1988; Osborne et al., 1989; Mathieu et al., 1991). The secretion of other lysosomal enzymes was also affected, but their biosynthesis was not increased (Capony et al., 1990). In normal, confluent rat hepatocytes, down-regulation of the IGF II/M6PR was observed (Scott et al., 1987), while IGF II/M6PR expression was found to be independent of cell density in hepatoma cells (Scott et al., 1988). Controversely, IGF II/M6PR expression was induced during liver regeneration (Burguera et al., 1990).

Disruption of the IGF II/M6PR causes lethality in mice. In mouse, the IGF II/M6PR has been mapped to the Tme locus (Barlow et al., 1991). It has been shown to be maternally imprinted and has a function in embryonic growth regulation. In mice inheritance of a maternal null allele or homozygosity for the null allele of the IGF II/M6PR gene is lethal at birth (Wang et al., 1994). Furthermore, the mutants are about 30% larger than their litter mates. Mutant mice have organ and skeletal abnormalities and missort mannose-6-phosphate-containing proteins. The pheno-type appears to result from a loss of function of the IGF II/M6PR that endocytoses extracellular IGF II from the circulation and functions as a receptor for the degradation of IGF II. The phenotype is probably caused by an excess of IGF II because the introduction of an IGF II/M6PR null allele rescued the IGF II/M6PR mutant mice. A few mutants that inherited a maternal null allele reactivated their paternal IGF II/M6PR allele in some tissues and survived to adulthood. However, no mice homozygous for the inactive allele survived, indicating a function for the reactivated paternal allele in postnatal survival. As the defect in M6P-recognition marker synthesis in man (I-cell disease) is lethal at the age of 4–10 years, the defect of the IGF II/M6PR binding of M6P-containing ligands does not appear to be the cause of embryonic death.

The targeted disruption of the MPR 46 gene in mice resulted in viable animals that had no discernable phenotype upon inspection or autopsy at the age of 6–9 weeks as shown in the studies of Köster et al. (1993) and Ludwig et al. (1993). No signs of lysosomal storage were detectable on light- and electron-microscopical examination of a variety of tissues. In contrast to whole animals, isolated cells from MPR 46-deficient mice (fibroblasts, splenocytes, thymocytes) displayed a three to five-fold increase in levels of M6P-polypeptides secreted into the medium. These results indicate a misrouting of M6P-polypeptides in MPR 46-deficient cells and strongly argue for an essential function of MPR 46 in targeting newly synthesized lysosomal enzymes from the secretory to the endocytic route. *In vivo*, re-uptake of secreted lysosomal enzymes via carbohydrate specific receptors like the IGF II/M6PR and mannose receptor and asialoglycoprotein receptor has been shown to correct the misrouting of newly synthesized lysosomal enzymes and to result in a normal phenotype (Köster et al., 1994). Under cell culture conditions with its grossly expanded extracellular space, the efficiency of the re-uptake is reduced and the increased secretion becomes obvious.

The MPR 46-deficient mice and mice double deficient for IGF II and IGF II/M6PR may help to define whether MPR 46 and IGF II/M6PR transport different groups of lysosomal enzymes, whether the two receptors feed different subpopulations of endosomes/lysosomes, and why the receptors are expressed in a tissue-specific manner. Complementary studies in mice or cell lines deficient in either MPR or both MPR will be required to answer these questions.

VII. CONCLUSION

MPR 46 and IGF II/M6PR are integral type I transmembrane glycoproteins that mediate the targeting of newly synthesized lysosomal proteins to lysosomes. Their structure is related and they appear to be derived from a common ancestor gene. The signals that direct their intracellular routing between TGN, endosomes, and plasma membrane are beginning to be defined. The IGF II/M6PR has been shown to be a multifunctional protein that in mammalian species additionally binds the insulin-like growth factor II and functions as a morphogenic receptor in mice.

The question why cells express two different MPR remains open. The two MPR may have different, or at least partly different functions, for example, binding of IGF II. It is not clear, whether the acquisition of IGF II binding or the need for other functional differences, such as endocytosis and secretion of M6P-containing ligands, has provided the evolutionary force for the development and expression of two MPR. Further information on specific functions of the two MPR is expected from the analysis of mice and cell lines deficient in MPR 46, IGF II/M6PR and IGF II, or both MPR and IGF II.

ACKNOWLEDGMENTS

The author would like to thank all colleagues who have made results of their studies available prior to publication and who supported her with reprints. The author would like to especially thank Angelika Thiel for her help in the preparation of this manuscript. The research done in the author's laboratory has been supported by the Deutsche Forschungsgemeinschaft (SFB 236/B 12 and grant Po 303/2-1).

REFERENCES

Bansal, A., & Gierasch, L. M. (1991). The NPXY internalization signal of the LDL receptor adopts a reverse-turn conformation. Cell 67, 1195–1201.

Baranski, T. J., Cantor, A. B., & Kornfeld, S. (1992). Lysosomal enzyme phosphorylation I. J. Biol. Chem. 267, 23342–23348.

Barlow, D. P., Stöger, R., Herrmann, B., Saito, K., & Schweifer, N. (1991). The mouse insulin-like growth factors type 2 receptor is imprinted and closely linked to the Tme locus. Nature 349, 84–87.

Beck, F., Samani, N. J., Penschow, J. D., Thorley, B., Tregear, G. W., & Coghlan, J. P. (1987). Histochemical localization of IGF I and II mRNA in the developing rat embryo. Development 101, 175–184.

Bleekemolen, J. E., Stein, M., von Figura, K., Slot, J. W., & Geuze, H. J. (1988). The two mannose 6-phosphate receptors have almost identical subcellular distributions in U 937 monocytes. Eur. J. Cell Biol. 47, 366–372.

Braulke, T., Gartung, C., Hasilik, A., & von Figura, K. (1987). Is movement of mannose 6-phosphate-specific receptor triggered by binding of lysosomal enzymes? J. Cell Biol. 104, 1735–1742.

Braulke, T., Causin, C., Waheed, A., Junghans, U., Hasilik, A., Maly, P., Humbel, R. E., & von Figura, K. (1988). Mannose 6-phosphate/insulin like growth factor II receptor: Distinct binding sites for mannose 6-phosphate and insulin like growth factor II. Biochem. Biophys. Res. Commun. 150, 1287–1293.

Braulke, T., Tippmer, S., Neher, E., & von Figura, K. (1989). Regulation of mannose 6-phosphate/IGF II receptor expression at the cell surface by mannose 6-phosphate, insulin like growth factors and epidermal growth factor. EMBO J. 8, 681–686.

Braulke, T., Tippmer, S., Chao, H. H. J., & von Figura, K. (1990a). Insulin like growth factor I and II stimulate the endocytosis but do not affect sorting of lysosomal enzymes in human fibroblasts. J. Biol. Chem. 265, 6650–6655.

Braulke, T., Tippmer, S., Chao, H. J., & von Figura, K. (1990b). Regulation of mannose 6-phosphate/insulin-like growth factor II receptor distribution by activators and inhibitors of protein kinase C. Eur. J. Biochem. 189, 609–616.

Braulke, T., & Mieskes, G. (1992). Role of protein phosphatases in insulin-like growth factor II stimulated mannose 6-phosphate/IGF II receptor distribution. J. Biol. Chem. 267, 17347–17353.

Brown, W. J., & Farquhar, M. G. (1984). Accumulation of control vesicles bearing mannose 6-phosphate receptors for lysosomal enzymes in the Golgi region of I-cell fibroblasts. Proc. Natl. Acad. Sci. USA 81, 5135–5139.

Brown, W. J., Constantinescu, E., & Farquhar, M. G. (1984). Redistribution of mannose 6-phosphate receptors induced by tunicamycin and chloroquine. J. Cell Biol. 99, 320–326.

Burguera, B., Werner, H., Sklar, M., Shen-Orr, Z., Stannard, B., Roberts, C. T., Nissley Jr., S. P., Vore, S. J., Caro, J. F., & Keröith, D. (1990). Liver regeneration is associated with increased expression of the insulin-like growth factor II/mannose 6-phosphate receptor. Mol. Endocrinol. 4, 1539–1544.

Canfield, W. M., & Kornfeld, S. (1989). The chicken liver cation-independent mannose 6-phosphate receptor lacks the high affinity binding site for insulin like growth factor II. J. Biol. Chem. 264, 7100–7103.

Canfield, W. M., Johnson, K. F., Ye, R. D., Gregory, W., & Kornfeld, S. (1991). Localization of the signal for rapid internalization of the bovine cation-independent mannose 6-phosphate/insulin-like growth factor II receptor to amino acids 24–29 of the cytoplasmic tail. J. Biol. Chem. 266, 5682–5688.

Cantor, A. B., & Kornfeld, S. (1992). Phosphorylation of Asn-linked oligosaccharides located at novel sites on the lysosomal enzyme cathepsin D. J. Biol. Chem. 267, 23357–23363.

Cantor, A. B., Baranski, T. J., & Kornfeld, S. (1992). Lysosomal enzyme phosphorylation II. J. Biol. Chem. 267, 23349–23356.

Capony, F., Rougeot, C., Cavailles, V., & Rochefort, H. (1990). Estradiol increases the secretion by MCF cells of several lysosomal proenzymes. Biochem. Biophys. Res. Commun. 171, 972–978.

Cavailles, V., Angeran, P., Garcia, M., & Rochefort, H. (1988). Estrogens and growth factors induce the mRNA of the 52 k procathepsin D secreted by breast cancer cells. Nucleic Acids Res. 16, 1903–1919.

Chao, H. H. J., Waheed, A., Pohlmann, R., Hille, A., & von Figura, K. (1990). Mannose 6-phosphate receptor dependent secretion of lysosomal enzymes. EMBO J. 9, 3507–3513.

Clairmont, K. B., & Czech, M. (1989). Chicken and xenopus mannose 6-phosphate receptors fail to bind insulin-like growth factor II. J. Biol. Chem. 264, 16390–16392.

Collawn, J. F., Stangel, M., Kuhn, L. A., Esekogwu, V., Jing, S. Q., Trowbridge, I. S., & Tainer, J. A. (1990). Transferrin receptor internalization sequence YXRF implicates a tight turn as the structural recognition motif for endocytosis. Cell 63, 1061–1072.

Collawn, J. F., Kuhn, L. A., Lin, L. F. S., Tainer, J. A., & Trowbridge, I. S. (1991). Transplanted LDL and mannose 6-phosphate receptor internalization signals promote high efficiency endocytosis of the transferrin receptor. EMBO J. 10, 3247–3253.

Conibear, E., & Pearse, B. M. (1994). A chimera of the cytoplasmic tail of the mannose-6-phosphate/IGF II receptor and lysozyme localizes to the TGN rather than prelysosomes where the bulk of the endogenous receptor is found. J. Cell Sci. 107, 923–932.

Conover, C. A., Misra, P., Hintz, R. L., & Rosenfeld, R. G. (1986). Effect of an anti-insulin-like growth factor I receptor antibody on insulin-like growth factor II stimulation of DNA synthesis in human fibroblasts. Biochem. Biophys. Res. Commun. 139, 501–508.

Corvera, S., Folander, K., Clairmont, K. B., & Czech, M. P. (1988a). A highly phosphorylated subpopulation of insulin-like growth factor II/mannose 6-phosphate receptors is concentrated in a clathrin-enriched plasma membrane fraction. Proc. Natl. Acad. Sci. USA 85, 7567–7571.

Corvera, S., Roach, P. J., De Paoli-Roach, A. A., & Czech, M. P. (1988b). Insulin action inhibits insulin-like growth factor II (IGF II) receptor phosphorylation in H35 hepatoma cells. IGF II-receptors isolated from insulin-treated cells exhibit enhanced in vitro phosphorylation by casein kinase II. J. Biol. Chem. 263, 3116–3122.

Cushman, S. W., & Wardzala, L. J. (1980). Potential mechanism of insulin action on glucose transport in the isolated rat adipose cell. J. Biol. Chem. 255, 4758–4762.

Dahms, N. M., Lobel, P., Breitmeyer, J., Chirgwin, J. M., & Kornfeld, S. (1987). 46 kDa mannose 6-phosphate receptor: Cloning, expression and homology to the 215 kD mannose 6-phosphate receptor. Cell 50, 181–192.

Dahms, N. M., & Kornfeld, S. (1989). The cation-dependent mannose 6-phosphate receptor. J. Biol. Chem. 264, 11458–11467.

Dahms, N. M., Lobel, P., & Kornfeld, S. (1989). Mannose 6-phosphate receptors and lysosomal enzyme targeting. J. Biol. Chem. 264, 12115–12118.

Dahms, N. M., Rose, P. A., Mokentin, J. D., Zhang, Y., & Brzycki, M. A. (1993). The bovine mannose 6-phosphate/insulin-like growth factor II receptor. J. Biol. Chem. 268, 54457–54463.

Damke, H., von Figura, K., & Braulke, T. (1992). Simultaeous redistribution of mannose 6-phosphate and transferrin receptors by insulin-like growth factors and phorbolester. Biochem. J. 281, 225–229.

Damke, H., Bouterfa, H., & Braulke, T. (1993). The insulin-like growth factor II does not induce the generation of inositol trisphosphate, diacylglycerol or cAMP in human fibroblasts. Mol. Cell. Endocrinology 99, 225–229.

Denzer, K., Weber, B., von Figura, K., & Pohlmann, R. (1996). Internalization signals in the cytoplasmic tail of the human Mr 46000 mannose 6-phosphate receptor. 6th Int. Congress on Cell Biology, San Francisco.

Dintzis, S. M., & Pfeffer, S. R. (1990a). The mannose 6-phosphate receptor cytoplasmic domain is not sufficient to alter the subcellular distribution of a chimeric EGF-receptor. EMBO J. 9, 77–84.

Dintzis, S. M., & Pfeffer, S. R. (1990b). Mannose 6-phosphate receptor ectodomain sequences influence the rate of receptor recycling from endosomes to the cell surface. J. Cell Biol. 111 (1126) 203a.

Distler, J. J., Guo, J., Jourdian, G. W., Srivastava, O. P., & Hindsgaul, O. (1991). The binding specificity of high and low molecular weight phosphomannosyl receptors from bovine testes. Inhibition studies with chemically synthesized 6-O-phosphorylated oligomannosides. J. Biol. Chem. 266, 21687–21692.

Duncan, J. R., & Kornfeld, S. (1988). Intracellular movement of two mannose 6-phosphate receptors: Return to the Golgi apparatus. J. Cell Biol. 106, 617–628.

Eberle, W., Sander, C., Klaus, W., Schmidt, B., von Figura, K., & Peters, C. (1991). The essential tyrosine of the internalization signal in lysosomal acid phosphatase is part of a β turn. Cell 67, 1203–1209.

Fukuda, M. (1991). Lysosomal membrane glycoproteins. J. Biol. Chem. 266, 21327–21330.

Gabel, C. A., Goldberg, D. E., & Kornfeld, S. (1983). Identification and characterization of cells deficient in the mannose 6-phosphate receptor: evidence for an alternate pathway for lysosomal enzyme targeting. Proc. Natl. Acad. Sci. USA 80, 775–779.

Glickman, J. N., Conibear, E., & Pearse, B. M. (1989). Specificity of binding of clathrin adaptors to signals on the mannose 6-phosphate/insulin-like growth factor II receptor. EMBO J. 8, 1041–1047.

Goda, Y., & Pfeffer, S. R. (1988). Selective recycling of the mannose 6-phosphate/IGF II receptor to the trans Golgi network in vitro. Cell 55, 309–320.

Green, S. A., & Kelly, R. B. (1992). Low density lipoprotein receptor and cation-independent mannose 6-phosphate receptor are transported from the cell surface to the Golgi apparatus at equal rates in PC12 cells. J. Cell Biol. 117, 47–55.

Guse, A. H., Kiess, W., Funk, B., Kessler, U., Berg, I., & Gercken, G. (1992). Identification and characterization of insulin-like growth factor receptors on adult rat cardiac myocytes: Linkage to inositol 1,4,5-trisphosphate formation. Endocrinology 130, 145–151.

Haft, C. R., Klausner, R. D., & Taylor, S. J. (1994). Involvement of di-leucine motifs in the internalization and degradation of the insulin receptor. J. Biol. Chem. 269, 26286–26294.

Haig, D., & Graham, C., (1991). Genomic imprinting and the strange case of the insulin-like growth factor II receptor. Cell 64, 1045–1064.

Hari, J., Pierce, S. B., Morgan, D. O., Sara, V., Smith, M. C., & Roth, R. A. (1987). The receptor for insulin-like growth factor II mediates an insulin-like response. EMBO J. 6, 3367–3371.

Harvey, M. B., & Kaye, P. L. (1991). IGF II receptors are first expressed at the 2-cell stage of mouse development. Development 111, 1057–1060.

Hemer, F., Körner, C., & Braulke, T. (1993). Phosphorylation of the human 46 kDa mannose 6-phosphate receptor in the cytoplasmic domain at serine 56. J. Biol. Chem. 268, 17108–17113.

Higashijima, T., Uzu, S., Nakajima, T., & Ross, E. M. (1988). Mastoparan, a peptide toxin from wasp venom, mimics receptors by activating GTP-binding regulatory proteins (G proteins). J. Cell Biol. 263, 6491–6494.

Hille, A., Waheed, A., & von Figura, K. (1989). The ligand-binding conformation of Mr 46.000 mannose 6-phosphate receptor. Acquisition of binding activity during *in vitro* synthesis. J. Biol. Chem. 264, 13460–13467.

Hille, A., Waheed, A., & von Figura, K. (1990). Assembly of the ligand-binding conformation of M_r 46,000 mannose 6-phosphate-specific receptor takes place before reaching the Golgi complex. J. Cell Biol. 110, 963–972.

Hille-Rehfeld, A. (1995). Mannose 6-phosphate receptors in sorting and transport of lysosomal enzymes. Bioch. Biophys. Acta 1241, 177–194.

Hoflack, B., & Kornfeld, S. (1985). Purification and characterization of a cation-dependent mannose 6-phosphate receptor from murine P388D1 macrophages and bovine liver. J. Biol. Chem. 260, 12008–12014.

Hoflack, B., Fujimoto, K., & Kornfeld, S. (1987). The interaction of phosphorylated oligosaccharides and lysosomal enzymes with bovine liver cation-dependent mannose 6-phosphate receptor. J. Biol. Chem. 262, 123–129.

Jadot, M., Canfield, W. M., Gregory, W., & Kornfeld, S. (1992). Characterization of the signal for rapid internalization of the bovine mannose 6-phosphate/insulin-like growth factor II receptor. J. Biol. Chem. 267, 11069–11077.

Jin, M., Sahagian, G. G., Jr., & Snider, M. D. (1989). Transport of surface mannose 6-phosphate receptor to the Golgi complex in cultured human cells. J. Biol. Chem. 264, 7675–7680.

Johnson, F. K., Chan, W., & Kornfeld, S. (1990). Cation-dependent mannose 6-phosphate receptor contains two internalization signals in its cytoplasmic domain. Proc. Natl. Acad. Sci. USA 87, 10010–10014.

Johnson, K. F., & Kornfeld, S. (1992a). The cytoplasmic tail of the mannose 6-phosphate IGF II receptor has two signals for lysosomal sorting in the Golgi. J. Cell Biol. 119, 249–257.

Johnson, K. F., & Kornfeld, S. (1992b). A His-Leu-Leu sequence near the carboxyl terminus of the cytoplasmic domain of the cation-dependent mouse 6-phosphate receptor is necessary for the lysosomal enzyme sorting function. J. Biol. Chem. 267, 17110–17115.

Kasper, D., Dittmer, F., von Figura, K., & Pohlmann, R. (1996). Neither type of mannose-6-phosphate receptor is sufficient for targeting of lysosomal enzymes along intracellular routes. J. Cell Biol. 134, 1–9.

Kiess, W., Haskell, J. F., Lee, L., Greenstein, L. A., Miller, B. E., Aarons, A. L., Rechler, M. M., & Nissley, S. P. (1987). An antibody that blocks insulin-like growth factor (IGF) binding to the type II IGF receptor is neither an agonist nor an inhibitor of IGF-stimulated biologic responses in L6 myoblasts. J. Biol. Chem. 262, 12745–12751.

Kiess, W., Blickenstaff, G. D., Sklar, M. M., Thomas, C. L., Nissley, S. P., & Sahagian, G. G. (1988). Biochemical evidence that the type II insulin-like growth factor receptor is identical to the cation-independent mannose 6-phosphate receptor. J. Biol. Chem. 263, 9339–9344.

Kiess, W., Thomas, C. L., Greenstein, L. A., Lee, L., Sklar, M. M., Rechler, M. M., Sahagian, G. G., & Nissley, S. P. (1989). Insulin-like growth factor II inhibits both the cellular uptake of β-galactosidase and the binding of β-galactosidase to purified IGF II/mannose 6-phosphate receptor. J. Biol. Chem. 264, 4710–4714.

Kiess, W., Thomas, C. L., Sklar, M. M., & Nissley, S. P. (1990). β-Galactosidase decreases the binding affinity of the insulin-like growth factor II/mannose 6-phosphate receptor for insulin-like growth factor II. Eur. J. Biochem. 190, 71–77.

Klier, H. J., von Figura, K., & Pohlmann, R. (1991). Isolation and analysis of the human Mr 46000 mannose 6-phosphate receptor gene. Eur. J. Biochem. 197, 23–28.

Klumperman, J., Hille, A., Veenendaal, T., Oorschot, V., Stoorvogel, W., von Figura, K., & Geuze, H. J. (1993). Differences in the endosomal distributions of the two mannose 6-phosphate receptors. J. Cell Biol. 121, 997–1010.

Kojima, I., Nishimoto, I., Iiri, T., Ogata, E., & Rosenfeld, R. (1988). Evidence that type II insulin-like growth factor receptor is coupled to calcium gating system. Biochem. Biophys. Res. Commun. 154, 9–19.

Kornfeld, S., & Mellman, I. (1989). The biogenesis of lysosomes. Annu. Rev. Cell Biol. 5, 307–330.

Kornfeld, S. (1992). Structure and function of the mannose 6-phosphate/insulin-like growth factor II receptors. Annu. Rev. Biochem. 61, 307–330.

Köster, A., Nagel, G., von Figura, K., & Pohlmann, R. (1991). Molecular cloning of the mouse 46 kDa mannose 6-phosphate receptor (MPR 46). Biol. Chem. Hoppe-Seyler 372, 297–300.

Köster, A., Saftig, P., Matzner, U., von Figura, K., Peters, C., & Pohlmann, R. (1993). Missorting of lysosomal proteins in mice with targeted disruption of the Mr 46000 mannose 6-phosphate receptor gene. EMBO J. 12, 5219–5223.

Köster, A., von Figura, K., & Pohlmann, R. (1994). Mistargeting of lysosomal enzymes in Mr 46000 mannose-6-phosphate receptor deficient mice is compensated by carbohydrate-specific endocytotic receptors. Eur. J. Bioch. 224, 685–689.

Kovacina, K. S., Steele-Perkins, G., Purchio, A. F., Lioubin, M., Miyazona, K., Heldin, C. H., & Roth, R. A. (1989). A role for the insulin-like growth factor II/mannose 6-phosphate receptor in the insulin-induced inhibition of protein catabolism. Biochem. Biophys. Res. Commun. 160, 393–403.

Kyle, J. W., Nolan, C. M., Oshima, A., & Sly, W. S. (1988). Expression of human cation-independent mannose 6-phosphate receptor cDNA in receptor negative mouse P388D$_1$ cells following gene transfer. J. Biol. Chem. 263, 16230–16235.

Laureys, G., Barton, D. E., Ullrich, A., & Francke, U. (1988). Chromosomal mapping of the gene for the type II insulin-like growth factor receptor/cation-independent mannose 6-phosphate receptor in man and mouse. Genomics 3, 224–229.

Lee, S. J., & Nathans, D. (1988). Proliferin secreted by cultured cells binds to mannose 6-phosphate receptors. J. Biol. Chem. 263, 3521–3527.

Lemansky, P., Gieselmann, V., Hasilik, A., & von Figura, K. (1985). Synthesis and transport of lysosomal acid phosphatase in normal and I-cell fibroblasts. J. Biol. Chem. 260, 9023–9030.

Letourneur, F., & Klausner, R. D. (1992). A novel di-leucine motif and a tyrosine-based motif independently mediate lysosomal targeting and endocytosis of CD 3 chanins. Cell 69, 1143–1157.

Li, M., Distler, J. J., & Jourdian, G. W. (1990). The aggregation and dissociation properties of a low molecular weight mannose 6-phosphate receptor from bovine testis. Arch. Biochem. Biophys. 283, 150–157.

Li, M., & Jourdian, G. W. (1991). Isolation and characterization of two glycosylation isoforms of low molecular weight mannose 6-phosphate receptor from bovine testis. J. Biol. Chem. 266, 17621–17630.

Lobel, P., Dahms, N. M., Breitmeyer, J., Chirgwin, J. M., & Kornfeld, S. (1987). Cloning of the bovine 215 kDa cation-independent mannose 6-phosphate receptor. Proc. Natl. Acad. Sci. USA 84, 2233–2237.

Lobel, P., Dahms, N. M., & Kornfeld, S. (1988). Cloning and sequence analysis of the cation-independent mannose 6-phosphate receptor. J. Biol. Chem. 263, 2563–2570.

Lobel, P., Fujimoto, K., Ye, R. D., Griffiths, G., & Kornfeld, S. (1989). Mutations in the cytoplasmic domain of the 275 kd mannose 6-phosphate receptor differentially alter lysosomal enzyme sorting and endocytosis. Cell 57, 787–796.

Ludwig, T., Rüther, U., Metzger, R., Copeland, N. G., Jenkins, N. A., Lobel, P., & Hoflack, B. (1992). Gene and Pseudogene of the mouse cation-dependent mannose 6-phosphate receptor. J. Biol. Chem. 267, 12211–12219.

Ludwig, T., Ovitt, C. E., Bauer, U., Hollinshead, M., Remmler, J., Lobel, P., Rüther, U., & Hoflack, B. (1993). Targeted disruption of the mouse cation-dependent mannose 6-phosphate receptor gene results in a phenotype similar to the human disease mucolipidosis III. EMBO J. 12, 5225–5235.

Ludwig, T., Munier-Lehmann, H., Bauer, U., Hollinshead, H., Ovitt, C., Lobel, P., & Hoflack, B. (1994). Differential sorting of lysosomal enzymes in mannose 6-phosphate receptor-deficient fibroblasts. EMBO J. 13, 3430–3437.

Ma, Z. M., Grubb, J. H., & Sly, W. S. (1991). Cloning, sequencing and functional characterization of the murine 46 kD mannose 6-phosphate receptor. J. Biol. Chem. 266, 10589–10595.

MacDonald, R. G., Pfeffer, S. R., Coussens, L., Tepper, M. A., Brocklebank, C. M., Mole, J. E., Anderson, J. K., Chen, E., Czech, M. P., & Ullrich, A. (1988). A single receptor binds both insulin-like growth factor II and mannose 6-phosphate. Science 239, 1134–1137.

Mainferme, F., Wattiaux, R., & von Figura, K. (1985). Synthesis, transport and processing of cathepsin C in Morris hepatoma 7777 cells and rat hepatocytes. Eur. J. Biochem. 153, 211–216.

Mathieu, M., Vignon, F., Capony, F., & Rochefort, H. (1991). Estradiol down regulated the mannose 6-phosphate/insulin-like growth factor II receptor gene and induces cathepsin-D in breast cancer cells: a receptor saturation mechanism to increase the secretion of lysosomal proenzymes. Mol. Endocrinology 5, 815–822.

Matzner, U., von Figura, K., & Pohlmann, R. (1992). Expression of the two mannose 6-phosphate receptors is spatially and temporally differential during mouse embryogenesis. Development 114, 965–972.

Matzner, U., von Figura, K., & Pohlmann, R. (1996). Expression of the two mannose-6-phosphate receptors in chicken. Developmental Dynamics, in press.

Mauxion, F., LeBorgne, R., Munier-Lehmann, H., & Hoflack, B. (1996). A casein kinase II phosphorylation site in the cytoplasmic domain of the cation-dependent mannose-6-phosphate receptor determines the high affinity interaction of the AP-1 Golgi assembly proteins with the membrane. J. Biol. Chem. 271, 2171–2178.

Mellas, J., Gavin, J. R., & Hammerman, M. R. (1986). Multiplication-stimulating activity-induced alkalinization of canine renal proximal tubular cells. J. Biol. Chem. 261, 14437–14442.

Meresse, S., Ludwig, T., Frank, R., & Hoflack, B. (1990). Phosphorylation of the cytoplasmic domain of the bovine cation-independent mannose 6-phosphate receptor. J. Biol. Chem. 265, 18833–18842.

Meresse, S., & Hoflack, B. (1993). Phosphorylation of the cation-independent mannose 6-phosphate receptor is closely associated with its exit from the trans-Golgi network. J. Cell Biol. 120, 67–75.

Morgan, D. O., Edman, J. C., Standring, D. N., Fried, V. A., Smith, M. C., Roth, R. A., & Rutter, W. J. (1987). Insulin-like growth factor II receptor as a multifunctional binding protein. Nature 329, 301–307.

Mottala, C., & Czech, M. P. (1984). The type II insulin-like growth factor receptor does not mediate increased DNA synthesis in H-35 hepatoma cells. J. Biol. Chem. 259, 12705–12713.

Murayama, Y., Okamoto, T., Ogata, E., Asano, T., Iiri, T., Katada, T., Ui, M., Grubb, J. H., Sly, W. S., & Nishimito, I. (1990). Distinctive regulation of the functional linkage between the human cation-independent mannose 6-phosphate receptor and GTP-binding proteins by insulin-like growth factor II and mannose 6-phosphate. J. Biol. Chem. 265, 17456–17462.

Nishimoto, I., Hata, Y., Ogata, E., & Kojima, I. (1987). Insulin-like growth factor II stimulates calcium influx in competent BALB/c 3T3 cells primed with epidermal growth factor. Characteristics of calcium influx and involvement of GTP-binding protein. J. Biol. Chem. 262, 12120–12126.

Nishimoto, I., Murayama, Y., Katada, T., Ui, M., & Ogata, E. (1989). Possible direct linkage of insulin-like growth factor II receptor with guanine nucleotide-binding proteins. J. Biol. Chem. 264, 14029–14038.

Nolan, C. M., & Sly, W. S. (1987). Intracellular traffic of the mannose 6-phosphate receptors and its ligands. Adv. Exp. Med. Biol. 225, 199–212.

Nolan, C. M., Kyle, J. W., Watanabe, H., & Sly, W. S. (1990). Binding of insulin-like growth factor II by human cation-independent mannose 6-phosphate receptor/IGF II receptor expressed in receptor deficient mouse L cells. Cell regulation 1, 197–213.

Ohlsson, R., Holmgren, L., Glaser, A., Szpecht, A., & Pfeifer-Ohlsson, B. (1989). Insulin-like growth factor II and short stimulatory loops in control of human placental growth. EMBO J. 8, 1993–1999.

Oka, Y., Mottola, C., Oppenheimer, C. L., & Czech, M. P. (1984). Insulin activates the appearance of insulin-like growth factor II receptors on the adipocyte cell surface. Proc. Natl. Acad. Sci. USA 81, 4028–4032.

Oka, Y., Rozek, L. M., & Czech, M. P. (1985). Direct demonstration of rapid insulin-like growth factor II receptor internalization and recycling in rat adipocytes. Insulin stimulates [^{125}I]insulin-like growth factor II degradation by modulating the IGF II receptor recycling process. J. Biol. Chem. 260, 9435–9442.

Oka, Y., & Czech, M. P. (1986). The type II insulin-like growth factor receptor is internalized and recycles in the absence of ligand. J. Biol. Chem. 261, 9090–9093.

Okamoto, T., Katada, T., Murayama, Y., Ui, M., Ogata, E., & Nishimoto, I. (1990a). A simple structure encodes G protein-activating function of the IGF II/mannose 6-phosphate receptor. Cell 62, 709–717.

Okamoto, T., Nishimoto, I., Murayama, Y., Ohkuni, Y., & Ogata, E. (1990b). Insulin-like growth factor II/mannose 6-phosphate receptor is incapable of activating GTP-binding proteins in response to mannose 6-phosphate but capable in response to insulin-like growth factor II. Biochem. Biophys. Res. Commun. 168, 1201–1210.

Osborne, C. K., Coronado, E. B., Kitten, L. J., Artega, C. I., Fuqua, S. A. W., Ramasharma, K., Marshall, M., & Li, C. H. (1989). IGF II: A potential autocrine/paracrine growth factor for human breast cancer acting via the IGF I receptor. Mol. Endocrinol. 3, 1701–1709.

Oshima, A., Nolan, C. M., Kyle, J., Grubb, H. J., & Sly, W. S. (1988). The human cation-independent mannose 6-phosphate receptor. Cloning and sequence of the full length cDNA and expression of the functional receptor in COS-cells. J. Biol. Chem. 263, 2553–2562.

Pearse, B. M. (1988). Receptors compete for adaptors found in plasma membrane coated pits. EMBO J. 7, 3331–3336.

Pearse, B. M. F., & Robinson, M. S. (1990). Clathrin, adaptors and sorting. Annu. Rev. Cell Biol. 6, 151–171.

Pelham, H. R. B. (1988). Evidence that luminal ER proteins are sorted from secreted proteins in a post ER compartment. EMBO J. 4, 2457–2460.

Perdue, J. F., Chan, J. K., Thibault, C., Radaj, P., Mills, B., & Daughaday, W. H. (1983). The biochemical characterization of detergent-solubilized insulin-like growth factor II receptors from rat placenta. J. Biol. Chem. 258, 7800–7811.

Pfeffer, S. R. (1988). Mannose 6-phosphate receptors and their role in targeting proteins to lysosomes. J. Membr. Biol. 103, 7–16.

Pohlmann, R., Waheed, A., Hasilik, A., & von Figura, K. (1982). Synthesis of phosphorylated recognition marker in lysosomal enzymes is located in the cis part of Golgi apparatus. J. Biol. Chem. 257, 5323–5325.

Pohlmann, R., Nagel, G., Schmidt, B., Stein, M., Lorkowski, G., Krentler, C., Cully, J., Meyer, H. E., Grzeschik, K. H., Mersmann, G., Hasilik, A., & von Figura, K. (1987). Cloning of a cDNA encoding the human cation-dependent mannose 6-phosphate specific receptor, Proc. Natl. Acad. Sci. USA 84, 5575–5579.

Pohlmann, R., Wendland, M., Böker, C., & von Figura, K. (1995). The two mannose-6-phosphate receptors transport distinct complements of lysosomal proteins. J. Biol. Chem. 270, 27311–27318.

Prence, E. M., Dong, J., & Sahagian, G. G. (1990). Modulation of the transport of a lysosomal enzyme by PDGF. J. Cell Biol. 110, 319–326.

Rechler, M. M., Zapf, J., Nissley, S. P., Froesch, E. R., & Moses, A. C. (1980). Interactions of insulin-like growth factor I and II and multiplication-stimulating activity with receptors and serum carrier proteins. Endocrinology 107, 1451–1459.

Rijnboutt, S., Aerts, H. M. F. G., Geuze, H. J., Tager, J. M., & Strous, G. J. (1990). Mannose 6-phosphate independent membrane association of cathepsin D, glucocerebrosidase and sphingolipid activating protein in HepG2 cells. J. Biol. Chem. 266, 4862–4868.

Rogers, S. A., & Hammerman, M. R. (1988). Insulin-like growth factor II stimulates production of inositol trisphosphate in proximal tubular basolateral membranes from canine kidney. Proc. Natl. Acad. Sci. USA 85, 4037–4041.

Rogers, S. A., & Hammerman, M. R. (1989). Mannose 6-phosphate potentiates insulin-like growth factor II-stimulated inositol trisphosphate production in proximal tubular basolateral membranes. J. Biol. Chem. 264, 4273–4276.

Rogers, S. A., Purchio, A. F., & Hammerman, M. R. (1990). Mannose 6-phosphate containing peptides activate phospholipase C in proximal tubular basolateral membranes from canine kidney. J. Biol. Chem. 265, 9722–9727.

Rohrer, J., Schweizer, A., Johnson, K. F., & Kornfeld, S. (1995). A determinant in the cytoplasmic tail of the cation-dependent mannose-6-phosphate receptor prevents trafficking to lysosomes. J. Cell. Biol. 130, 1297–1306.

Rosorius, O., Mieskes, G., Issinger, O. G., Körner, C., Schmidt, B., von Figura, K., & Braulke, T. (1993a). Characterization of phosphorylation sites in the cytoplasmic domain of the 300 kDa mannose 6-phosphate receptor. Biochem. J. 292, 833–838.

Rosorius, O., Issinger, O. G., & Braulke, T. (1993b). Phosphorylation of the cytoplasmic tail of the 300 kDa mannose 6-phosphate receptor is required for the interaction with a cytosolic protein. J. Biol. Chem. 268, 21470–21473.

Roth, R. A., Stover, C., Hari, J., Morgan, D. O., Smith, M. C., Sara, V., & Fried, V. A. (1987). Interaction of the receptor for insulin-like growth factor II with mannose 6-phosphate and antibodies to the mannose 6-phosphate receptor. Biochem. Biophys. Res. Commun. 149, 600–606.

Roth, R. A. (1988). Structure of the receptor for insulin-like growth factor II: The puzzle amplified. Science 239, 1269–1271.

Sahagian, G. G., Distler, J., & Jourdian, G. W. (1981). Characterization of a membrane associated receptor from bovine liver that binds phosphomannosyl residues of bovine testicular β-galactosidase. Proc. Natl. Acad. Sci. USA 78, 4289–4293.

Sahagian, G. G., & Neufeld, E. F. (1983). Biosynthesis and turnover of the mannose 6-phosphate receptor in Chinese hamster ovary cells. J. Biol. Chem. 258, 7121–7128.

Sakano, K., Enjoh, T., Numata, F., Fujiwara, H., Marumoto, Y., Higashihashi, N., Sato, Y., Perdue, J. F., & Fujita-Yamaguchi, Y. (1991). The design, expression and characterization of human insulin-like growth factor II (IGF-II) mutants specific for either the IGF-II/cation-independent mannose 6-phosphate receptor of IGF-I receptor. J. Biol. Chem. 266, 20626–20635.

Schmidt, B., Kieke-Siemsen, C., Waheed, A., Braulke, T., & von Figura, K. (1995). Localization of the insulin-like growth factor II binding site to amino acids 1508-1566 in repeat 11 of the mannose-6-phosphate insulin like growth factor receptor. J. Biol. Chem. 270, 14975–14982.

Schulze-Garg, C., Böker, C., Nadimpalli, S., von Figura, K., & Hille-Rehfeld, A. (1993). Tail-specific antibodies that block return of Mr 46,000 mannose 6-phosphate receptor to the trans Golgi network. J. Cell Biol. 122, 541–551.

Schweizer, A., Kornfeld, S., & Rohrer, J. (1996). Cysteine 34 of the cytoplasmic tail of the cation dependent mannose-6-phosphate receptor is reversibly palmitoylated and required for normal trafficking and lysosomal enzyme sorting. J. Cell. Biol. 132, 577–584.

Scott, C. D., & Baxter, R. C. (1987). Insulin-like growth factor II receptors in cultured rat hepatocytes: Regulation by cell density. J. Cell Physiol. 133, 532–538.

Scott, C. D., Taylor, J. E., & Baxter, R. C. (1988). Differential regulation of insulin-like growth factor II-receptors in rat hepatocytes and hepatoma cells. Biochem. Biophys. Res. Commun. 151, 815–821.

Senior, P. V., Byrne, S., Brammar, W. J., & Beck, F. (1990). Expression of the IGF II/mannose 6-phosphate receptor mRNA and protein in the developing rat. Development 109, 67–73.

Sklar, M. M., Kiess, W., Thomas, C. L., & Nissley, S. P. (1989). Developmental expression of the tissue insulin-like growth factor II/mannose 6-phosphate receptor in the rat. J. Biol. Chem. 264, 16733–16738.

Sosa, M. A., Schmidt, B., von Figura, K., & Hille–Rehfeld, A. (1993). In vitro binding of plasma membrane coated vesicle adaptors to the cytoplasmic domain of lysosomal acid phosphatase. J. Biol. Chem. 268, 12537–12543.

Stein, M., Meyer, H. E., Hasilik, A., & von Figura, K. (1987a). 46-kDa mannose 6-phosphate-specific receptor: Purification, subunit composition, chemical modification. Biol. Chem. Hoppe-Seyler 368, 927–936.

Stein, M., Braulke, T., Krentler, C., Hasilik, A., & von Figura, K. (1987b). 46-kDa mannose 6-phosphate-specific receptor: Biosynthesis, processing, subcellular location and topology. Biol. Chem. Hoppe-Seyler 368, 937–947.

Stein, M., Zijderhand-Bleekemolen, J. E., Geuze, H., Hasilik, A., & von Figura, K. (1987c). M_r 46000 Mannose 6-phosphate specific receptor: Its role in targeting of lysosomal enzymes. EMBO J. 6, 2677–2681.

Stöger, R., Kubicka, P., Liu, C. G., Kafri, T., Razin, A., Cedar, H., & Barlow, D. (1993). Maternal-specific methylation of the imprinted mouse IGF 2r locus identifies the expressed locus as carrying the imprinting signal. Cell 73, 61–71.

Stylianopoulou, F., Efstratiadis, A., Herbert, J., & Pintar, J. (1988). Pattern of the insulin-like growth factor II gene expression during rat embryogenesis. Development 103, 497–506.

Szebenyi, G., & Rotwein, P. (1994). The mouse insulin like growth factor II/cation-independent mannose-6-phosphate receptor gene: Molecular cloning and genomic organisation. Genomics 19, 120–129.

Thakker, J. K., DiMarchi, R., MacDonald, K., & Carot, J. F. (1989). Effect of insulin-like growth factors I and II on phosphatidylinositol and phosphatidylinositol 4,5 bisphosphate break down in liver from humans with and without type II diabetes. J. Biol. Chem. 264, 7169–7175.

Tollefsen, S. E., Sadow, J. L., & Rotwein, P. (1989). Coordinate expression of insulin-like growth factor II and its receptor during muscle differentiation. Proc. Natl. Acad. Sci. USA 86, 1543–1547.

Tong, P. Y., Tollefsen, S. E., & Kornfeld, S. (1988). The cation-independent mannose 6-phosphate receptor binds insulin-like growth factor II. J. Biol. Chem. 263, 2585–2588.

Tong, P. Y., Gregory, W., & Kornfeld, S. (1989a). Ligand interactions of the cation-independent mannose 6-phosphate receptor. J. Biol. Chem. 264, 7962–7969.

Tong, P. Y., & Kornfeld, S. (1989b). Ligand interactions of the cation-dependent mannose 6-phosphate receptor. J. Biol. Chem. 264, 7970–7975.

van Dongh, J. M., Barneveld, R. A., Geuze, J. J., & Galjaard, H. (1984). Immunocytochemistry of lysosomal hydrolases and their precursor forms in normal and mutant human cells. Histochem. J. 16, 941–954.

von Figura, K., & Hasilik, A. (1986). Lysosomal enzymes and their receptors. Annu. Rev. Biochem. 55, 167–193.

von Figura, K. (1991). Molecular recognition and targeting of lysosomal proteins. Curr. Opinion in Cell Biology 3, 642–646.

Waheed, A., Braulke, T., Junghans, U., & von Figura, K. (1988). Mannose 6-phosphate/insulin like growth factor II receptor: The two types of ligands bind simultaneously to one receptor at different sites. Biochem. Biophys. Res. Commun. 152, 1248–1254.

Waheed, A., & von Figura, K. (1990a). Rapid equilibrium between monomeric, dimeric and tetrameric forms of the Mr 46000 mannose 6-phosphate receptor at 37°C. Possible relation to the function of the receptor. Eur. J. Biochem. 193, 47–54.

Waheed, A., Hille, A., Junghans, U., & von Figura, K. (1990b). Quaternary structure of the Mr 46000 mannose 6-phosphate specific receptor: Effect of ligand, pH, and receptor concentration on equilibrium between dimeric and tetrameric receptor forms. Biochemistry 29, 2449–2455.

Wang, Z. Q., Fung, M. R., Barlow, D. P., & Wagner, E. F. (1994). Regulation of embryonic growth and lysosomal enzyme targeting by the imprinted IGF II/M6PR gene. Nature 372, 464–467.

Wardzala, L. J., Simpson, I. A., Rechler, M. M., & Cushman, S. W. (1984). Potential mechanism of the stimulatory action of insulin on insulin-like growth factor II binding to the isolated rat adipose cell. Apparent redistribution of receptors cycling between a large intracellular pool and the plasma membrane. J. Biol. Chem. 259, 8378–8383.

Watanabe, H., Grubb, J. H., & Sly, W. S. (1990). The overexpressed human 46 kDa mannose 6-phosphate receptor mediates endocytosis and sorting of beta-glucuronidase. Proc. Natl. Acad. Sci. USA 87, 8036–8040.

Wendland, M., Hille, A., Nagel, G., Waheed, A., von Figura, K., & Pohlmann, R. (1989). Synthesis of a truncated Mr 46000 mannose 6-phosphate receptor that is secreted and retains ligand binding. Biochem. J. 260, 201–206.

Wendland, M., Waheed, A., von Figura, K., & Pohlmann, R. (1991a). Mr 46000 mannose 6-phosphate receptor: The role of histidine and arginine residues for binding of ligands. J. Biol. Chem. 266, 2917–2923.

Wendland, M., Waheed, A., Schmidt, B., Hille, A., Nagel, G., von Figura, K., & Pohlmann, R. (1991b). Glycosylation of the Mr 46000 mannose 6-phosphate receptor: Effect on ligand binding and conformation. J. Biol. Chem. 266, 4598–4604.

Wendland, M., von Figura, K., & Pohlmann, R. (1991c). Mutational analysis of disulfide bridges in the Mr 46000 mannose 6-phosphate receptor: Localization and role for ligand binding. J. Biol. Chem. 266, 7132–7136.

Wenk, J., Hille, A., & von Figura, K. (1991). Quantitation of MR 46000 and Mr 300000 mannose 6-phosphate receptors in human cells and tissues. Biochemistry International 23, 723–732.

Westcott, K. R., & Rome, L. H. (1988). Cation-independent mannose 6-phosphate receptor contains covalently bound fatty acid. J. Cell. Biochem. 38, 23–33.

Yang, Y. W. H., Robbins, A. R., Nissley, S. P., & Rechler, M. M. (1991). The chick embryo fibroblast cation-independent mannose-6-phosphate receptor is functional and immunologically related to the mammalian insulin like growth factor II (IGF II/Man6P) receptor but does not bind IGF II. Endocrinology 128, 1177–1189.

Yee, D., Cullen, K. J., Paik, S., Purdue, J. F., Hampton, B., Schwartz, A., Lippman, M. E., & Rosen, N. (1988). Insulin-like growth factor II mRNA expression in human breast cancer. Cancer Res. 48, 6691–6696.

Zhou, M., Ma, Z., & Sly, W. (1995). Cloning and expression of the cDNA of the chicken cation independent mannose-6-phosphate receptor. Proc. Natl. Acad. Sci. USA 92, 9762–9766.

Zorzano, A., Wilkinson, W., Kotliar, N., Thoidis, G., Wadzinski, B. E., Ruoho, A. E., & Pilch, P. F. (1989). Insulin-regulated glucose uptake in rat adipocytes is mediated by two transporter isoforms present in at least two vesicle populations. J. Biol. Chem. 264, 12358–12363.

THE TRANSFERRIN RECEPTOR

Caroline A. Enns, Elizabeth A. Rutledge, and
Anthony M. Williams

Biomembranes
Volume 4, pages 255–287.
Copyright © 1996 by JAI Press Inc.
All rights of reproduction in any form reserved.
ISBN: 1-55938-661-4.

I. INTRODUCTION

Receptor-mediated endocytosis is an essential process in eukaryotic cells. It is the predominant pathway by which proteins are taken into cells. This mode of uptake has been implicated in such diverse events as: (1) the transport of nutrients into cells; (2) the degradation of serum proteins; (3) the down-regulation of signaling receptors and their ligands; (4) signal transduction events when the activated receptor is translocated to another location in the cell; and (5) the processing of antigens.

The transferrin receptor (TfR) is an excellent model system to study ligand–receptor interactions and receptor-mediated endocytosis, as well as being important in the delivery of iron to cells. Iron is essential to many processes within the cell. Transferrin (Tf) is the major iron-transport protein in the blood of a wide variety of vertebrates and invertebrates. The uptake of Tf into cells that are rapidly proliferating, or that have a specialized requirement for iron, is mediated by a specific plasma membrane receptor, the TfR.

II. STRUCTURE AND FUNCTION OF THE HUMAN TfR

A. Endocytosis and Recycling

Overview

The endocytic cycle of the TfR has been extensively characterized. Surface TfR binds the iron-bound form of Tf present in the extracellular medium. The binding affinity is approximately 10^8–10^9 M^{-1} at neutral pH, and in experimental systems reaches equilibrium in 15–90 minutes at 4°C (Morgan, 1964a; Wada et al., 1979; Karin and Mintz, 1981; Ward et al., 1982; Klausner et al., 1983a). The receptor undergoes endocytosis in the presence or absence of bound Tf (Schulman et al., 1981; Klausner et al., 1983b), although the kinetics of cycling may be accelerated in the presence of bound Tf (Gironès and Davis, 1989). The receptor is internalized into clathrin-coated pits, which bud off from the cell membrane and become clathrin-coated vesicles. After losing their coats, the vesicles fuse with other vesicles or an existing endosome. As the pH is lowered to 5.5–6.5, Fe^{3+} dissociates from Tf; the apo-Tf remains associated with the receptor (Morgan, 1979). The mechanism by which Fe^{3+} is transported from endosomes to cytoplasm has not been identified. Acidification is essential for the dissociation of Fe^{3+} from Tf because

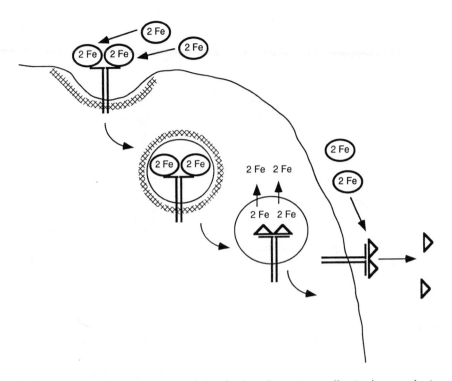

Figure 1. A schematic diagram of the uptake of iron into cells via the transferrin receptor.

Fe^{3+} uptake into cells is inhibited by agents that raise endosomal pH, such as chloroquine and monensin (Harding and Stahl, 1983; Paterson et al., 1984). Upon recycling of the apo-Tf–TfR complex to the cell surface, the affinity of apo-Tf for receptor is reduced at neutral pH and apo-Tf is released (Morgan, 1981; Young and Aisen, 1981; Morgan, 1983; Tsunoo and Sussman, 1983). This pathway is summarized in Figure 1.

Details of the Intracellular Pathway

TfR is endocytosed through clathrin-coated pits. The TfR may act to nucleate sites of coated pit formation. Iacopetta et al. (1988) found that in mouse cells expressing high levels of human TfR, 3- to 4-fold more coated pits were present on the surface as on nontransfected mouse cells. In contrast, Miller et al. (1991) found no increase in clathrin-coated pits in chick embryo fibroblasts transfected with human TfR, but there was a 3-fold increase in flat clathrin lattices associated with the plasma membrane. The difference may be due to the different conditions utilized by the two groups. In any case, an increased recruitment of clathrin to the plasma membrane was seen by both groups.

The TfR concentrates in coated pits. Early estimates of the proportion of TfR in these structures ran as high as 70% (Hopkins, 1983), but more recent studies indicate a much lower proportion, perhaps 10–15% (Iacopetta et al., 1988; Hansen et al., 1992). Coated pits only occupy 1–2% of the cell surface (Hansen et al., 1992). Thus, the TfR is concentrated 5–15 fold in coated pits. The internalization rates of Tf for a number of cell lines are about 10% of surface-bound ligand per minute. The coated pit containing the Tf–TfR complex buds off from the plasma membrane, becomes uncoated, and enters the tubulo-vesicular endosomal system (Geuze et al., 1983; Gruenberg et al., 1989; Schmid, 1992). An H^+-adenosine triphosphatase acidifies the endosome (Yamashiro and Maxfield, 1983), the Tf–TfR complex releases its iron, is sorted away from the co-endocytosed proteins that are not recycled, and returns to the cell surface as an apo-Tf–TfR complex.

Studies of endosomal sorting using density-shift techniques with Tf-horseradish peroxidase conjugates and diaminobenzidine in the presence of ^{125}I-labeled TfR, asialoglycoprotein receptor (ASGPR), and mannose-6-phosphate receptor (MPR), indicate that all three receptors are at least transiently present in the same compartment (Stoorvogel et al., 1989). TfR and ASGPR are rapidly recycled back to the surface from early endosomes, while MPR is largely transported to the *trans* Golgi network, and from there shuttles back and forth to late endosomes. The kinetics of Tf cycling from the cell surface through the endosomes and back to the cell surface is identical to that of three membrane lipids, N-[N-(7-nitro-2,1,3-benzoxadiazol-4-yl)-ε-amino-hexanoyl]-sphingosylphosphorylcholine, C_6-NBD-phosphatidyl-choline, and galactosylceramide. These results indicate that the receptor complex moves through the endocytic pathway by a bulk flow mechanism, and that proteins which sort to the lysosome are selectively sequestered (Mayor et al., 1993).

Neuraminidase treatment of cells indicates that some TfRs pass through the *trans* Golgi network where they can be partially resialylated with a half-life of 1.5 hours (Snider and Rogers, 1985). The extent of this pathway has been a matter of some controversy (Snider and Rogers, 1985; Neefjes et al., 1988; Robertson et al., 1992) and may correspond to the long recycling pathways described previously (Hopkins and Trowbridge, 1983; Stein and Sussman, 1986).

The cytoplasmic domain of the receptor possesses the signal for endocytosis. Removal of the cytoplasmic domain of the TfR reduces the rate of endocytosis 10 to 20-fold and reduces its concentration in coated pits (Rothenberger et al., 1987; Iacopetta et al., 1988; and reviewed in Trowbridge et al., 1993). The endocytic signal has been more precisely localized, by site-directed mutagenesis, to the YTRF sequence at positions 20–23, and the tyrosine at position 20 appears to be particularly critical for endocytosis (Jing et al., 1990). The type of amino acid at position 21 does not appear to be critical for endocytosis, leading to a generalized motif of YXRF for internalization. A tyrosine is also a key ingredient in the internalization motif (NPVY) of the LDL receptor (Davis et al., 1986; Chen et al., 1990). Similar motifs have been identified in several other receptors (reviewed in Trowbridge et al., 1993). Mutations of phenylalanines at positions 13 and 23 of the TfR have a negative effect on endocytosis, suggesting that sequence information extending beyond tyrosine 20 contributes to efficient endocytosis. Also, substitution of tyrosine at position 34 for the normal serine causes reversion of an internalization-defective receptor (McGraw et al., 1991). Therefore, cryptic internalization signals in addition to the major motif may be present.

Comparison of the major internalization motifs of TfR (YXRF) and LDL receptor (NPXY) with known crystallographic structures showed a tendency of similar tetrapeptides to adopt a "reverse beta-turn" structure (Collawn et al., 1990). Chemically synthesized peptides containing internalization motifs for lysosomal acid phosphatase and LDL receptor examined by nuclear magnetic resonance displayed the reverse beta-turn in aqueous solution (Bansal and Gierasch, 1991; Eberle et al., 1991).

In spite of extensive characterization of the internalization motif for endocytosis, an interaction of the TfR with any known components of the endocytic apparatus, has not been demonstrated (Turkewitz and Harrison, 1989 and unpublished results). Peptides comprising portions the cytoplasmic domains of the LDL receptor and the mannose-6-phosphate receptor cytoplasmic domains showed weak interactions with adaptins (AP-2 proteins) *in vitro* (Pearse, 1988; Glickman et al., 1989; Chang et al., 1993), but no such interaction has been shown for the TfR cytoplasmic domain. The recent cloning of a new protein, enigma, which interacts specifically with the human insulin receptor raises the possibility of multiple adaptor-like proteins involved in the endocytic process (Wu and Gill, 1993).

B. TfR Sorting in Polarized Cells

The TfR, like many plasma membrane proteins, is sorted preferentially to either the basolateral- or apical-side of polarized cells. This phenomenon has been studied extensively in MDCK cells where at least 95% of the TfR is sorted to the basolateral portion of the cell membrane (Fuller and Simons, 1986). Polarity of the TfR is maintained by the accurate recycling of the TfR to the basolateral membrane (Fuller and Simons, 1986). The cytoplasmic domain of the TfR was examined to determine whether it played a role in sorting. Like the poly-Ig receptor, proper targetting of the TfR to the basolateral surface did not require phosphorylation (Dargemont et al., 1993). Deletion of 36 amino acids including the internalization sequence (Δ 6–41) resulted in 20% of the TfR being redirected to the apical membrane (Dargemont et al., 1993). Thus, basolateral targetting was not solely dependent on the internalization signal. Therefore, the TfR falls into the class of receptors with distinct sorting and internalization signals. The TfR is also found on the basolateral side of hepatocytes and CaCo-2 cells, an intestinal cell line, but the sorting in these cells has not been as extensively studied.

C. Biosynthesis of the TfR

Biosynthesis of the TfR has been examined both *in vivo* and *in vitro* (in the presence of rabbit reticulocyte lysate and a microsomal system) (Omary and Trowbridge, 1981a; Schneider et al., 1983a; Enns et al., 1991). The receptor is a type II transmembrane protein, that is, the N-terminus is cytoplasmic and the C-terminus is extracellular. The sequence that signals translocation through the endoplasmic reticulum membrane is within the putative transmembrane domain (Zerial et al., 1986, 1987). Deletion of the transmembrane region results in the failure of the TfR to undergo translocation and glycosylation *in vitro*. Attachment of the transmembrane domain of the TfR onto the N-terminus of mouse dihydro-folate reductase and chimpanzee alpha-globulin, results in the insertion of these normally soluble proteins into the microsomal membrane in the same orientation as the TfR (Zerial et al., 1986, 1987). Asn-linked glycosylation of the TfR occurs co-translationally (Omary and Trowbridge, 1981; Enns et al., 1991).

The newly synthesized TfR is unable to bind Tf and is a monomer (Enns et al., 1991). Dimer formation and the ability to bind Tf occurs rapidly. Disulfide bond formation between subunits is a slower process which occurs with a $t_{1/2}$ of approximately 30 minutes. The presence of intersubunit disulfide bonds, however, is not required for dimer formation or for binding of Tf, because a mutated form of the TfR lacking disulfide bonds binds Tf with normal affinity (Jing and Trowbridge, 1987; Alvarez et al., 1989).

D. Post-translational Modifications

The human TfR undergoes several post-translational modifications: (1) intermo-lecular disulfide bond formation, (2) phosphorylation, (3) fatty-acylation with palmitate, (4) addition of three Asn-linked oligosaccharides, (5) addition of one Ser/Thr-linked oligosaccharide, and (6) proteolytic cleavage (Figure 2). The func-tion of many of these modifications is not known, however.

Disulfide bonds at Cys 89 and Cys 98 form intermolecular bonds, joining two monomers together, though it is not clear whether all the TfRs form both disulfide bonds. When only Cys 98 is mutated to Ser, only about half the TfRs have intersubunit disulfide bonds, suggesting that in this mutant, Cys 89 does not always form disulfide bonds. Mutation of both Cys 89 and Cys 98 to Ser eliminates the disulfide bonds (Jing and Trowbridge, 1987). The disulfide bonds are not necessary for normal endocytosis, recycling, iron accumulation, or the dimeric state. The dimeric state is maintained by the 70kD external domain of the TfR (obtained by trypsin treatment) and is disrupted in the presence of non-ionic detergent (Turkewitz et al., 1988). The full length Cys 89–98 double mutant in nonionic detergent forms a dimer only when Tf is present (Alvarez et al., 1989).

Ser 24 is the only known phosphorylation site, and it is phosphorylated both *in vivo* and *in vitro* by protein kinase C (May et al., 1985; Davis et al., 1986;

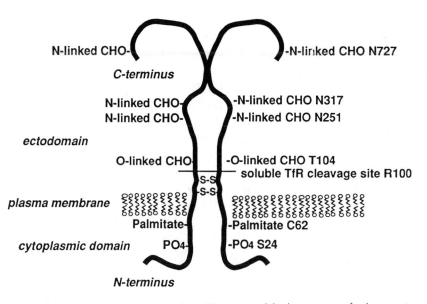

Figure 2. Co- and post-translational modifications of the human transferrin receptor.

Rothenberger et al., 1987). Phosphorylation of human TfR is not important for endocytosis in mouse L cells (Davis and Meisner, 1987; Rothenberger et al., 1987; McGraw et al., 1988). Though the function of the phosphorylation is not clear, it could have a role in transyctosis of receptors in polarized cells, as is the case with the poly-Ig receptor (Casanova et al., 1990).

The fatty acid, palmitate, was shown to be attached to TfR (Omary and Trowbridge, 1981; Adam et al., 1984) at Cys 62 (Jing and Trowbridge, 1987) and/or Cys 67 (Alvarez et al., 1990). Acylation does not appear to be necessary for efficient endocytosis of human TfR expressed in chicken embryo fibroblasts (Jing and Trowbridge, 1990), but its presence may cause a 20% decrease in endocytosis rate in Chinese hamster ovary (CHO) cells (Alvarez et al., 1990). The palmitate of TfR has a higher turnover rate than the TfR and can be added to mature TfR as long as 48 hours after synthesis (Omary and Trowbridge, 1981b). The fraction of the mature TfRs that are modified with palmitate is not known. The turnover of the palmitate is too slow for it to be involved in the endocytic cycle, although slower pathways, such as to the *trans*-Golgi network are not excluded (Omary and Trowbridge, 1981b).

The Asn-linked carbohydrate groups are added to TfR during synthesis and are further processed in the Golgi apparatus to yield one complex type and two high mannose oligosaccharides. In human TfR, the modified sites are at Asn 251, Asn 317, and Asn 727 (Omary and Trowbridge, 1981a; Schneider et al., 1982). When the human TfR is either expressed in mouse 3T3 cells or isolated from human placentae, the complex type oligosaccharide is at Asn 251, and the other two Asn residues at positions 317 and 727 are of the high mannose type (Williams, Hayes, Lucas, and Enns, unpublished observations). Mutation of each of these sites revealed that glycosylation of Asn 727 is essential for TfR transport to the cell surface. Without this oligosaccharide, the TfR remains localized to the endoplasmic reticulum and is associated with an endoplasmic reticulum chaperone protein, binding immunoglobulin protein (BiP; Williams and Enns, 1991, 1993). Transport of the TfR lacking the Asn 727 glycosylation site to the plasma membrane can be partially recovered by generation of a glycosylation site at Asn 722 (Williams and Enns, 1993). Mutation of the other two glycosylation sites had little effect on surface localization (Williams and Enns, 1991, 1993). A different mutation of the carbohydrate attachment site at Asn 251 resulted in a TfR that was retained and cleaved in the endoplasmic reticulum (Hoe and Hunt, 1992). These results are in contrast to those mentioned above, in which elimination of the carbohydrate at Asn 251 did not significantly affect transport of TfR to the cell surface, and imply that the amino acid substitution rather than the lack of glycosylation resulted in a misfolded TfR and created a cleavage site.

The effect of Asn-linked glycosylation on TfR transport is the most dramatic of any of the post-translational modifications thus far observed. The structure of the Asn-linked carbohydrates of the TfR shows considerable heterogeneity and varies with the cell line (Do et al., 1990) and with individual people (Orberger et al., 1992), and displays blood group antigens (Do et al., 1990). The possible physiological consequence of this heterogeneity is not known.

A single Ser/Thr (O-linked) carbohydrate of the human TfR has been characterized recently (Do et al., 1990) and localized to Thr 104 (Do and Cummings, 1992; Hayes et al., 1992). As with the variable structure of Asn-linked oligosaccharides, the structure of the O-linked carbohydrate from different human cell lines varies also (Do et al., 1990). The O-linked carbohydrate at Thr 104 protects the TfR from efficient cellular proteolytic cleavage near the plasma membrane (Rutledge et al., 1994). When Thr 104 is mutated to Asp, no O-linked oligosaccharide is attached, and a 78kD soluble TfR form accumulates in the growth medium. Sequencing of the soluble TfR at the cleavage site revealed that cleavage occurs between residues Arg 100 and Leu 101, only four amino acids from the normal site of O-linked carbohydrate attachment at Thr 104, near the membrane (Rutledge et al., 1994a). These results suggest that the carbohydrate inhibits access of a protease to this site. The cleavage of the TfR in cell culture is enhanced by treatment with neuraminidase (Rutledge and Enns, 1996), implying that removal of sialic acid from the O-linked carbohydrate is sufficient for efficient cleavage of the TfR. Since a soluble form of TfR has been observed in human blood with the same cleavage site (Shih et al., 1990), cleavage of the TfR may be an additional way to down-regulate the TfR on the surface of cells.

E. Tf–TfR Interactions

The interaction of Tf with its receptor plays an important role in the delivery of iron to cells. At neutral pH, Tf has an extremely high avidity for two ferric ions ($K_a \sim 10^{20}$–10^{23} M^{-1}; Aisen and Leibman, 1978; Evans and Williams, 1978), whereas, at pH 5, the binding constant falls to a K_a of $\sim 10^7$ M^{-1}. During its intracellular transit, the receptor cycles from the plasma membrane to an acidic compartment of pH ~ 5.4–6.5 depending on the cell type (Dautry-Varsat et al., 1983; Klausner et al., 1983a; Yamashiro et al., 1983; Paterson et al., 1984). This low pH and the rapid recycling of the receptor back to the cell surface are not sufficient to explain the release of iron from Tf inside the endosomal compartment. Recent evidence from two separate laboratories shows that the binding of Tf to its receptor facilitates the loss of iron at endosomal pH (Bali et al., 1991; Sipe and Murphy, 1991). In contrast, at pH 6.2 or higher, the receptor actually stabilizes the interaction between the ferric ions and Tf, making it more difficult to remove iron from Tf bound to the receptor than from unbound Tf (Bali et al., 1991). The cycle is completed when the apo-Tf–TfR complex returns to the cell surface. At neutral pH, the binding constant of apo-Tf for the receptor is ~ 2000-fold lower than that of di-ferric Tf (Tsunoo and Sussman, 1983) and consequently, the apo-Tf readily dissociates from the receptor.

The structural features of the TfR that are involved in the binding of Tf are not known. Glycosylation of the TFR is important for TF binding. Initial studies using tunicamycin-treated cells indicated that blocking Asn-linked glycosylation resulted in a form of the TfR that was unable to bind Tf (Reckhow and Enns, 1988; Hunt et al., 1989) and greatly inhibited the ability of both the murine and human TfR to reach the cell surface (Omary and Trowbridge, 1981a; Reckhow and Enns, 1988; Hunt et al., 1989; Ralton et al., 1989). Site-directed mutagenesis of all the consensus sequences for Asn-linked glycosylation confirmed the initial observations using tunicamycin (Williams and Enns, 1991; Yang et al., 1993). The third glycosylation at Asn-727 appears to be the most critical site for Tf binding and transport to the cell surface (Williams and Enns, 1993). The TfR which lacks a glycosylation site at 727 is still predominantly a disulfide bonded dimer, and therefore retains much of its quaternary structure (Williams and Enns, 1993). Inhibition of Asn-linked processing of the core oligosaccharides of the human and murine TfRs by treatment of cells with swainsonine or deoxynojiramycin does not affect the ability of the TfR to reach the cell surface (Ralton et al., 1989; Enns et al., 1991) and only lowered the binding affinity of Tf ~ 3-fold (Enns et al., 1991). Abnormal processing of the Asn-linked carbohydrates in an insect cell line, Sf9, using a baculovirus expression system, results in over a 10-fold lower binding affinity for Tf (Domingo and Trowbridge, 1988).

The structural features of Tf required for TfR binding have been partially defined. As just discussed, apo-Tf is easily displaced by di-ferric Tf at neutral pH. Vertebrate Tfs have arisen from a gene duplication (reviewed in Williams, 1982). Each lobe binds a single ferric ion with similar affinities. Tf can be cleaved by mild protease digestion into two peptides (Williams, 1974). Each peptide retains its ability to bind its ferric ion, but only the C-terminal fragment of Tf is capable of binding to either the human (Zak et al., 1994) or bacterial TfR (Alcantara et al., 1993). The C-terminal domain of Tf contains the two Asn-linked oligosaccharides (reviewed in De Jong and Van Eijk, 1989), which are not necessary for binding to the receptor. Deglycosylation of Tf with glycosidases or elimination of the Asn-linked glycosyl groups by site-directed mutagenesis does not diminish the binding of Tf to the human (Tsunoo and Sussman, 1993; Mason et al., 1993) or bacterial TfRs (Alcantara et al., 1993). An N-terminal fragment of Tf generated by site-directed mutagenesis has been generated and it could be tested for binding to the human TfR (Chow et al., 1991).

III. EXPRESSION OF THE TfR

The human TfR was first cloned independently by two different groups in 1984 (McClelland et al., 1984; Schneider et al., 1984). In one case, a monoclonal antibody, OKT9, was used to detect transfected mouse cells expressing the human TfR using fluorescence-activated cell sorting (Kühn et al., 1984; McClelland et al., 1984). The entire gene was isolated and a probe to a portion of the 5' end was used to isolate cDNA coding for the TfR. The other technique employed immunoselection of polyribosomes with a polyclonal antibody to the TfR (Schneider et al., 1983, 1984). A cDNA library was constructed from this enriched fraction and screened.

A. Structure of the Gene

The TfR gene is 31 Kb and contains at least 19 introns (McClelland et al., 1984). The mRNA is 4.9 Kb, 2.3 Kb of which codes for the TfR and the remaining 2.5 Kb is the 3' untranslated region. Chromosome mapping of the TfR revealed that it is located on human chromosome 3 (Enns et al., 1982; Goodfellow et al., 1982). More detailed mapping indicated that it is located on q26.2→ qter (Rabin et al., 1985). Interestingly, two other iron proteins, Tf and p97, are also located on chromosome 3 in the same region (Plowman et al., 1983; Yang et al., 1984).

B. Regulation of TfR Expression by Iron

The TfR is highly regulated as the cell tightly controls the uptake of iron. Too little iron prevents proliferation and too much iron can result in oxidative injury to cells as free ferric ions catalyze free radical formation. In early work, chelators that deprive the cell of iron were found to up-regulate the number of TfRs, whereas exogenous iron sources, such as hemin or iron salts, have the opposite effect (Ward et al., 1982; Bridges and Cudkowicz, 1984; Louache et al., 1984). The regulation of the TfR by iron appears to be modulated by controlling the stability of the TfR mRNA (reviewed in Kühn, 1989; Klausner et al., 1993). Under low iron conditions, the mRNA becomes more stable and more TfR is synthesized (Owen and Kühn, 1987; Casey et al., 1988b; Müllner et al., 1989). The opposite happens under high iron conditions. Deletional analysis of the 3′ untranslated region of the TfR indicates that this portion is responsible for the iron sensitivity of the mRNA (Casey et al., 1988; Müllner and Kühn, 1988). Intensive investigation of this region indicates that the mRNA can form multiple stem–loop structures called IREs, iron-responsive elements, and these stem–loop structures convey sensitivity to degradation (reviewed in Kühn, 1989; Klausner et al., 1993). An iron-responsive element binding protein, IRE-BP, or iron-regulatory factor, IRF, has been identified, isolated, cloned, and sequenced (Rouault et al., 1988; Koeller et al., 1989; Müllner et al., 1989). Sequence analysis and enzyme kinetics of the purified protein indicate that this protein is cytoplasmic aconitase (Rouault et al., 1991). Cytoplasmic aconitase is an iron–sulfur protein which binds four iron ions/molecule. One of these iron binding sites is more labile than the others. Loss of one iron inactivates the enzyme and allows the protein to bind to the TfR mRNA thereby protecting this region from rapid degradation (Rouault et al., 1991). Thus, under low iron conditions, the TfR mRNA is more stable and more TfR is synthesized. Under high iron conditions, the replete aconitase does not bind to the IRE and the TfR mRNA is less stable. Phylogenetic comparisons indicate that the IRE-stem–loops are found across many species indicating that this mechanism of iron regulation is commonly used (Rothenberger et al., 1990).

C. Regulation of TfR Expression by Mitogens

Early Tf binding studies indicated that the level of TfR was increased when cells were stimulated to divide by a host of mitogens or by serum, and decreased when cells were induced to differentiate and cease proliferating (Nunez et al., 1977; Hamilton et al., 1979; Larrick and Cresswell, 1979; Yeoh and Morgan, 1979). The monoclonal antibodies, OKT9 and B3/25, were termed proliferation specific markers before they were known to be specific for the TfR (Judd et al., 1980; Omary et al., 1980; Goding and Burns, 1981; Trowbridge and Omary, 1981).

Cloning of the TfR enabled investigators to determine that the TfR can be regulated at the transcriptional level (Ho et al., 1986, 1989; Miskimins et al., 1986). Deletional analysis of the promoter indicated a critical region 70 base pairs upstream of the initiation site (Casey et al., 1988a; Ouyang et al., 1993). Footprinting analysis of the 5′ untranslated portion revealed two regions of the TfR DNA protected from DNase1, which indicates protein(s) bind to these areas (Miskimins, 1992; Ouyang et al., 1993). The first element is a GC-rich area just upstream from the TATA box (-34 to -55) and a second element 5′ to the first (-60 to -78; Ouyang et al., 1993). The first region contains a consensus sequence for SP1 and the second has a site that is similar to AP1 and CREB binding sites. Recent studies using deletional analysis of the upstream region of DNA coupled to a CAT gene indicate the second site (-78 to -55) is necessary for the response to serum. Both elements are required for the full response to serum or mitogens (a sixfold increase in CAT activity; Ouyang et al., 1993).

The regulation of TfR levels by interleukin-2 (IL-2) in IL-2 dependent cell lines appears to be more complex. IL-2 induction of cells increases steady-state levels of TfR mRNA 50-fold (Seiser et al., 1993). In addition, IL-2 appears to stabilize TfR mRNA by IRF/IRE-BP in the cytoplasm and prevent TfR mRNA degradation (Seiser et al., 1993).

IV. ROLE OF TfR IN CELL PROLIFERATION

Iron is essential for many aspects of cell metabolism: it is a component of cytochromes, the glycolytic enzyme aconitase, ribonucleotide diphosphate reductase, and peroxisomal enzymes (Cammack et al., 1990). The synthesis of DNA requires deoxyribonucleotides, which are produced in all known organisms through the reduction of ribonucleotides by the activity of ribonucleotide diphosphate reductase. Iron is an integral component of the active form of this enzyme, and consequently, ability of cells to synthesize DNA may be expected to be dependent on a supply of iron (Thelander, 1990). Robbins and Pederson showed that the iron chelator, deferoxamine, inhibited DNA synthesis in HeLa cells (Robbins and Pederson, 1970). Cells depleted of iron show evidence of reduced ribonucleotide reductase activity and impoverishment of nucleotide pools (Lederman et al., 1984; Chitambar et al., 1988; Hedley et al., 1988). Thymidine incorporation into cells of iron-deficient patients is also reduced (Hershko et al., 1970). Growth arrest of normal cells resulting from the absence of iron or the presence of deferoxamine generally occurs during G1 phase, while growth arrest of transformed cells occurs predominantly in S phase (compiled in Kühn et al., 1990).

How critical is the TfR for adequate iron supply? Cells that undergo continuous proliferation, such as crypt cells of the intestine, express higher levels of TfR than associated nonproliferative tissues (Levine and Seligman, 1984; Anderson et al., 1991). Many malignant cell lines show increased numbers of TfR in comparison to cells of benign lesions or differentiated cell lines (Faulk et al., 1980; Shindelman et al., 1981; Gatter et al., 1983). On the other hand, toxin-conjugated Tf has been used to select for mutant cell lines which show no detectable Tf binding, yet continue to propagate at a rate comparable to the parent cell line in culture, providing that the medium contains a source of iron (McGraw et al., 1987; Alvarez et al., 1989). These results imply that cells in culture can obtain iron via non-Tf sources and that the TfR itself is not required for cell proliferation.

There may be alternative iron sources for proliferating cells. L1210 leukemia cells have been shown to possess a specific, saturable non-Tf iron uptake system (Basset et al., 1986). Several ferric complexes of acyl-hydrazones can support erythroid cell proliferation in the absence of Tf (Ponka et al., 1982; Ponka and Schulman, 1985). While normal lymphocytes, normal marrow progenitors, and most malignant blood cell lines are inhibited in their growth by antibodies to TfR

(Mendelsohn et al., 1983; Taetle et al., 1983; Lesley and Schulte, 1985), this may be due to iron limitation, since the presence of soluble iron counteracts the growth inhibition (for a review of the literature on TfR in cell proliferation, see Kühn et al., 1990; for the role of TfR in proliferation of hematopoietic cell proliferation, see Taetle, 1990).

Is Tf itself a mitogen? Tf has been shown to have growth-promoting activity for murine T-lymphoma cells, chicken nerves, chick embryo extract, rat bladder carcinoma cells, and human prostate cancer cell lines (Beach et al., 1983; Kitada and Hays, 1985; Hayashi et al., 1987; Chackal-Roy et al., 1989; Chackal-Rossi and Zetter, 1992). Elevated levels of TfR were found to be present in prostate cancer cell lines *in vitro* and *in vivo* (Keer et al., 1990), and in human breast carcinoma (Shindelman et al., 1981). Although there have been observations that apo-Tf can support the proliferation of lymphocytes (Dillner-Centerlind et al., 1979), the general consensus is that the apo-Tf so readily acquires iron from even minor impurities in its environment that it is not feasible to obtain truly iron-free Tf in the absence of iron chelators.

Iron is required for cell proliferation, and *in vivo*, TfR is the main conduit for iron acquisition by proliferating cells. Consequently, TfR is expressed at increased levels in cells that are actively proliferating. The mechanism of this increase, whether transcriptional or post-transcriptional, has not been clearly established. While it has been suggested that Tf is itself a "growth factor," it is most likely that the Tf requirement in certain malignant cell lines is due to iron limitation in the absence of Tf.

V. PHYSIOLOGY—IRON AND TISSUE DISTRIBUTION OF THE TfR

A. The Role of TfRs in Transport of Iron Across Barriers in the Body

The TfR appears to play a role in the transport of iron across several barriers in the body including the maternal blood–placenta, the blood–brain, and the blood–testis barriers.

Early studies using ^{125}I-labeled ^{59}Fe-Tf injected into pregnant animals indicated that iron, but not Tf, was transported across the placenta (Gitlin et al., 1964; Morgan, 1964b; Wong and Morgan, 1973). The placenta is especially rich in TfRs. Immunolocalization studies indicate that the majority of the TfRs are found on the microvillus membrane, which is bathed in maternal blood, although a significant amount is also on the basolateral side. Cultured choriocarcinoma (BeWo and JAR) cell lines have been used as model systems to study the cycling of the TfR from apical and basolateral membranes. These studies indicate that there is bidirectional transcytosis of the TfR across these cells (van der Ende et al., 1987; Cerneus and van der Ende, 1991; Cerneus et al., 1993). Some transcytosis of ^{125}I-Tf was detected, but the relative amounts were not determined. The extent of ^{59}Fe transport with respect to Tf transport remains to be determined.

The TfR-mediated transport of iron across the blood–testes and the blood–brain barriers have also been studied. In the case of the blood–testes barrier, iron, but not Tf, crosses the barrier where it is incorporated into Tf synthesized by the Sertoli cells (Wauben-Penris et al., 1988). In the brain, ^{59}Fe crosses the barrier and Tf does not, as discussed in greater detail in the section on TfR in the brain. In addition, the TfR has been used to transport various drugs and growth factors across the blood–brain barrier. For example, nerve growth factor covalently bound to a monoclonal antibody to the TfR was able to cross the blood–brain interface (van der Ende et al., 1987; Cerneus and van der Ende, 1991; Friden et al., 1991, 1992; Cerneus et al., 1993; Friden et al., 1993). The compartments through which these compounds and iron are transported and whether the TfR is transcytosed has not been elucidated. Further studies are required to answer these questions.

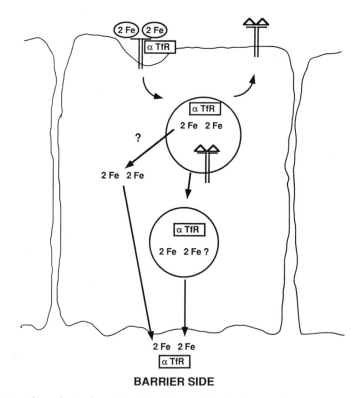

Figure 3. A hypothetical model of how iron and antibodies to the transferrin receptor could be transported across epithelial and endothelial barriers.

One problem with the transcytosis of Tf is that to carry iron across the cell, Tf must not enter an acidic organelle, such as an endosome, or it would lose its iron. Either the transcytotic pathway involves a special neutral route, or the cell has a special way to keep iron in Tf under acidic conditions. Another dilemma is how the iron-rich Tf is released from the TfR at the cell surface, because under neutral pH, Fe-Tf has a high affinity for TfR. One possible model for iron transport across epithelial layers is based on what is known about iron internalization in non-polarized cells. Iron could be uncoupled from Tf in a low pH endosome and be transported either by vesicular traffic or though the cytoplasm to be delivered to the opposite side (Figure 3). Alternatively, the release of Tf from the TfR could be due to a proteolytic cleavage of the TfR from the membrane (see section on soluble TfR). Cleavage of TfR after transcytosis would insure one way delivery of Fe-Tf.

In summary, the TfR appears to play a role in the transport of iron across various barriers in the body. In all three of the cases discussed, the majority of the TfR appears to face the blood side of the barrier, whether it be the apical side in the case of blood–brain and maternal–fetal transport, or the basolateral side in the case of blood–testes transport.

B. TfR in the Brain

TfR, along with Tf and ferritin, provides an important function in regulating the iron balance in the brain. An imbalance of iron could have important consequences and may play a role in Alzheimer's disease and in myelin disorders. In addition, TfR and Tf have a role in developing (Beach et al., 1983) and regenerating (Graeber et al., 1989) neurons and may also act as neuro-modulators (Hyndman et al., 1991).

Location of TfR in Brain

Tf binding in the brain to TfR has the same K_d, about 1–10 nM, as in other tissues and cells. Also it appears that there is a single class of TfR (Pardridge et al., 1987; Mash et al., 1990; Kalaria et al., 1992; Roskams and Connor, 1992).

TfR is expressed in endothelial cells, neurons, and oligodendrocytes, though the reported expression levels vary widely. TfR expression in endothelial cells of the cerebral blood capillaries is 6 to 10-fold higher in rats and about three-fold higher in humans as compared to the cerebral cortex (Kalaria et al., 1992). However, Mash et al. report that in crude membrane fractions, capillary vessels account for only 2.5% of total bound ^{125}I-Tf in the brain (Mash et al., 1990). In the rat brain, more than 80% of Tf is in oligodendrocytes (Connor et al., 1987). In myelin deficient rats (Simons and Riordan, 1990), the level of Tf in neural tissue is decreased by more than 90%, and there is no TfR expression in the myelin deficient oligodendrocytes (which die prematurely). However, the TfR expression in the capillary endothelial cells in these rats is normal (Connor and Benkovic, 1992). These results suggest that the majority of Tf and TfR in neural tissue is in oligodendrocytes.

Immunohistochemical staining shows TfR on cortical neurons (Jefferies et al., 1984), oligodendrocytes (Lin and Connor, 1989; Giometto et al., 1990) and the choroid plexus (which produces cerebral spinal fluid) (Giometto et al., 1990). TfR is expressed 2–3 times higher in the cerebellum than the cerebral cortex in rats (Roskams and Connor, 1992) as determined by ^{125}I-Tf binding. Using the same technique, however, Mash et al. found high TfR expression throughout the rat cerebral cortex and dentate gyrus of the hippocampus (Mash et al., 1990). The hippocampus is involved in learning and memory, and is also affected in Alzheimer's disease. TfR is also present in areas of motor functions in the basal ganglia (the caudate-putamen, nucleus accumbens, substantia nigra, red nucleus, pontine nuclei) and cerebellum. By avidin–biotin immunocytochemistry Giometto et al. used anti-TfR monoclonal antibody to detect TfR in rat brain slices after trypsin treatment, and found TfR staining on the choroid plexus and strong TfR staining in cerebral cortical and brain stem pontine neurons, and weak TfR staining in Purkinje cells of the cerebellum and in spinal cord neurons (Giometto et al., 1990).

The significance of localized TfR concentrations in the brain is not yet clear. The TfR distribution pattern in rat brain is markedly different from the iron distribution, and suggests that iron is transported by TfR to sites where it accumulates. TfR distribution is similar to neuropeptide receptors which are concentrated in "nodal points" in the brain, often in areas associated with mood. Similar to neuropeptides, Hill and co-workers have suggested that Tf may modulate behavior (Hill et al., 1985).

The above results suggest the intriguing possibility that Tf and TfR could function as neuromodulators. This is supported by the observations of Hyndman et al. (1991) who measured the effect of Tf on the intracellular calcium levels and membrane potential of isolated chick embryo retinal neurons. Tf caused rapid membrane depolarization and increased calcium levels, an effect that is also caused by glutamate. Regenerating rat facial motor neurons increase TfR and iron uptake (Graeber et al., 1989), and Tf has been purified as a neurotrophic factor (Beach et al., 1983). The necessity of iron in cell growth as a nutrient is clear. However, more work is necessary to determine whether Tf and TfR have effects on neural tissue that are separate from the necessity of iron.

TfR from chicken forebrain neurons has been described (Roberts et al., 1992). This TfR appears to be related to the estrogen inducible oviduct TfR, but not the chicken embryonic red cell TfR as determined by antibody reactivity (Poola et al., 1990; Fuernkranz et al., 1991). This is the first instance of two different TfRs in the same organism, however, these receptors have not been sequenced, and more work is needed to characterize their differences.

Blood–Brain Barrier

Identifying what types of molecules are transported from the blood into the brain and understanding how this process occurs in the highly selective capillary endothelial cells is of fundamental significance. Brain cells have a high requirement for iron, which is supplied as Fe–Tf from the blood, and endothelial cells express a significant number of TfRs (Jefferies et al., 1984). There are conflicting reports in the literature about whether Tf crosses the blood–brain barrier, namely whether it is transcytosed from the apical (blood) side of endothelial cells to the basolateral (brain) side. Some of the conflict is related to the technical difficulties of these types of experiments. Two approaches have been used to determine whether Tf crosses the blood–brain barrier: (1) fractionation of the brain by isolating brain microvessels and (2) brain perfusion.

Isolation of brain microvessels after perfusing rat brain with ^{125}I-Tf showed that 50% of the Tf that entered endothelial cells was transcytosed into the brain (Fishman et al., 1987). However, the technical steps involved in separating the brain tissue from the vascular elements may contaminate the brain fraction with ^{125}I-Tf (Pardridge et al., 1987). In contrast, perfusion of rat brain with Tf linked to horseradish peroxidase and subsequent electron microscopy of brain sections showed no evidence of Tf transcytosis into the brain (Roberts et al., 1992). Perfusion with ^{125}I-Tf also showed a very low amount of transcytosis (Banks et al., 1988).

Drug Delivery to the Brain by TfR

Methods are just recently becoming available that are able to deliver drugs into the brain (reviewed in Friden et al., 1992; Pardridge, 1992). TfR has been one of the best vehicles for drug delivery and has been extensively studied in rat brain with the use of a mouse anti-TfR monoclonal antibody, MRC OX-26 (Jefferies et al., 1984). The strategy is to covalently couple the desired drug to the antibody and inject the conjugate into the blood circulation. This monoclonal antibody has been shown to react mainly with liver and brain, and so this system has good potential for drug delivery to the brain (Jefferies et al., 1984).

Recently, drug conjugates to MRC OX-26 have been shown to produce measured activity in the rat brain. Nerve growth factor conjugated to this antibody stimulated the growth of a section of medial forebrain tissue that had been transplanted to the rat eye (Friden et al., 1993). This work has important implications for Alzheimer's disease. A vasoactive intestinal peptide analog conjugated to MRC OX-26 produced peptide activity in the brain, causing cerebral blood flow to increase 65% (Bickel et al., 1993). The use of drug delivery by anti-TfR antibodies appears promising as potential treatments for many disorders of the brain.

Aluminum Uptake into the Brain

Aluminum is carried by Tf in the blood and binds Tf with about 10^{10}-fold lower affinity than iron (Martin et al., 1987). The affinity of Tf for the TfR is the same whether iron or aluminum is bound to Tf (Roskams and Connor, 1990).

In chronic renal dialysis patients, aluminum accumulation in the brain corresponded to regions of high TfR density, and also had a distribution pattern similar to aluminum accumulation in people with Alzheimer's disease. Aluminum was elevated in the frontal cortex in the cortical and limbic areas. The regions of high aluminum did not correspond to regions of high iron, again suggesting that there are other mechanisms that govern iron distribution in the brain (Morris et al., 1989). The highest region of focal iron accumulation in dialysis patients was found in the globus pallidus, where the level of TfR was low (Morris et al., 1989).

Examination of the brains from Alzheimer's patients revealed decreased TfR in the temporal cortex, occipital cortex, and the hippocampus. The decrease of TfR in the hippocampus is probably due to the degeneration of the neuronal cells in this region. No changes in TfR were seen in parietal and frontal cortices, which are regions usually affected in Alzheimer's disease. Thus, the TfR is not markedly affected in Alzheimer's disease, and other mechanisms likely cause the accumulation of aluminum (Kalaria et al., 1992).

VI. TfR AS A DIAGNOSTIC MARKER OF DISEASE

A.   The Soluble TfR

In 1986, a soluble form of the TfR in human serum was described (Kohgo et al., 1986) and has been found to be increasingly useful as diagnostic marker of certain clinical disorders (Cook et al., 1993; Thorstensen and Romslo, 1993). Initially, normal levels of serum TfR were reported to be 0.26 mg/liter of serum (Kohgo et al., 1986, 1988). Subsequently, values of 5.6 mg/liter (Klemow et al., 1990) and 8.3 mg/liter (Huebers et al., 1990) were reported. The differences between these values are probably due to different TfR standards used in the immunological assays. Flowers et al. (1989) showed that the differences between the low value of 0.26 mg/liter and the 5.6 mg/liter were due to a combination of the monoclonal antibodies used and the standard TfR used. Independent of which standard is used, normal values are not significantly different for men and women (Kohgo et al., 1988; Cook et al., 1990) and there are no differences with age of adults (Flowers et al., 1989). The level of serum TfR is increased in the sera of patients with conditions such as iron deficiency anemia (Kohgo et al., 1988) and diseases such as sickle cell disease (Flowers et al., 1989; Singhal et al., 1993; Thorstensen and Romslo, 1993), systemic lupus erythematosus (Woith et al., 1993), myeloproliferative disorders such as polycythemia vera and idiopathic myelofibrosis (Kohgo et al., 1988; Klemow et al., 1990), autoimmune hemolytic anemia (Kohgo et al., 1988), and chronic lymphocytic leukemia (Klemow et al., 1990). Serum TfR levels have been reported as high as 143 mg/liter in a patient with β-thalassemia/HbE (Huebers et al., 1990). There are a few diseases in which the level of serum TfR is slightly reduced (Thorstensen and Romslo, 1993). With some diseases, the progression of the disease can be monitored by the change in serum TfR levels, as is the case for chronic lymphocytic leukemia (Klemow et al., 1990) and megaloblastic anemia (Carmel and Skikne, 1992).

The level of serum TfR is believed to reflect the number of erythroid cells. In general, the increased levels of serum TfR reflect increased erythropoiesis. Diseases that cause increased production of reticulocytes show increased serum TfR, such as in iron deficiency anemia and sickle cell anemia. Decreased serum TfR is thought to be a result of decreased reticulocyte numbers. However, the exact relationship between erythropoiesis and serum TfR is unknown. The origin of serum TfR in patients with chronic lymphocytic leukemia may be other than erythroid, since serum TfR is increased independently of erythroid mass and may be derived from increased lymphoid proliferation (Klemow et al., 1990).

One important use of the measurement of serum TfR is that it can distinguish the anemia of chronic disease from iron deficiency anemia. This distinction had previously been possible only with bone marrow examinations. Currently, the use of serum TfR provides a reliable measure and is much less invasive. Measurement of serum TfR, combined with that of serum ferritin allows the differentiation between several types of anemia (Ferguson et al., 1992; Means and Krantz, 1992; Cook et al., 1993). With a less invasive method for determining iron deficiency, it should be easier to determine the iron status of a large number of people and thus provide more effective diagnosis and treatment.

The source of serum TfR is unknown. It appears to reflect erythroid mass, since about 50% of TfRs are in blood cells (Klemow et al., 1990). One model for the generation of serum TfR involves remodeling of the plasma membrane as reticulocytes mature into erythrocytes. During this process, TfRs are lost from the cell, possibly by cleavage as part of this maturation process (Ahn and Johnstone, 1993). However, other body tissues are likely to contribute to serum TfR levels, because when marrow is ablated, or in patients with aplastic anemia, the serum TfR level decreases only about 50% (Flowers et al., 1989; Huebers et al., 1990; Klemow et al., 1990). In patients with chronic lymphocytic leukemia, elevated serum TfR levels may be due to increased lymphocyte proliferation (Klemow et al., 1990). Also, TfR has been detected in cerebral spinal fluid at a level 1/1,000 of serum TfR (Thorstensen and Romslo, 1993). In addition, several cell lines not of erythroid origin, including human HL60 (Chitambar and Zivkovic, 1989) and Chinese hamster ovary transfected with human TfR (Rutledge et al., 1994a), have been shown to produce a cleaved soluble form of TfR. The soluble TfR appears to be very stable, both in the body and in serum-containing culture medium from cell lines. After bone marrow ablation, the mean-time to half the initial serum TfR level was 15 days (Klemow et al., 1990). If the serum TfR level is below normal, this could be due to diminished erythropoiesis, diminished production from other tissues, or increased degradation of the serum TfR.

Only trace amounts (0.6%) of full-length TfR are detected in serum from normal subjects. The level of full-length TfR is increased slightly to 3.9% in patients with sickle cell anemia (Shih et al., 1993). The serum TfR is derived from proteolytic cleavage of full-length TfR, and cleavage is between residues Arg 100 and Leu 101 (Shih et al., 1990).

The protease that cleaves TfR has not been identified, however, it appears to be a membrane protease and does not require acidic endosomes for cleavage to occur (Baynes et al., 1993; Rutledge et al., 1994b).

B. ^{67}Ga-Tf

TfR has been the vehicle for many tumor-specific agents for both detecting and killing cancer cells. Gallium binds to Tf with an affinity about 250 times less than iron (Martin et al., 1987), and it is the Ga-Tf that binds to TfR and interferes with iron uptake and thus cell proliferation (Larson et al., 1980; Chitambar et al., 1989; Seligman et al., 1992). Uptake of gallium has been shown to occur via Tf and TfR in human HL60 cells (Chitambar and Zivkovic, 1987) and in human melanomas introduced into mice (Chan et al., 1987).

^{67}Gallium was first used as a tumor imaging agent in 1969 (Edwards and Hayes, 1969) and has become widely used to detect tumors (Pinsky and Henkin, 1976). Uptake of Ga is greatest in the colon, then the liver, and faintly in bones (especially the pelvis and spine, probably due to the bone marrow). Gallium is useful for detecting tumors, both initially and in follow-up treatment, of lymph nodes (Hodgkin's disease; Johnston et al., 1974), lung cancer (Edwards and Hayes, 1970; Siemsen et al., 1976), liver tumors, brain tumors, testicular tumors, and melanoma. Gallium scanning is also useful for detecting non-tumor conditions such as inflammation (Lavender et al., 1971), rheumatoid arthritis, and bone fractures.

VII. HETEROGENEITY OF THE TfR AND TfR IN OTHER ORGANISMS

A.   Heterogeneity of the TfR

A tissue dependent epitopic variation in the TfR has been defined by a variety of monoclonal antibodies generated to the TfR. Three different monoclonal antibodies displayed different reactivities to TfR on a single cell line (RPMI 8402 leukemia cells, for example, showed high reactivity with the 5E9 antibody, but no reactivity with HuLy-m9; other human cell lines showed different patterns of reactivity with these antibodies; Panaccio et al., 1987). TfR antibodies have been developed which discriminate between high-grade and low-grade lymphoma tumors from the same patient; the Trump antibody which recognizes the TfR in high-grade lymphomas, reacts with only a subpopulation of OKT9-positive cells (Takahashi et al., 1991). This suggests that an epitope is exposed on some cells that is unavailable for antibody binding on other cells. The reason for this heterogeneity has not been established. As far as is known, in humans there is only a single TfR gene. The heterogeneity may arise from differential splicing, differential post-translational modifications, or steric hindrance from other cell surface components.

B. Vertebrates and Invertebrates

Virtually nothing is known about the evolution or origin of the TfR. To date, the cDNA has been cloned from four vertebrate species: chicken, mouse, hamster, and man (McClelland et al., 1984; Schneider et al., 1984; Trowbridge et al., 1988; Gerhardt et al., 1991; Collawn et al., 1993). Among these forms, a number of important features are conserved. The YTRF putative internalization motif for endocytosis is strictly conserved, as is the phosphorylated serine immediately following the internalization signal. The chicken receptor lacks a cysteine corresponding to human position 89, which is thought to be the main site of intersubunit disulfide bond formation, but possesses the alternate cysteine at position 98, and another at position 105. The O-glycosylated Thr at position 104 is also absent from chicken receptor, but there are several Thr residues downstream from this site, which could serve the same function. The three potential Asn-linked glycosylation sites of the human TfR are conserved, although in the chicken, the third site (human position 727) is shifted slightly N-terminal, and the rodent sequences possess two potential Asn-linked glycosylation sequences in the C-terminus, at positions 722 and 727. Glycosylation in this region is important for transport of the TfR to the cell surface; failure of transport to the cell surface of the mutant human receptor lacking the 727 site can be corrected by introducing a new site at 722 (Williams and Enns, 1993; Yang et al., 1993).

The taxonomic distribution of Tf may provide clues to the evolutionary origin of the receptor. Tf itself has been cloned from several invertebrates — all insects — as well as a variety of chordates. The sequences from vertebrate Tf suggest that, in higher vertebrates, the complete gene has resulted from gene duplication and fusion (Williams, 1982). Indirect evidence suggests that in primitive chordates, e.g., the tunicate *Pyura*, the gene duplication has not yet occurred (Bowman et al., 1988). Tf from the moth *Manduca sexta* and the cockroach have been sequenced and found to be approximately 30% identical in amino acid sequence to vertebrate Tf and 46% identical to each other, and to have independently undergone sequence duplication (Bartfield and Law, 1990; Jamroz et al., 1993). It would, therefore, appear that Tf, and presumably its receptor, was present in an ancestor common to arthropods and vertebrates, more than 500 million years ago, and that the duplication/fusion event had not yet occurred.

C. Microbials

Tf binding proteins unrelated to the TfR have been found in several eukaryotic and prokaryotic pathogens, apparently having evolved independently as an adaptation to the host environment. The bacteria *Haemophilus influenzae*, *Neisseria gonorrheae*, *N. meningitidis*, *Bordetella pertussis*, and *Pasteurella haemolytica* all possess Tf binding proteins (Redhead et al., 1987; Blanton et al., 1990; Ogunnariwo and Schryvers, 1990; Stevenson et al., 1992; Ala'aldeen et al., 1993). Bacteria, in general, produce small soluble iron binding molecules referred to as siderophores, which are capable of competing with Tf for iron association, but with slow kinetics, such that bacterial acquisition of iron is retarded. It has been suggested that Tfs evolved as a way of sequestering iron from invading microbes. *Neisseria* lacks a siderophore system and can acquire iron from Tf. It may also be able to utilize the TfR, via its Tf binding capability, as a portal of passage through transcytosing membranes. One of the gonococcal Tf-binding proteins has been cloned and shows homology to B-dependent outer membrane receptors of *E. coli* and *Pseudomonas putida*. Therefore, the Tf-binding proteins of bacteria were not appropriated from eukaryotes, but evolved independently (Cornelissen et al., 1992).

Tf binding proteins have also been reported from the protozoan parasites *Trypanosoma brucei, T. cruzi*, and *Leishmania* (Coppens et al., 1987; Lima and Villalta, 1990; Voyiatzaki and Soteriadou, 1990, 1992; Schell et al., 1991). Like the bacterial TfRs, they show no homology to mammalian TfRs and have probably evolved independently.

ACKNOWLEDGMENT

The authors would like to thank Robin Warren for careful reading of the manuscript. The work was supported by a grant from the National Institute of Health (DK 40608).

REFERENCES

Adam, M., Rodriguez, A., Turbide, C., Larrick, J., Meighen, E., & Johnstone, R. M. (1984). *In vitro* acylation of the transferrin receptor. J. Biol. Chem. 259, 15460–15463.

Ahn, J., & Johnstone, R. M. (1993). Origin of a soluble truncated transferrin receptor. Blood 81, 2442–2451.

Aisen, P., & Leibman, A. (1978). Transport by proteins. Berlin, de Gruyter.

Ala'aldeen, D. A. A., Powell, N. B. L., Wall, A. R., & Borriello, S. P. (1993). Localization of the meningococcal receptors for human transferrin. Infect. and Immun. 61, 751–754.

Alcantara, J., Yu, R. H., & Schryvers, A. B. (1993). The region of human transferrin involved in binding to bacterial transferrin receptors is localized in the C-lobe. Molec. Micro. 8, 1135–1143.

Alvarez, E., Gironès, N., & Davis, R. J. (1989). Intermolecular disulfide bonds are not required for the expression of the dimeric state and functional activity of the transferrin receptor. EMBO J. 8, 2231–2240.

Alvarez, E., Gironès, N., & Davis, R. J. (1990). Inhibition of the receptor-mediated endocytosis of diferric transferrin is associated with the covalent modification of the transferrin receptor with palmitic acid. J. Biol. Chem. 265, 16644–16655.

Anderson, G. J., Walsh, M. D., Powell, L. W., & Halliday, J. W. (1991). Intestinal transferrin receptors and iron absorption in the neonatal rat. Br. J. Haematol. 77, 229–236.

Bali, P. K., Zak, O., & Aizen, P. (1991). A new role for the transferrin receptor in the release of iron from transferrin. Biochemistry 30, 324–328.

Banks, W. A., Kastin, A. J., Fasold, M. B., Barrera, C. M., & Augereau, G. (1988). Studies of the slow bidirectional transport of iron and transferrin across the blood-brain barrier. Brain Research Bulletin 21, 881–885.

Bansal, A., & Gierasch, L. M. (1991). The NPXY internalization signal of the LDL receptor adopts a reverse-turn conformation. Cell 67, 1195–1201.

Bartfield, N. S., & Law, J. H. (1990). Isolation and molecular cloning of transferrin from the tobacco hornworm, *Manduca sexta*. J. Biol. Chem. 265, 21684–21691.

Basset, P., Quesneau, Y., & Zwiller, J. (1986). Iron-induced L1210 cell growth: Evidence of a transferrin-independent iron transport. Cancer Research 1644–1647.

Baynes, R. D., Shih, Y. J., Hudson, B. G., & Cook, J. D. (1993). Production of the serum form of the transferrin receptor by a cell membrane-associated serine protease. Proc. Soc. Exp. Biol. Med. 204, 65–69.

Beach, R. L., Popiela, H., & Festoff, B. W. (1983). The identification of neurotrophic factor as transferrin. FEBS Lett. 156, 151–156.

Bickel, U., Yoshikawa, T., Landaw, E. M., Faull, K. F., & Pardridge, W. M. (1993). Pharmacologic effects *in vivo* in brain by vector-mediated peptide drug delivery. Proc. Natl. Acad. Sci. USA 90, 2618–2622.

Blanton, K. J., Biswas, G. D., Tsai, J., Adams, J., Dyer, D. W., Davis, S. M., Koch, G. G., Sen, P. K., & Sparling, F. (1990). Genetic evidence that *Neisseria gonorrheae* produces specific receptors for transferrin and lactoferrin. J. Bact. 172, 5225–5235.

Bowman, B. H., Yang, F., & Adrian, G. S. (1988). Transferrin: Evolution and genetic regulation. Adv. Genetics 25, 1–38.

Bridges, K. R., & Cudkowicz, A. (1984). Effect of iron chelators on the transferrin receptor in K562 cells. J. Biol. Chem. 259, 12970–12977.

Cammack, R., Wrigglesworth, J. M., & Baum, H. (1990). Iron-dependent enzymes in mammalian systems. 17–41 Iron Transport and Storage Boston, CRC Press.

Carmel, R., & Skikne, B. S. (1992). Serum transferrin receptor in megaloblastic anemia of cobalamin deficiency. Eur. J. Haem. 49, 246–250.

Casanova, J. E., Breitfeld, P. P., Ross, S. A., & Mostov, K. E. (1990). Phosphorylation of the polymeric immunoglobulin receptor required for its efficient transcytosis. Science 248, 742–745.

Casey, J. L., Di Jeso, B., Rao, K. K., Rouault, T. A., Klausner, R. D., & Harford, J. B. (1988a). Deletional analysis of the promoter region of the human transferrin receptor gene. Nucleic Acids Res. 16, 629–646.

Casey, J. L., Hentze, M. W., Koeller, D. M., Caughman, S. W., Rouault, T. A., Harford, J., & Klausner, R. (1988b). Iron-responsive elements: Regulatory RNA sequences that control mRNA levels and translation. Science 240, 924–928.

Cerneus, D. P., & van der Ende, A. (1991). Apical and basolateral transferrin receptors in polarized BeWo cells recycle through separate endosomes. J. Cell Biol. 114, 1149–1158.

Cerneus, D. P., Strous, G. J., & van der Ende, A. (1993). Bidirectional transcytosis determines the steady state distribution of the transferrin receptor at opposite plasma membrane domains of BeWo cells. J. Cell Biol. 122, 1223–1230.

Chackal-Roy, M., & Zetter, B. R. (1992). Selective stimulation of prostatic carcinoma cell proliferation by transferrin. Proc. Natl. Acad. Sci. USA 89, 6197–6201.

Chackal-Roy, M., Niemeyer, C., Moore, M., & Zetter, B. R. (1989). Stimulation of human prostatic carcinoma cell growth by factors present in human bone marrow. J. Clin. Invest. 84, 43–50.

Chan, S. M., Hoffer, P. B., Maric, N., & Duray, P. (1987). Inhibition of gallium-67 uptake in melanoma by an anti-human transferrin receptor monoclonal antibody. J. Nuc. Med. 28, 1303–1307.

Chang, M. P., Mallett, W. G., Mostov, K. E., & Brodsky, F. M. (1993). Adaptor self-aggregation, adaptor-receptor recognition and binding of alpha-adaptin subunits to the plasma membrane contribute to recruitment of adaptor (AP2) components of clathrin-coated pits. EMBO J. 12, 2169–2180.

Chen, W. J., Goldstein, J. L., & Brown, M. S. (1990). NPXY, a sequence often found in cytoplasmic tails, is required for coated pit-mediated internalization of the low density lipoprotein receptor. J. Biol. Chem. 265, 3116–3123.

Chitambar, C. R., & Zivkovic, Z. (1987). Uptake of gallium-67 by human leukemic cells: Demonstration of transferrin receptor-dependent and transferrin-independent mechanisms. Cancer Res. 47, 3929–3934.

Chitambar, C. R., Matthaeus, W. G., Antholine, W. E., Graff, K., & O'Brien, W. J. (1988). Inhibition of leukemic HL60 cell growth by transferrin-gallium: Effects on ribonucleotide reductase and demonstration of drug synergy with hydroxyurea. Blood 72, 1930–1936.

Chitambar, C. R., Craig, A., & Ash, R. C. (1989). Transferrin receptor-mediated suppression of in vitro hematopoiesis by transferrin-gallium. Exper. Hem. 17, 418–422.

Chitambar, C. R., & Zivkovic, Z. (1989). Release of soluble transferrin receptor from the surface of human leukemic HL60 cells. Blood 74, 602–608.

Chow, B. K., Funk, W. D., Banfield, D. K., Lineback, J. A., Mason, A. B., Woodworth, R. C., & MacGillivray, R. T. A. (1991). Structural-functional studies of human transferrin by using in vitro mutagenesis. Cur. Stud. Hem. Blood Transfus. 132–138.

Collawn, J. F., Stangel, M., Kuhn, L. A., Esekogwu, V., Jing, S., Trowbridge, I. S., & Tainer, J. A. (1990). Transferrin receptor internalization sequence YXRF implicates a tight turn as the structural recognition motif for endocytosis. Cell 63, 1061–1072.

Collawn, J. F., Lai, A., Domingo, D., Fitch, M., Hatton, S., & Trowbridge, I. S. (1993). YTRF is the conserved internalization signal of the transferrin receptor, and a second YTRF signal at position 31-34 enhances endocytosis. J. Biol. Chem. 268, 21686–21692.

Connor, J. R., & Benkovic, S. A. (1992). Iron regulation in the brain: Histochemical, biochemical, and molecular considerations. Ann. Neurol. 32, S51–S61.

Connor, J. R., Phillips, T. M., Lakshman, M. R., Barron, K. D., Fine, R. E., & Csiza, C. K. (1987). Regional variation in the levels of transferrin in the CNS of normal and myelin-deficient rats. J. Neurochem. 49, 1523–1529.

Cook, J. D., Dassenko, S., & Skikne, B. S. (1990). Serum transferrin receptor as an index of iron absorption. Br. J. Haematol. 75, 603–609.

Cook, J. D., Skikne, M. S., & Baynes, R. D. (1993). Serum transferrin receptor. Annu. Rev. Med. 44, 63–74.

Coppens, I., Opperdoes, F. R., Courtoy, P. J., & Baudhuin, P. (1987). Receptor-mediated endocytosis in the bloodstream form of trypanosoma brucei. J. Protozool. 34, 465–473.

Cornelissen, C. N., Biswas, G. D., Tsai, J., Paruchuri, D. K., Thompson, S. A., & Sparling, P. F. (1992). Gonococcal transferrin-binding protein 1 is required for transferrin utilization and is homologous to TonB-dependent outer membrane receptors. J. Bact. 174, 5788–5797.

Dargemont, C., Le Bivic, A., Rothenberger, S., Iacopetta, B., & Kühn, L. C. (1993). The internalization signal and the phosphorylation site of transferrin receptor are distinct from the main basolateral sorting information. EMBO J. 12, 1713–1721.

Dautry-Varsat, A., Ciechanover, A., & Lodish, H. F. (1983). pH and the recycling of transferrin during receptor-mediated endocytosis. Proc. Natl. Acad. Sci. USA 80, 2258–2262.

Davis, C..G., Lehrman, M. A., Russell, D. W., Anderson, R. G. W., Brown, M. S., & Goldstein, J. L. (1986). The J.D. mutation in familial hypercholesterolemia: Amino acid substitution in the cytoplasmic domain impedes internalization of LDL receptors. Cell 45, 15–24.

Davis, R. J., Johnson, G. L., Kelleher, D. J., Anderson, J. K., Mole, J. E., & Czech, M. P. (1986). Identification of serine 24 as the unique site on the transferrin receptor phosphorylated by protein kinase C. J. Biol. Chem. 261, 9034–9041.

Davis, R. J., & Meisner, H. (1987). Regulation of transferrin receptor cycling by protein kinase C is independent of receptor phosphorylation at serine 24 in Swiss 3T3 fibroblasts. J. Biol. Chem. 262, 16041–16047.

De Jong, G., & Van Eijk, H. G. (1989). Functional properties of the carbohydrate moiety of human transferrin. Int. J. Biochem. 21, 353–363.

Dillner-Centerlind, M. L., Hammmarstrom, S., & Perlmann, P. (1979). Transferrin can replace serum for *in vitro* growth of mitogen-stimulated T lymphocytes. Eur. J. Immunol. 9, 942–948.

Do, S.-I., & Cummings, R. D. (1992). Presence of O-linked oligosaccharide on a threonine residue in the human transferrin receptor. Glycobiol. 2, 345–353.

Do, S.-I., Enns, C. A., & Cummings, R. D. (1990). Human transferrin receptor contains O-linked oligosaccharides. J. Biol. Chem. 265, 114–125.

Domingo, D. L., & Trowbridge, I. S. (1988). Characterization of the human transferrin receptor produced in a baculovirus expression system. J. Biol. Chem. 263, 13386–13392.

Eberle, W., Sander, C., Klaus, W., Schmidt, B., von Figura, K., & Peters, C. (1991). The essential tyrosine of the internalization signal in lysosomal acid phosphatase is part of a β turn. Cell 67, 1203–1209.

Edwards, C. L., & Hayes, R. L. (1969). Tumor scanning with ^{67}Ga citrate. J. Nuc. Med. 10, 103–105.

Edwards, C. L., & Hayes, R. L. (1970). Scanning malignant neoplasms with gallium 67. J. Amer. Med. Assoc. 212, 1182–1190.

Enns, C. A., Clinton, E. M., Reckhow, C. L., Root, B. J., Do, S.-I., & Cook, C. (1991). Acquisition of the functional properties of the transferrin receptor during its biosynthesis. J. Biol. Chem. 266, 13272–13277.

Enns, C. A., Suomalainen, H. A., Gebhardt, J. E., Schroder, J., & Sussman, H. H. (1982). Human transferrin receptor: Expression of the receptor is assigned to chromosome 3. Proc. Natl. Acad. Sci. USA 79, 3241–3245.

Evans, R. W., & Williams, J. (1978). Studies of the binding of different iron donors to human serum transferrin and isolation of iron-binding fragments from the N- and C-terminal regions of the protein. Biochem. J. 173, 5442–5452.

Faulk, W. P., Stevens, P. J., & Hsi, B.-L. (1980). Transferrin and transferrin receptors in carcinoma of the breast. Lancet Aug. 23, 390–392.

Ferguson, B. J., Skikne, B. S., Simpson, K. M., Baynes, R. D., & Cook, J. D. (1992). Serum transferrin receptor distinguishes the anemia of chronic disease from iron deficiency anemia. J. Lab. Clin. Med. 19, 385–390.

Fishman, J. B., Rubin, J. B., Handrahan, J. V., Connor, J. R., & Fine, R. E. (1987). Receptor-mediated transcytosis of transferrin across the blood-brain barrier. J. Neurosci. Res. 18, 299–304.

Flowers, C. H., Skikne, B. S., Covell, A. M., & Cook, J. D. (1989). The clinical measurement of serum transferrin receptor. J. Lab. Clin. Med. 114, 368–377.

Friden, P. M., Walus, L. R., Musso, G. F., Taylor, M. A., Malfroy, B., & Starzyk, R. M. (1991). Anti-transferrin receptor antibody and antibody-drug conjugates cross the blood brain barrier. Proc. Natl. Acad. Sci. USA 88, 4771–4775.

Friden, P. M., Walus, L., Taylor, M., Musso, G. F., Abelleira, S. A., Malfroy, B., Tehrani, F., Eckman, J. B., III, Morrow, A. R., & Starzyk, R. M. (1992). Drug delivery to the brain using an anti-transferrin receptor antibody. National Institute on Drug Abuse Research Monograph Series.

Friden, P. M., Walus, L. R., Watson, P., Doctrow, S. R., Kozrich, J. W., Backman, C., Bergman, H., Hoffer, B., Bloom, F., & Granholm, A.-C. (1993). Blood-brain barrier penetration and *in vivo* activity of an NGF conjugate. Science 259, 373–377.

Fuernkranz, H. A., Schwob, J. E., & Lucas, J. J. (1991). Differential tissue localization of oviduct and erythroid transferrin receptors. Proc. Natl. Acad. Sci. USA 88, 7505–7508.

Fuller, S. D., & Simons, K. (1986). Transferrin receptor polarity and recycling accuracy in "tight" and "leaky" strains of Madin-Darby canine kidney cells. J. Cell Biol. 103, 1767–1779.

Gatter, K. C., Brown, G., Trowbridge, I. S., Woolston, R.-E., & Mason, D. Y. (1983). Transferrin receptors in human tissues: their distribution and possible clinical relevance. J. Clin. Path. 539–545.

Gerhardt, E. M., Chan, L. L., Jing, S., Meiying, Q., & Trowbridge, I. S. (1991). The cDNA sequence and primary structure of the chicken transferrin receptor. Gene 102, 249–254.

Geuze, H. J., Slot, J. W., Strous, G. J. A. M., Lodish, H. F., & Schwartz, A. L. (1983). Intracellular site of asialoglycoprotein receptor-ligand uncoupling: double-label immunoelectron microscopy during receptor-mediated endocytosis. Cell 32, 277–287.

Giometto, B., Bozza, F., Argentiero, V., Gallo, P., Pagni, S., Piccinno, M. G., & Tavolato, B. (1990). Transferrin receptors in rat central nervous system: An immunocytochemical study. J. Neuro. Sci. 98, 81–90.

Gironès, N., & Davis, R. J. (1989). Comparison of the kinetics of cycling of the transferrin receptor in the presence or absence of bound diferric transferrin. Biochem. J. 264, 35–46.

Gitlin, S. R., Kumate, J., Urrusti, J., & Morales, C. (1964). The selectivity of the human placenta in the transfer of plasma proteins from mother to fetus. J. Clin. Invest. 43, 1938–1951.

Glickman, J. N., Conibear, E., & Pearse, B. M. F. (1989). Specificity of binding of clathrin adaptors to signals on the mannose-6-phosphate/insulin-like growth factor II receptor. EMBO J. 8, 1041–1047.

Goding, J. W., & Burns, G. F. (1981). Monoclonal antibody OKT-9 recognized the receptor for transferrin on human acute lymphocytic leukemia cells. J. Immun. 127, 1256–1273.

Goodfellow, P. N., Banting, G., Sutherland, R., Greaves, M., Solomon, E., & Povey, S. (1982). Expression of human transferrin receptor is controlled by a gene on chromosome 3: Assignment using species specificity of a monoclonal antibody. Som. Cell Genetics 8, 197–206.

Graeber, M. B., Raivich, G., & Kreutzberg, G. W. (1989). Increase of transferrin receptors and iron uptake in regenerating motor neurons. J. Neurosci. Res. 23, 342–345.

Gruenberg, J., Griffiths, G., & Howell, K. E. (1989). Characterization of the early endosome and putative endocytic carrier vesicles in vivo and with an assay of vesicle fusion in vitro. J. Cell Biol. 108, 1301–1316.

Hamilton, T. A., Wada, H. G., & Sussman, H. H. (1979). Identification of transferrin receptors on the surface of human cultured cells. Proc. Natl. Acad. Sci. USA 76, 6406–6410.

Hansen, S. H., Sandvig, K., & van Deurs, B. (1992). Internalization efficiency of the transferrin receptor. Exp. Cell Res. 199, 19–28.

Harding, C., & Stahl, P. (1983). Transferrin recycling in reticulocytes: pH and iron are important determinants of ligand binding and processing. Biochem. Biophys. Res. Comm. 113, 650–658.

Hayashi, O., Noguchi, S., & Oyasu, R. (1987). Transferrin as a growth factor for rat bladder carcinoma cells in culture. Cancer Res. 47, 4560–4564.

Hayes, G. R., Enns, C. A., & Lucas, J. J. (1992). Identification of the O-linked glycosylation site of the human transferrin receptor. Glycobiol. 2, 355–359.

Hedley, D. W., Tripp, E. H., Slowiaczek, P., & Mann, G. J. (1988). Effect of gallium on DNA synthesis by human T-cell lymphoblasts. Cancer Res. 48, 3014–3018.

Hershko, C., Karsai, A., Eylon, L., & Izak, G. (1970). The effect of chronic iron deficiency on some biochemical functions of the human hemopoietic tissue. Blood 36, 321–329.

Hill, J. M., Ruff, M. R., Weber, R. J., & Pert, C. B. (1985). Transferrin receptors in rat brain: Neuropeptide-like pattern and relationship to iron distribution. Proc. Natl. Acad. Sci. USA 85, 4553–4557.

Ho, P. T., King, I., & Sartorelli, A. C. (1986). Transcriptional regulation of the transferrin receptor in differentiating HL-60 leukemic cells. Biochem. Biophys. Res. Comm. 138, 995–1000.

Ho, P. T. C., Ishiguro, K., & Sartorelli, A. C. (1989). Regulation of transferrin receptor in myeloid and monocytic differentiation of HL-60 leukemia cells. Cancer Research 49, 1989–1995.

Hoe, M. H., & Hunt, R. C. (1992). Loss of one asparagine-linked oligosaccharide from human transferrin receptors results in specific cleavage and association with the endoplasmic reticulum. J. Biol. Chem. 267, 4916–4929.

Hopkins, C., & Trowbridge, I. S. (1983). Internalization and processing of transferrin and the transferrin receptor in human carcinoma A431 cells. J. Cell Biol. 97, 508–521.

Hopkins, C. R. (1983). Intracellular routing of transferrin and transferrin receptors in epidermoid carcinoma A431 cells. Cell 35, 321–330.

Huebers, H. A., Beguin, Y., Pootrakul, P., Einspahr, D., & Finch, C. A. (1990). Intact transferrin receptors in human plasma and their relation to erythropoiesis. Blood 75, 102–107.

Hunt, R. C., Riegler, R., & Davis, A. A. (1989). Changes in glycosylation alter the affinity of the human transferrin receptor for its ligand. J. Biol. Chem. 264, 9643–9648.

Hyndman, A. G., Hockberger, P. E., Zeevalk, G. D., & Connor, J. A. (1991). Transferrin can alter physiological properties of retinal neurons. Brain Res. 561, 318–323.

Iacopetta, B. J., Rothenberger, S., & Kühn, L. C. (1988). A role for the cytoplasmic domain in transferrin receptor sorting and coated pit formation during endocytosis. Cell 54, 485–489.

Jamroz, R. C., Gasdaska, J. R., Bradfield, J. Y., & Law, J. H. (1993). Transferrin in a cockroach: Molecular cloning, characterization, and suppression by juvenile hormone. Proc. Natl. Acad. Sci. USA 90, 1320–1324.

Jefferies, W. A., Brandon, M. R., Hunt, S. V., Williams, A. F., Gatter, K. C., & Mason, D. Y. (1984). Transferrin receptor on endothelium of brain capillaries. Nature 312, 162–163.

Jing, S., Spencer, T., Miller, K., Hopkins, C., & Trowbridge, I. S. (1990). Role of the human transferrin receptor cytoplasmic domain in endocytosis: Localization of a specific signal sequence for internalization. J. Cell Biol. 110, 283–294.

Jing, S., & Trowbridge, I. S. (1987). Identification of the intermolecular disulfide bonds of the human transferrin receptor and its lipid attachment site. EMBO J. 6, 327–331.

Jing, S., & Trowbridge, I. S. (1990). Nonacylated human transferrin receptors are rapidly internalized and mediate iron uptake. J. Biol. Chem. 265, 11555–11559.

Johnston, G., Benua, R. S., Teates, C. D., Edwards, C. L., & Kniseley, R. M. (1974). ^{67}Ga-citrate imaging in untreated Hodgkin's disease: Preliminary report of cooperative group. J. Nuc. Med. 15, 399–403.

Judd, W. C., Poodry, C. A., & Strominger, J. L. (1980). Novel surface antigen expressed on dividing cells but absent on non-dividing cells. J. Exp. Med. 152, 1430–1435.

Kalaria, R. N., Sromek, S. M., Grahovac, I., & Harik, S. I. (1992). Transferrin receptors of rat and human brain cerebral microvessels and their status in Alzheimer's disease. Brain Res. 585, 87–93.

Karin, M., & Mintz, B. (1981). Receptor-mediated endocytosis of transferrin in developmentally totipotent mouse teratocarcinoma stem cells. J. Biol. Chem. 256, 3245–3252.

Keer, H. N., Kozlowski, J. M., Tsai, Y. C., Lee, C., McEwan, R. N., & Grayhack, J. T. (1990). Elevated transferrin receptor content in human prostate cancer cell lines assessed *in vitro* and *in vivo*. J. Urology 143, 381–385.

Kitada, S., & Hays, E. F. (1985). Transferrin-like activity produced by murine malignant T-lymphoma cell lines. Cancer Res. 45, 3537–3540.

Klausner, R. D., Ashwell, G., van Renswoude, J., Harford, J. B., & Bridges, K. R. (1983a). Binding of apotransferrin to K562 cells: Explanation of the transferrin cycle. Proc. Natl. Acad. Sci. USA 80, 2263–2266.

Klausner, R. D., van Renswoude, J., Ashwell, G., Kempf, C., Schechter, A. N., Dean, A., & Bridges, K. R. (1983b). Receptor-mediated endocytosis of transferrin in K562 cells. J. Biol. Chem. 258, 4715–4724.

Klausner, R. D., Rouault, T. A., & Harford, J. B. (1993). Regulating the fate of mRNA: The control of cellular iron metabolism. Cell 72, 19–28.

Klemow, D., Einsphar, D., Brown, T. A., Flowers, C. H., & Skikne, B. S. (1990). Serum transferrin receptor measurements in hematologic malignancies. Am. J. Hematol. 34, 193–198.

Koeller, D. M., Dasey, J. L., Hentze, M. W., Gerhardt, E. M., Chan, L.-N. L., Klausner, R. D., & Harford, J. B. (1989). A cytosolic protein binds to structural elements within the iron regulatory region of the transferrin receptor mRNA. Proc. Natl. Acad. Sci. USA 86, 3574–3578.

Kohgo, Y., Nishisato, T., Kondo, H., Tsushima, N., Niitsu, Y., & Urushizaki, I. (1986). Circulating transferrin receptor in human serum. Br. J. Haematol. 64, 277–281.

Kohgo, T., Nitsu, Y., Nishisato, T., Kato, J., Kondo, H., Sasaki, K., & Urushizaki, I. (1988). Quantitation and characterization of serum transferrin receptor in patients with anemias and polycythemias. Jap. J. Med. 27, 54–70.

Kühn, L. C. (1989). The transferrin receptor: A key function in iron metabolism. Scheiz. Med. Wschr. 119, 1319–1326.

Kühn, L. C., McClelland, A., & Ruddle, F. H. (1984). Gene transfer, expression, and molecular cloning of the human transferrin receptor gene. Cell 37, 95–103.

Kühn, L. C., Schulman, H. M., & Ponka, P. (1990). Iron-transferrin requirements and transferrin receptor expression in proliferating cells. 149–192 Iron Transport and Storage, Boston, CRC.

Larrick, J. W., & Cresswell, P. (1979). Transferrin receptors on human B and T lymphoblastoid cell lines. Biochim. et Biophys. 583, 483–490.

Larson, S. M., Rasey, J. S., Allen, D. R., Nelson, N. J., Grunbaum, Z., Harp, G. D., & Williams, D. L. (1980). Common pathway for tumor cell uptake of gallium-67 and iron-59 via a transferrin receptor. J. Nat. Can. Instit. 64, 41–53.

Lavender, J. P., Lowe, J., Barker, J. R., Burn, J. I., & Chaudhri, M. A. (1971). Gallium 67 citrate scanning in neoplastic and inflammatory lesions. Br. J. Radiol. 44, 361–366.

Lederman, H. M., Cohen, A., Lee, J. W. W., Freedman, M. H., & Gelfand, E. W. (1984). Desferoxamine: A reversible S-phase inhibitor of human lymphocyte proliferation. Blood 64, 748–753.

Lesley, J. F., & Schulte, R. J. (1985). Inhibition of cell growth by monoclonal anti-transferrin receptor antibodies. Mol. Cell. Biol. 5, 1814–1821.

Levine, J. S., & Seligman, P. A. (1984). The ultrastructural immunocytochemical localization of transferrin receptor and transferrin in the gastrointestinal tract of man. Gastroent. 86, 1161.

Lima, M. F., & Villalta, F. (1990). *Trypanosoma cruzi* receptors for human transferrin and their role. Mol. Biochem. Paras. 38, 245–252.

Lin, H. H., & Connor, J. R. (1989). The development of the transferrin-transferrin receptor system in relation to astrocytes, MBP and galactocerebroside in normal and myelin-deficient rat optic nerves. Dev. Brain Res. 49, 281–293.

Louache, F., Testa, U., Pelicci, P., Thomopoulos, P., Titeux, M., & Rochant, H. (1984). Regulation of transferrin receptors in human hematopoietic cell lines. J. Biol. Chem. 259, 11576–11582.

Martin, R. B., Savory, J., Brown, S., Bertholf, R. L., & Wills, M. R. (1987). Transferrin binding of Al^{3+} and Fe^{3+}. Clin. Chem. 33, 405–407.

Mash, D. C., Pablo, J., Flynn, D. D., Efange, S. M. N., & Weiner, W. J. (1990). Characterization and distribution of transferrin receptors in the rat brain. J. Neurochem. 55, 1972–1979.

Mason, A., Miller, M., Funk, W., Banfield, D., Savage, K., Oliver, R., & Green, B. N. (1993). Expression of glycosylated and nonglycosylated human transferrin in mammalian cells. Characterization of the recombinant proteins with comparison to three commercially available transferrins. Biochemistry 32, 5472–5479.

May, W. S., Sahyoun, N., Jacobs, S., Wolf, M., & Cuatrecasas, P. (1985). Mechanism of phorbol diester-induced regulation of surface transferrin receptor involves the action of activated protein kinase C and an intact cytoskeleton. J. Biol. Chem. 260, 9419–9426.

Mayor, S., Presley, J. F., & Maxfield, F. R. (1993). Sorting of membrane components from endosomes and subsequent recycling to the cell surface occurs by a bulk flow process. J. Cell Biol. 121, 1257–1269.

McClelland, A., Kühn, L. C., & Ruddle, F. H. (1984). The human transferrin receptor gene: Genomic organization, and the complete primary structure of the receptor deduced from a cDNA sequence. Cell 39, 267–274.

McGraw, T., Greenfield, L., & Maxfield, F. R. (1987). Functional expression of the human transferrin receptor cDNA in Chinese hamster ovary cells deficient in endogenous transferrin receptor. J. Cell Biol. 105, 207–214.

McGraw, T. E., Dunn, K. W., & Maxfield, F. R. (1988). Phorbol ester treatment increases the exocytic rate of the transferrin receptor recycling pathway independent of serine-24 phosphorylation. J. Cell Biol. 106, 1061–1066.

McGraw, T. E., Pytowski, B., Arzt, J., & Ferrone, C. (1991). Mutagenesis of the human transferrin receptor: Two cytoplasmic phenylalanines are required for efficient internalization and a second-site mutation is capable of reverting an internalization-defective phenotype. J. Cell Biol. 112, 853–861.

Means, R. T., & Krantz, S. B. (1992). Progress in understanding the pathogenesis of the anemia of chronic disease. Blood 80, 1639–1647.

Mendelsohn, J., Trowbridge, I., & Castagnola, J. (1983). Inhibition of human lymphocyte proliferation by monoclonal antibody to transferrin receptor. Blood 62, 821–826.

Miller, K., Shipman, M., Trowbridge, I. S., & Hopkins, C. R. (1991). Transferrin receptors promote the formation of clathrin lattices. Cell 65, 621–632.

Miskimins, K. W. (1992). Interaction of multiple factors with a GC-rich element within the mitogen responsive region of the human transferrin receptor gene. J. Cell. Biochem. 49, 349–356.

Miskimins, W. K., McClelland, A., Roberts, M. P., & Ruddle, F. H. (1986). Cell proliferation and expression of the transferrin receptor gene: Promoter sequence homologies and protein interactions. J. Cell Biol. 103, 1781–1788.

Morgan, E. H. (1964a). The interaction between rabbit, human, and rat transferrin and reticulocytes. Br. J. Haematol. 10, 442–452.

Morgan, E. H. (1964b). Passage of transferrin, albumin and gamma globulin from maternal plasma to foetus in the rat and rabbit. J. Physiol. 171, 26–41.

Morgan, E. H. (1979). Studies on the mechanism of iron release from transferrin. Biochim. Biophys. Acta 58, 312–326.

Morgan, E. H. (1981). Transferrin: Biochemistry, physiology, and clinical significance. Mol. Aspects Med. 4, 1–123.

Morgan, E. H. (1983). Effect of pH and iron content of transferrin on its binding to reticulocyte receptors. Biochim. Biophys. Acta 62, 498–502.

Morris, C. M., Candy, J. M., Oakley, A. E., Taylor, G. A., Mountfort, S., Bishop, H., Ward, M. K., Bloxham, C. A., & Edwardson, J. A. (1989). Comparison of the regional distribution of transferrin receptors and aluminum in the forebrain of chronic renal dialysis patients. J. Neuro. Sci. 94, 295–306.

Müllner, E. W., & Kühn, L. C. (1988). A stem-loop in the 3' untranslated region mediates iron dependent regulation of transferrin receptor mRNA stability in the cytoplasm. Cell 53, 815–825.

Müllner, E. W., Neupert, B., & Kühn, L. C. (1989). A specific mRNA binding factor regulates the iron-dependent stability of cytoplasmic transferrin receptor mRNA. Cell 58, 373–382.

Neefjes, J. J., Verkerk, J. M. G., van der Marel, G. A., van Boom, J. H., & Ploegh, H. L. (1988). Recyling glycoproteins do not return to the cis-Golgi. J. Cell Biol. 107, 79–87.

Nunez, M. T., Glass, J., Fischer, S., Lavidor, L. M., Lenk, E. M., & Robinson, S. H. (1977). Transferrin receptors in developing murine erythroid cells. Br. J. Haem. 36, 519–526.

Ogunnariwo, J. A., & Schryvers, A. B. (1990). Iron acquisition in *Pasteurella haemolytica*: Expression and identification of a bovine-specific transferrin receptor. Infect. and Immun. 58, 2091–2097.

Omary, M. B., & Trowbridge, I. S. (1981a). Biosynthesis of the human transferrin receptor in cultured cells. J. Biol. Chem. 256, 12888–12892.

Omary, M. B., & Trowbridge, I. S. (1981b). Covalent binding of fatty acid to the transferrin receptor in cultured human cells. J. Biol. Chem. 256, 4715–4718.

Omary, M. B., Trowbridge, I. S., & Minowada, J. (1980). Human cell-surface glycoprotein with unusual properties. Nature 286, 888–891.

Orberger, G., Geyer, R., Stirm, S., & Tauber, R. (1992). Structure of the N-linked oligosaccharides of the human transferrin receptor. Eur. J. Biochem. 205, 257–267.

Ouyang, Q., Bommakanti, M., & Miskimins, W. K. (1993). A mitogen-responsive promoter region that is synergistically activated through multiple signalling pathways. Mol. Cell. Biol. 13, 1796–1804.

Owen, D., & Kühn, L. C. (1987). Noncoding 3' sequences of the transferrin receptor gene are required for mRNA regulation by iron. EMBO J. 6, 1287–1293.

Panaccio, M., Zalcberg, J. R., Thompson, C. H., Leyden, M. J., Sullivan, J. R., Lichtenstein, M., & McKenzie, I. F. C. (1987). Heterogeneity of the human transferrin receptor and use of anti-transferrin receptor antibodies to detect tumors in vivo. Immunol. Cell Biol. 65, 461–472.

Pardridge, W. M. (1992). Recent developments in peptide drug delivery to the brain. Pharmacol. Toxicol. 71, 3–10.

Pardridge, W. M., Eisenberg, J., & Yang, J. (1987). Human blood-brain barrier transferrin receptor. Metabolism 36, 892–895.

Paterson, S., Armstrong, N. J., Iacopetta, B. J., McArdle, H. F., & Morgan, E. H. (1984). Intravesicular pH and iron uptake by immature erythroid cells. J. Cell. Phys. 120, 225–232.

Pearse, B. M. F. (1988). Receptors compete for adaptors found in plasma membrane coated pits. EMBO J. 7, 3331–3336.

Pinsky, S. M., & Henkin, R. E. (1976). Gallium-67 tumor scanning. Sem. Nuc. Med. 6, 397–409.

Plowman, G. D., Brown, J. P., Enns, C. A., Schroder, J., Nikinmaa, B., Sussman, H., Hellstrom, K. E., & Hellstrom, I. (1983). Assignment of the gene for human melanoma-associated antigen p97 to chromosome 3. Nature 303, 70–72.

Ponka, P., & Schulman, H. M. (1985). Acquisition of iron from transferrin regulates reticulocyte heme synthesis. J. Biol. Chem. 260, 14717–14721.

Ponka, P., Schulman, H. M., & Wilczynska, A. (1982). Ferric pyridoxal isonicotinoyl hydrazone can provide iron for heme synthesis in reticulocytes. Biochim. Biophys. Acta 718, 151–156.

Poola, I., Mason, A. B., & Lucas, J. J. (1990). The chicken oviduct and embryonic red blood cell transferrin receptors are distinct molecules. Biochem. Biophys. Res. Comm. 171, 26–32.

Rabin, M., McClelland, A., Kuhn, L., & Ruddle, F. H. (1985). Regional localization of the human transferrin receptor gene to 3q26.2→qter. Am. J. Hum. Genetics 37, 1112–1116.

Ralton, J. E., Jackson, H. J., Zanoni, M., & Gleeson, P. A. (1989). Effect of glycosylation inhibitors on the structure and function of the murine transferrin receptor. Eur. J. Biochem. 186, 637–647.

Reckhow, C. L., & Enns, C. A. (1988). Characterization of the transferrin receptor in tunicamycin-treated A431 cells. J. Biol. Chem. 263, 7297–7301.

Redhead, K., Hill, T., & Chart, H. (1987). Interaction of lactoferrin and transferrins with the outer membrane of Bordetella pertussis. J. Gen. Microbiol. 133, 891–898.

Robbins, E., & Pederson, P. (1970). Iron: Its intracellular localization and possible role in cell division. Proc. Natl. Acad. Sci. USA 66, 1244–1251.

Roberts, R., Sandra, A., Siek, G. C., Lucas, J. J., & Fine, R. E. (1992). Studies of the mechanism of iron transport across the blood-brain barrier. Ann. Neurol. 32, S43–S50.

Robertson, B. J., Park, R. D., & Snider, M. (1992). Role of vesicular traffic in the transport of surface transferrin receptor to the Golgi complex in cultured human cells. Arch. Biochem. Biophys. 292, 190–198.

Roskams, A. J., & Connor, J. R. (1990). Aluminum access to the brain: A role for transferrin and its receptor. Proc. Natl. Acad. Sci. USA 87, 9024–9027.

Roskams, A. J., & Connor, J. R. (1992). Transferrin receptor expression in myelin deficient (md) rats. J. Neuro. Res. 31, 421–427.

Rothenberger, S., Iacopetta, B. J., & Kühn, L. C. (1987). Endocytosis of the transferrin receptor requires the cytoplasmic domain but not its phosphorylation site. Cell 49, 423–431.

Rothenberger, S., Müllner, E. W., & Kühn, L. C. (1990). The mRNA-binding protein which controls ferritin and transferrin receptor expression is conserved during evolution. Nucl. Acids Res. 18, 1175–1179.

Rouault, T. A., Hentze, M. W., Caughman, S. W., Harford, J. B., & Klausner, R. D. (1988). Binding of a cytosolic protein to the iron-responsive element of human ferritin messenger RNA. Science 241, 1207–1210.

Rouault, T. A., Stout, D. C., Kaptain, S., Harford, J. B., & Klausner, R. D. (1991). Structural relationship between an iron-regulated RNA-binding protein (IRE-BP) and aconitase: functional implications. Cell 64, 881–883.

Rutledge, E. A., Root, B. J., Lucas, J. J., & Enns, C. A. (1994a). Elimination of the O-linked glycosylation site at Thr 104 results in the generation of a soluble human transferrin receptor. Blood 83, 580–586.

Rutledge, E. A., Green, F. A., & Enns, C. A. (1994b). Generation of the soluble transferrin receptor requires cycling through an endosomal compartment. J. Biol. Chem. 269, 31864–31868.

Rutledge, E. A., & Enns, C. A. (1996). Cleavage of the transferrin receptor is influenced by the composition of the O-linked carbohydrate at position 104. J. Cell. Physiol. In press.

Schell, D., Evers, R., Preis, D., Ziegelbauer, K., Kiefer, H., Lottspeich, F., Cornelissen, A. W. C. A., & Overath, P. (1991). A transferrin-binding protein of *Trypanosoma brucei* is encoded by one of the genes in the variant surface glycoprotein gene expression site. EMBO J. 10, 1061–1066.

Schmid, S. L. (1992). The mechanism of receptor-mediated endocytosis; more questions than answers. Bioessays 14, 589–596.

Schneider, C., Sutherland, R., Newman, R., & Greaves, M. (1982). Structural features of the cell surface receptor for transferrin that is recognized by the monoclonal antibody OKT9. J. Biol. Chem. 257, 8516–8522.

Schneider, C., Asser, U., Sutherland, D. R., & Greaves, M. F. (1983a). *In vitro* biosynthesis of the human cell surface receptor for transferrin. FEBS Lett. 158, 259–264.

Schneider, C., Kurkinen, M., & Greaves, M. (1983b). Isolation of cDNA clones for the human transferrin receptor. EMBO J. 2, 2259–2263.

Schneider, C., Owen, M. J., Banville, D., & Williams, J. G. (1984). Primary structure of human transferrin receptor deduced from the mRNA sequence. Nature 311, 675–678.

Schulman, H. M., Wilczynska, A., & Ponka, P. (1981). Transferrin and iron uptake by human lymphoblastoid and K562 cells. Biochem. Biophys. Res. Comm. 100, 1523–1530.

Seiser, C., Teixeira, S., & Kuhn, L. C. (1993). Interleukin-2-dependent transcriptional and posttranscriptional regulation of transferrin receptor mRNA. J. Biol. Chem. 268, 13074–13080.

Seligman, P. A., Morgan, P. L., Schleicher, R. B., & Crawford, E. D. (1992). Treatment with gallium nitrate: Evidence for interference with iron metabolism *in vivo*. Am. J. Hem. 41, 232–240.

Shih, Y. J., Baynes, R. D., Hudson, B. G., Flowers, C. H., Skikne, B. S., & Cook, J. D. (1990). Serum transferrin receptor is a truncated form of tissue receptor. J. Biol. Chem. 265, 19077–19081.

Shih, Y. J., Baynes, R. D., Hudson, B. G., & Cook, J. D. (1993). Characterization and quantitation of the circulating forms of serum transferrin receptor using domain-specific antibodies. Blood 81, 234–238.

Shindelman, J. E., Ortmeyer, A. E., & Sussman, H. H. (1981). Demonstration of the transferrin receptor in human breast cancer tissue. Potential marker for identifying dividing cells. Int. J. Cancer 27, 329–334.

Siemsen, J. K., Grebe, S. F., Sargent, E. N., & Wentz, D. (1976). Gallium-67 scintography of pulmonary diseases as a complement to radiography. Radiology 118, 371–375.

Simons, R., & Riordan, J. R. (1990). The myelin-deficient rat has a single base substitution in the third exon of the myelin proteolipid protein gene. J. Neurochem. 54, 1079–1081.

Singhal, A., Cook, J. D., Skikne, B. S., Thomas, P., Serjeant, B., & Serjeant, G. (1993). The clinical significance of serum transferrin receptor levels in sickle cell disease. Br. J. Haematol. 84, 301–304.

Sipe, D. M., & Murphy, R. F. (1991). Binding to cellular receptor results in increased iron release from transferrin at mildly acidic pH. J. Biol. Chem. 266, 8002–8007.

Snider, M. D., & Rogers, O. C. (1985). Intracellular movement of cell surface receptors after endocytosis: Resialylation of asialo-transferrin receptor in human erythroleukemia cells. J. Cell Biol. 100, 826–834.

Stein, B. S., & Sussman, H. H. (1986). Demonstration of two distinct transferrin receptor recycling pathways and transferrin-independent receptor internalization in K562 cells. J. Biol. Chem. 261, 10319–10331.

Stevenson, P., Williams, P., & Griffiths, E. (1992). Common antigenic domains in transferrin-binding protein 2 of *Neisseria gonorrheae*, and *Haemophilus influenzae* type b. Infect. and Immun. 60, 2391–2396.

Stoorvogel, W., Geuze, H. J., Griffith, J. M., Schwartz, A. L., & Strous, G. J. (1989). Relations between the intracellular pathways of the receptors for transferrin, asialoglycoprotein, and mannose 6-phosphate in human hepatoma cells. J. Cell Biol. 108, 2137–2148.

Taetle, R. (1990). The role of transferrin receptors in hemopoietic cell growth. Exp. Hematol. 18, 360–365.

Taetle, R., Honeysett, J. M., & Trowbridge, I. (1983). Effects of anti-transferrin receptor antibodies on growth of normal and malignant myeloid cells. Int. J. Cancer 32, 343–349.

Takahashi, S., Esserman, L., & Levy, R. (1991). An epitope on the transferrin receptor preferentially exposed during tumor progression in human lymphoma is close to the ligand binding site. Blood 77, 826–832.

Thelander, L. (1990). Ribonucleotide reductase. In: *Iron Transport and Storage.* CRC Press, Boston, pp. 193–299.

Thorstensen, K., & Romslo, I. (1993). The transferrin receptor: Its diagnostic value and its potential as therapeutic target. Scand. J. Clin. Lab. Invest. 53, 113–120.

Trowbridge, I. S., & Omary, M. B. (1981). Human cell surface glycoprotein related to cell proliferation is the receptor for transferrin. Proc. Natl. Acad. Sci. USA 78, 3039–3043.

Trowbridge, I. S., Domingo, D. L., Thomas, M. L., & Chain, A. (1988). Cell surface molecules of the hematopoietic system: T200 glycoprotein and the transferrin receptor as models for proteins involved in growth and differentiation. Inflam. Bowel Disease 441–447.

Trowbridge, I. S., Collawn, J. F., & Hopkins, C. R. (1993). Signal-dependent membrane protein trafficking in the endocytic pathway. Ann. Rev. Cell Biol. 9, 129–161.

Tsunoo, H., & Sussman, H. H. (1983). Characterization of transferrin binding and specificity of the placental transferrin receptor. Arch. Biochem. Biophys. 225, 42–54.

Turkewitz, A. P., Amatruda, J. F., Borhani, D., Harrison, S. C., & Schwartz, A. L. (1988). A high yield purification of the human transferrin receptor and properties of its major extracellular fragment. J. Biol. Chem. 263, 8318–8325.

Turkewitz, A. P., & Harrison, S. C. (1989). Concentration of transferrin receptor in human placental coated vesicles. J. Cell Biol. 108, 2127–2135.

van der Ende, A., du Maine, A., Simmons, C. F., Schwartz, A. L., & Strous, G. J. (1987). Iron metabolism in BeWo chorion carcinoma cells. J. Biol. Chem. 262, 8910–8916.

Voyiatzaki, C. S., & Soteriadou, K. P. (1990). Evidence of transferrin binding sites on the surface of *Lieishmania* Promastigotes. J. Biol. Chem. 265, 22380–22385.

Voyiatzaki, C. S., & Soteriadou, K. P. (1992). Identification and isolation of the *Leishmania* transferrin receptor. J. Biol. Chem. 267, 9112–9117.

Wada, H. G., Hass, P. E., & Sussman, H. H. (1979). Transferrin receptor in human placental brush border membranes: Studies on the binding of transferrin to placental membrane vesicles and the identification of a placental brush border glycoprotein with high affinity for transferrin. J. Biol. Chem. 254, 12629–12635.

Ward, J. H., Kushner, J. P., & Kaplan, J. (1982). Regulation of HeLa cell transferrin receptors. J. Biol. Chem. 257, 10317–10323.

Wauben-Penris, P. J., Strous, G. J., & van der Donk, H. A. (1988). Kinetics of transferrin endocytosis and iron uptake by intact isolated rat seminiferous tubules and Sertoli cells in culture. Biol. of Reprod. 38, 853–861.

Williams, A. M., & Enns, C. A. (1991). A mutated transferrin receptor lacking asparagine-linked glycosylation sites shows reduced functionality and an association with binding immunoglobulin protein. J. Biol. Chem. 266, 17648–17654.

Williams, A. M., & Enns, C. A. (1993). A region of the C-terminal portion of the human transferrin receptor contains an asparagine-linked glycosylation site critical for receptor structure and function. J. Biol. Chem. 268, 12780–12786.

Williams, J. (1974). The formation of iron-binding fragments of hen ovotransferrin by limited proteolysis. Biochem. J. 141, 745–752.

Williams, J. (1982). The evolution of transferrin. Trends in Biochem. Sci. 7, 394–397.

Woith, W., Nusslein, I., Antoni, C., Dejica, D. I., Winkler, T. H., Herrmann, M., Pirner, K., Kalden, J. R., & Manager, B. (1993). A soluble form of the human transferrin receptor is released by activated lymphocytes *in vitro*. Clin. Exp. Immunol. 92, 537–542.

Wong, C. T., & Morgan, E. H. (1973). Placental transfer of iron in the guinea pig. Quarterly J. Exper. Physi. 58, 27–58.

Wu, R., & Gill, G. N. (1993). A protein that interacts specifically with the endocytic code of the human insulin receptor. Mol. Biol. Cell, suppl. 4, 117a.

Yamashiro, D. J., & Maxfield, F. R. (1983). Acidification of endocytic compartments and the intracellular pathways of ligands and receptors. J. Cell. Biol. 231–245.

Yamashiro, D. J., Fluss, S. R., & Maxfield, F. R. (1983). Acidification of endocytic vesicles by an ATP-dependent proton pump. J. Cell Biol. 97, 929–934.

Yang, F., Yum, J. B., McGill, F. R., Moore, C. M., Naylor, S. L., vanBragt, P. H., Baldwin, W. S., & Bowman, B. H. (1984). Human transferrin: cDNA characterization and chromosomal localization. Proc. Natl. Acad. Sci. USA 81, 2752–2756.

Yang, B., Hoe, M. H., Black, P., & Hunt, R. C. (1993). Role of oligosaccharides in the processing and function of human transferrin receptors. J. Biol. Chem. 286, 7435–7441.

Yeoh, G. C. T., & Morgan, E. H. (1979). Dimethyl sulphoxide induction of transferrin receptors on Friend erythroleukemia cells. Cell Different. 8, 331–343.

Young, S. P., & Aisen, P. (1981). Transferrin receptors and the uptake and release of iron by isolated hepatocytes. Hepatology 1, 114–119.

Zak, O., Trinder, D., & Aisen, P. (1994). Primary receptor-recognition site of human transferrin is in the C-terminal. J. Biol. Chem. 269, 7110–7114.

Zerial, M., Melancon, P., Schneider, C., & Garoff, H. (1986). The transmembrane segment of the human transferrin receptor functions as a signal peptide. EMBO J. 5, 1543–1550.

Zerial, M., Huylebroeck, D., & Garoff, H. (1987). Foreign transmembrane peptides replacing the internal signal sequence of transferrin receptor allow its translocation and membrane binding. Cell 48, 147–155.

INDEX

Biomembranes
A Multi-Volume Treatise

Edited by **A.G. Lee**, *Department of Biochemistry, University of Southampton*

"Progress in understanding the nature of the biological membrane has been very rapid over a broad front, but still pockets of ignorance remain. Application of the techniques of molecular biology has provided the sequences of a very large number of membrane proteins, and led to the discovery of superfamilies of membrane proteins of related structure. In turn, the identification of these superfamilies has led to new ways of thinking about membrane processes. Many of these processes can now be discussed in molecular terms, and unexpected relationships between apparently unrelated phenomena are bringing a new unity to the study of biological membranes.

The quantity of information available about membrane proteins is now too large for any one person to be familiar with anything but a small part of the primary literature. A series of volumes concentrating on molecular aspects of biological membranes therefore seems timely. The hope is that, when complete, these volumes will provide a convenient introduction to the study of a wide range of membrane functions."

— From the Preface

Volume 3, Receptors of Cell Adhesion and Cellular Recognition
1996, 330 pp. $128.50
ISBN 1-55938-660-6

CONTENTS: Molecules of Cell Adhesion and Recognition: An Overview, *R. Marsh and R. Brackenbury.* Cell Recognition Molecules of the Immunoglobulin Superfamily in the Nervous System, *G. Gegelashvili and E. Bock.* T-cell Antogen Receptors, *C. Horgan and J.D. Fraser.* The Major Presentation and Histocompatibility Complex, *J. Colombani.* Cadherins: A Review of Structure and Function, *J. Wallis, R. Moore, P. Smith and F.S. Walsh.* The Integrin Family, *R.D. Bowditch and R.J. Faull.* The Selectin Family, *M.A. Jutila.* The CD44 Family of Cell Adhesion Molecules: Functional Aspects, *C. B. Underhill.* Membrane-Associated Mucins, *H.L. Vos, J. Wesseling and J. Hilkens.* Platelet Membrane Glycoproteins, *K. J. Clemetson.* Immunoglobulin Fc Receptors: Diversity, Structure and Function, *P.M. Hogarth and M.D. Hulett.*

Also Available:
Volume 1 (1995) $128.50
  Volume 2 (2 Part Set) $257.00

**J
A
I

P
R
E
S
S**

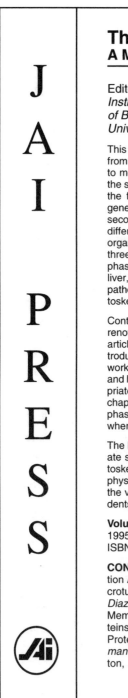

J
A
I

P
R
E
S
S

The Cytoskeleton
A Multi-Volume Treatise

Edited by **J.E. Hesketh,** *Rowett Research Institute, Aberdeen* and **Ian Pryme,** *Department of Biochemistry and Molecular Biology, University of Bergen*

This three volume treatise will cover aspects of the cytoskeleton from basic biochemistry and cell biology to those of relevance to medical science. The first volume of the treatise deals with the structural aspects of the cytoskeleton: the characteristics of the filaments and their components, the organization of the genes, motor proteins, and interactions with membranes. The second volume deals with the functions of the cytoskeleton in-different cellular processes such as cell compartmentation and organanelle transport, secretion and cell attachment. In volume three the functional theme is continued but in this case the emphasis is on the cytoskeleton in different tissues such as bone, liver, and intestine. Lastly, in the fourth volume a selection of pathological situations which are related to defects in the cytoskeleton are discussed.

Contributions from leading scientists working in internationally renowned research centres have been commissioned. These articles will not only describe the background necessary to introduce the topical area and present sufficient details of recent work so as to give an up to date account of the various concepts and hypotheses involved, but the authors will also where appropriate give their own research perspective on the subject. The chapters will thus represent the cutting edge of research, emphasizing directions in which the subjects are developing and where future progress will prove fruitful.

The Editors envisage the books will be of interest to postgraduate student/post-doctoral workers both actively involved in cytoskeleton research and other areas of cell biology. By including physiological and clinical aspects of the subject, the hope is that the volumes will also be of use in the teaching of medical students.

Volume 1, Structure and Assembly
1995, 300 pp. $109.50
ISBN 1-55938-687-8

CONTENTS: Preface. Introduction. Microfilament Organization Actin-Binding Proteins, *Sutherland McIvor.* Control of Microtubule Polymerization and Stablity, *Jesus Avila and Javier Diaz Nido.* Motor Proteins in Mitosis and Meiosis, *Tim J. Yen.* Membrane Cytoskeleton, *Verena Niggli.* Actin-Binding Proteins-Lipid Interactions, *G. Isenberg and W.H. Goldman.* The Proteins of Intermediate Filament Systems, *Robert I. Shoeman and Peter Traub.* Nuclear Lamins and the Nucleoskeleton, *Reimer Stick.* Index.

J A I   P R E S S

Advances In Cell and Molecular Biology of Membranes and Organelles

Edited by **Alan M. Tartakoff,** *Institute of Pathology, Case Western Reserve University*

REVIEW: "This is one of the books which clearly recognizes its full appurtenance to the cellular and molecular biology since it considers by so doing that there is no more possibility to investigate the biological problems without integrating the molecular aspects."

— *Cellular and Molecular Biology*

Volume 3 Signal Transduction Through Growth Factor Receptors
1994, 223 pp. $109.50
ISBN 1-55938-344-5

Edited by: **Yasuo Kitagawa,** *BioSciences Center, Laboratory of Organogenesis, Nagoya University* and **Ryuzo Sasaki,** *Faculty of Agriculture, Department of Food Science and Technology, Kyoto University,*

CONTENTS: The Hepatocyte Growth Factor/c-MET Signaling Pathway, *D. P. Bottario, A. M.-L. Chan, J. S. Rubin, E. Gak, E. Fortney, J. Schindler, M. Chedid, and S. A. Aaronson.* Insulin Receptor, *Y. Ebina, H. Hayashi, F. Kanai, S. Kamohara, and Y. Nishioka.* Interleukin-3 Receptor: Structure and Signal Transduction, *T. Kitamura, and A. Miyajima.* Interleukin-5 Receptor, *K. Takatsu.* Interleukin-6 Receptor and Signal Transduction, *T. Matsuda, T. Nakajima, T. Kaisho, K. Nakajima, and T. Hirano.* Receptor for Granulocyte Colony-Stimulating Factor, *S. Nagata and R. Fukunaga.* Receptor for Granulocyte/Macrophage Colony-stimulating Factor, *K. Kurata, T. Yokota, A. Miyajima and K. Arai.* Perspectives on the Structure and Mechanisms of Signal Transduction by the Erythropoietin Receptor, *S. S. Jones.* Interleukin-1 Signal Transduction, *J. E. Sims, T. A. Bird, J. G. Giri, and K. S. Dower.*